AF361706

Forming Processes of Modern Metallic Materials

Forming Processes of Modern Metallic Materials

Editor

Tomasz Trzepiecinski

MDPI • Basel • Beijing • Wuhan • Barcelona • Belgrade • Manchester • Tokyo • Cluj • Tianjin

Editor
Tomasz Trzepiecinski
Rzeszow University of Technology
Poland

Editorial Office
MDPI
St. Alban-Anlage 66
4052 Basel, Switzerland

This is a reprint of articles from the Special Issue published online in the open access journal *Metals* (ISSN 2075-4701) (available at: https://www.mdpi.com/journal/metals/special_issues/metallic_forming).

For citation purposes, cite each article independently as indicated on the article page online and as indicated below:

LastName, A.A.; LastName, B.B.; LastName, C.C. Article Title. *Journal Name* **Year**, *Article Number*, Page Range.

ISBN 978-3-03943-156-4 (Hbk)
ISBN 978-3-03943-157-1 (PDF)

Contents

About the Editor

Tomasz Trzepiecinski is currently an associate professor in the Department of Materials Forming and Processing of the Rzeszow University of Technology. After he finalized his Ph.D. thesis, he received an assistant professor position in the same department. In 2013, Dr. Trzepieciński competed his habilitation thesis, "Selected Aspects of Tribological Evaluation of Plastically Deformed Sheets". His research experience abroad included a three stay of about four months at the Stavanger University (Norway). Tomasz Trzepieciński has many years research experience in metal forming processes, plasticity, tribological behavior of metallic materials, numerical modeling of plastic working processes, and the application of artificial neural networks and genetic algorithms to optimize forming processes. His research areas include advanced manufacturing technology, processing of advanced materials, and analysis of contact phenomena in sheet metal forming. He is reviewer for more than 70 scientific journals in materials science and engineering. He has co-authored 7 monographs and 14 scientific book chapters. He authored and coauthored more than 200 papers in national and international peer review journals, 96 in ISI Web of Knowledge, for an h-index of 14 (Scopus).

Preface to "Forming Processes of Modern Metallic Materials"

The aim of this Special Issue is to present the latest achievements in various modern metal forming processes and the latest research related to the computational methods used for metal forming technologies. The subjects of the research articles published in this Special Issue are multidisciplinary, including friction and lubrication in sheet metal forming (SPF), hot strip rolling and tandem strip rolling, application of numerical methods to simulating metal forming processes, development of new creep performance materials, the single point incremental forming (SPIF) process, and the fatigue fracture characteristics of Alclad 7075-T6 aluminum alloy sheets joined by refill friction stir spot welding (RFSSW). Two review articles summarize the approaches on the innovative numeric algorithms, experimental methods, and theoretical contributions that have recently been proposed for SMF by researchers and business research centers.

Tomasz Trzepiecinski
Editor

Editorial

Forming Processes of Modern Metallic Materials

Tomasz Trzepieciński

Department of Materials Forming and Processing, Rzeszow University of Technology, al. Powst. Warszawy 8, 35-959 Rzeszów, Poland; tomtrz@prz.edu.pl; Tel.: +48-17743-2527

Received: 2 July 2020; Accepted: 16 July 2020; Published: 18 July 2020

1. Introduction and Scope

The plastic working of metallic materials is one of the most efficient and important manufacturing technologies used in industry today. Lightweight materials, such as titanium alloys, aluminium alloys, and ultra-high-strength steels are used extensively in the automotive, aerospace, transportation, and construction industries, leading to increasing demand for advanced innovative forming technologies. Numerical simulation is currently highly focused and provides a better understanding of innovative forming processes. Computational methods and numerical analysis, coupled with the modelling of structural change, allow us to reduce the time taken and eliminate experimental tests. The aim of this Special Issue is to present the latest achievements in various modern metal forming processes and the latest research related to the computational methods used for metal forming technologies. Research articles focusing on new developments in the forming of metallic materials are welcome for consideration for publication. I truly believe that this Special Issue will help the metals research community enhance their understanding of the present status and trends of the forming processes of modern metallic materials.

The modernisation of methods of plastic forming and the development of new techniques is mainly conditioned by the provision of innovative processing methods for shaping new types of materials, primarily in the automotive and aircraft industries. In parallel, a comprehensive approach is required to reduce the consumption of energy and materials in the context of the implementation of systems linked to the fourth industrial revolution (Industry 4.0). The search for high-performance unconventional methods of plastic working is reflected in the topics of the articles published in this Special Issue.

2. Contributions

Two review articles and ten research papers have been published in this Special Issue of *Metals*. The subjects are multidisciplinary, including (i) friction and lubrication in sheet metal forming (SPF), (ii) hot strip rolling and tandem strip rolling, (iii) application of numerical methods to simulate metal forming processes, (iv) development of new creep performance materials, (v) the Single Point Incremental Forming (SPIF) process, and (vi) the fatigue fracture characteristics of Alclad 7075-T6 aluminium alloy sheets joined by refill friction stir spot welding (RFSSW).

Firstly, two articles summarise the approaches on the innovative numerical algorithms, experimental methods, and theoretical contributions that have recently been proposed for SMF by researchers and business research centres. In the paper by Trzepiecinski and Lemu [1], the methods used to represent the friction conditions in conventional SMF and incremental sheet forming (ISF) have been summarised. Furthermore, the main disadvantages and limitations of the modelling methods of friction phenomena in specific areas of the material to be formed have been discussed. It has been concluded that although most of the friction tests were developed many decades ago, they are still valid and used by many authors. Although the methods to determine friction resistance in conventional SMF at different temperature conditions have been fairly well understood, there are a very limited number of experimental tests modelling conditions in ISF. This is due to the difficulty in separating the frictional resistance from the plastic deformation resistance in the process. The article of Trzepieciński [2] is intended to summarise recent development trends in both the numerical and experimental fields of conventional deep-drawing, spinning, flexible-die forming, electromagnetic forming, and computer-controlled forming methods, like incremental sheet forming. The review is limited to the considerable changes that have occurred in the SMF sector in the last decade, with special attention given to the 2015–2020 period. The progress observed in the last decade in the area of SMF mainly concerns the development of nonconventional methods of forming difficult-to-form lightweight materials for automotive and aircraft applications. In evaluating the ecological convenience of SMF processes, the tribological aspects have also become the subject of great attention.

Secondly, the modelling of tribological phenomena in SMF was discussed in three articles. Thus, one objective of the work reported in the paper by Trzepiecinski and Lemu [3] was to better understand the contact pressure acting on the rounded edge of the cylindrical counter-sample in the bending under tension (BUT) test by assuming the anisotropic properties of the metallic sheet. Another objective of this article was to calculate the coefficient of friction (COF) that acts during the BUT test for different strains of sheet metal. It was found that the effectiveness of the lubrication depends on the balance between two mechanisms accompanied by the friction process: roughening of workpiece asperities and adhesion of the contacting surfaces. The used lubricants were able to reduce the value of the friction coefficient approximately by 3–52% in relation to the surface roughness of the rolls. Trzepiecinski and Fejkiel [4] carried out both experimental and numerical finite element method-based studies of the frictional phenomena and material deformation during the flow of sheet metal through a drawbead. Different set-up conditions, including drawbead height and specimen widths, have been considered. Tribological tests have been carried out on DC04 steel sheets commonly used in the automotive industry using a special tribological drawbead simulator. Trzepiecinski [5] has compared the tribological properties of a deep drawing quality steel sheet using the three commonly used friction tests, i.e., the strip drawing test, draw bead test, and bending under tension test. Friction tests have been conducted under different pressure and lubrication conditions, surface roughness of tools represented by counter-samples, and orientations of the specimens according to the direction of the sheet rolling. The effectiveness of the reduction in the coefficient of friction by the lubricant clearly depended on the specimen orientation and surface roughness of the counter-samples. Differences in the coefficient of friction obtained from various tribological tests indicate a strong dependence of the kinematic conditions of the process on parameters directly related to the characteristics of the friction process.

Thirdly, the analysis of the rolling process was discussed in three articles. Li et al. [6] provided the basic principles of the six models; they verified the predictability of the bending force using these models based on a real-life dataset of a 1580-mm hot rolling process in a steel factory. In this paper, six machine learning models, including Artificial Neural Network (ANN), Support Vector Machine (SVR), Classification and Regression Tree (CART), Bagging Regression Tree (BRT), Least Absolute Shrinkage and Selection operator (LASSO), and Gaussian Process Regression (GPR), were applied to predict the bending force in the hot strip rolling (HSR) process. A comparative experiment was carried out based on a real-life dataset, and the predicted performance of the six models was analysed on the basis of prediction accuracy, stability, and computational cost. The results show that the Gaussian process regression model is considered the optimal model for the prediction of bending force with the best prediction accuracy, better stability, and acceptable computational cost. Kraner et al. [7] studied the influence of asymmetric rolling in comparison with symmetric rolling, with respect to the technological, mechanical, and metallographic perspectives. The interactions between the measured rolling forces, the achieved strains, the tensile and yield strength, the hardness, the indicators of planar anisotropy, the microstructures with an average size of grains and crystallographic textures with volume fractions of different rolling, shear, and recrystallization texture components were investigated. The impact of asymmetric cold rolling was quantitatively assessed for an industrial 5754 aluminium alloy. The main advantage of asymmetric rolling is the presence of significant shear strain, leading to a gradient structure. It was found that the increased isotropy of the deformed and annealed aluminium sheet is a product of the texture heterogeneity and reduced volume fractions of the separate texture components. Mao et al. [8] proposed tandem skew rolling (TSR) for the production of seamless tubes. In order to study the deformation characteristics and mechanism on tubes obtained by the TSR process, a numerical simulation of the process was analysed using Deform-3D software. Based on numerical analysis, experiments were carried out with carbon steel 1045, high-strength steel (HSS) 42CrMo, and magnesium alloy AZ31 on the TSR testing mill to verify the feasibility of the process. Based on the studies carried out, it was found that rolling on a TSR testing mill allowed TSR to be run with high efficiency in the deformation of seamless tubes pierced and rolled at the same time, which makes continuous skew rolling more economical and practical.

Fourthly, as far as electromagnetic forming (EMF) is concerned, Cui et al. [9] established a three-dimensional (3D) finite element model (FEM), including quasi-static stamping, sequential coupling for electromagnetic forming (EMF), and springback, to analyse springback calibration by electromagnetic force. It was found that the tangential stress in the sheet bending region is reduced, and even the direction of tangential stress in the bending region is changed after EMF. Forming with a higher discharge voltage significantly reduced the springback phenomenon.

Fifthly, the development of new materials with good creep performance is a topic of the paper of Heo et al. [10]. A modified 9Cr-2W (alloy B) steel was developed in this study for use as fuel cladding with good creep performance for sodium-cooled fast reactors. The causes of fracture during the manufacturing process of alloy B steels with low production yield were investigated in order to develop a novel manufacturing process for alloy B steels with high formability.

Sixthly, the dimensional accuracy and mechanical properties of metal components formed by SPIF have been extensively studied by Slota et al. [11]. An X-ray diffraction method has been applied to achieve an understanding of the residual stress formation caused by the SPIF process on DC04 steel sheet drawpieces. It was found that step size had a very large impact on the value of residual stresses. An increase in step size causes an increase in the absolute values of axial and tangential residual stresses. X-ray analysis of the walls of truncated cones led to the conclusion that residual stresses measured in a tangential direction decrease with increasing values of the surface roughness parameters Sa and Sq.

Finally, Kubit et al. [12] have shown that the RFSSW technique has great potential to be a replacement for single-lap joining techniques such as riveting or resistance spot welding used in the aircraft industry. The RFSSW technique is used to join Alclad 7075-T6 aluminium alloy plates in a single-lap configuration, which were subjected to a high-cycle fatigue test. Paris' law for crack propagation has been successfully adopted to predict the fatigue crack growth of lap-shear RFSSW specimens. Although some assumptions have been made, the comparison of the analytical and experimental fatigue crack growth rate confirms the potential of Paris' law to analyse the crack growth in RFSSW joints.

3. Conclusions and Outlook

The emergence of new plastic forming methods forces the modernisation of the existing machine park and the economically justified implementation of automation systems and robotisation of the machining process. In the SMF area, there is a tendency to adapt machines for flexible production, as well as to improve material flow through the production lines. Striving to improve the functionality of machines and extend the period of their trouble-free operation are the most important challenges facing the creators and propagators of the concept of the so called fourth industrial revolution (Industry 4.0). The implementation of this concept creates an increased demand for innovation in the field of new materials and technologies, which are one of the leading topics presented in this Special Issue.

As guest editor of this Special Issue, I am very happy with the final result and hope that the papers presented will be useful to academic researchers and industrial designers working in the plastic working of metallic materials sector. I also hope that all the scientific results in this Special Issue will contribute to the future development of research on the plastic working of metallic materials. I would like to thank all the authors for their contributions and all of the anonymous reviewers who assisted me in the reviewing process. I would also like to give special thanks to staff at the *Metals* Editorial Office, especially to Ms. Betty Jin, Assistant Editor, who managed and facilitated this Special Issue.

Conflicts of Interest: The author declares no conflicts of interest.

References

1. Trzepiecinski, T.; Lemu, H.G. Recent Developments and Trends in the Friction Testing for Conventional Sheet Metal Forming and Incremental Sheet Forming. *Metals* **2020**, *10*, 47. [CrossRef]
2. Trzepieciński, T. Recent Developments and Trends in Sheet Metal Forming. *Metals* **2020**, *10*, 779. [CrossRef]
3. Trzepiecinski, T.; Lemu, H.G. Effect of Lubrication on Friction in Bending under Tension Test-Experimental and Numerical Approach. *Metals* **2020**, *10*, 544. [CrossRef]
4. Trzepiecinski, T.; Fejkiel, R. A 3D FEM-Based Numerical Analysis of the Sheet Metal Strip Flowing Through Drawbead Simulator. *Metals* **2020**, *10*, 45. [CrossRef]
5. Trzepiecinski, T. A Study of the Coefficient of Friction in Steel Sheets Forming. *Metals* **2019**, *9*, 988. [CrossRef]
6. Li, X.; Luan, F.; Wu, Y. A Comparative Assessment of Six Machine Learning Models for Prediction of Bending Force in Hot Strip Rolling Process. *Metals* **2020**, *10*, 685. [CrossRef]
7. Kraner, J.; Fajfar, P.; Palkowski, H.; Kugler, G.; Godec, M.; Paulin, I. Microstructure and Texture Evolution with Relation to Mechanical Properties of Compared Symmetrically and Asymmetrically Cold Rolled Aluminum Alloy. *Metals* **2020**, *10*, 156. [CrossRef]
8. Mao, F.; Wang, F.; Shuang, Y.; Hu, J.; Chen, J. Deformation Behavior and Experiments on a Light Alloy Seamless Tube via a Tandem Skew Rolling Process. *Metals* **2020**, *10*, 59. [CrossRef]
9. Cui, X.; Zhang, Z.; Yu, H.; Xiao, X.; Cheng, Y. Springback Calibration of a U-Shaped Electromagnetic Impulse Forming Process. *Metals* **2019**, *9*, 603. [CrossRef]
10. Heo, H.M.; Kim, J.H.; Kim, S.H.; Kim, J.R.; Moon, W.J. Influence of Heat Treatment on the Workability of Modified 9Cr-2W Steel with Higher B Content. *Metals* **2019**, *9*, 904. [CrossRef]
11. Slota, J.; Krasowski, B.; Kubit, A.; Trzepiecinski, T.; Bochnowski, W.; Dudek, K.; Neslušan, M. Residual Stresses and Surface Roughness Analysis of Truncated Cones of Steel Sheet Made by Single Point Incremental Forming. *Metals* **2020**, *10*, 237. [CrossRef]
12. Kubit, A.; Drabczyk, M.; Trzepiecinski, T.; Bochnowski, W.; Kaščák, Ľ.; Slota, J. Fatigue Life Assessment of Refill Friction Stir Spot Welded Alclad 7075-T6 Aluminium Alloy Joints. *Metals* **2020**, *10*, 633. [CrossRef]

Article

A Study of the Coefficient of Friction in Steel Sheets Forming

Tomasz Trzepiecinski

Department of Materials Forming and Processing, Rzeszow University of Technology, al. Powst. Warszawy 8, 35-959 Rzeszów, Poland; tomtrz@prz.edu.pl

Received: 22 August 2019; Accepted: 4 September 2019; Published: 6 September 2019

Abstract: The aim of this paper was to compare the tribological properties of a deep drawing quality steel sheet using the three commonly used friction tests, i.e., the strip drawing test, draw bead test, and bending under tension test. All tests have been carried out using a specially designed friction simulator. The test material was a 0.8-mm-thick DC04 steel sheet, commonly used in the automotive industry. Uniaxial tensile tests have been carried out to characterise the mechanical properties of the specimens. Furthermore, measurements of the sheet surface topography have been carried out to characterise the tribological properties of the specimens. The friction tests have been conducted under different pressure and lubrication conditions, surface roughnesses of tools represented by counter-samples, and orientations of the specimens according to the direction of the sheet rolling. A comparative analysis of the results of the friction tests revealed different values of friction. In the strip drawing test, the value of the coefficient of friction decreases as the contact pressure increases for both dry and lubricated conditions. In the draw bead test, the specimens oriented along the rolling direction demonstrated a higher value of the coefficient of friction compared to the samples cut transverse to the rolling direction. In contrast to the strip drawing test, the specimens tested in the bending under tension test exhibit a tendency to an increase in the value of the coefficient of friction with the increasing contact pressure.

Keywords: coefficient of friction; deep drawing; draw bead; material properties; sheet metal forming; surface properties

---

1. Introduction

One of the most important technological phenomena, which largely influences the flow characteristics of a deformed material, is friction. The amount of friction generated in metal forming is mainly determined by the shear strength of the friction connections [1]. The type of friction joint created mainly depends on the materials of the friction pair and the surface roughness of the bodies in contact. Friction depends on several parameters and phenomena such as lubrication, normal pressure, surface roughness of the tool and sheet, sliding speed, materials of the contact pair, and temperature [2]. Under warm and hot forming conditions the coefficient of friction is usually higher than in the cold forming due to the increased adhesion between the bodies in contact. Under lubricated conditions, it is known [3,4] that friction changes as the material plastically deforms and surface roughness increases, causing a transition from a hydrodynamic to a mixed friction regime where the metal to metal contact is increased.

In sheet metal forming, the value of the drawing force, friction conditions occurring in the various areas of the workpiece and the temperature are changing. The non-uniformity of the deformation of the drawpiece is mainly determined by the occurrence of friction forces at the interface of the deformed material and the tool. Many factors affect the contact phenomena: The amount of normal pressure, material grade, topography both of the sheet and tool surfaces, type of lubricant, and temperature [5].

The unfavourable effects of friction include [6,7]: Non-uniformity in the workpiece deformation, increase of the surface roughness of the workpiece, and the intensification of the tool wear. The advantageous effect of the friction between the workpiece and punch on the stamping process is the increasing of the maximum allowable drawing force. Moreover, Menezes et al. [8] concluded that the variations in the coefficient of friction between the tool surfaces and the workpiece directly affect the stress distribution and shape of the workpiece, having implications on the microstructure of the material being processed.

To represent the friction phenomena in sheet metal forming, simulation tests have been developed. The friction tests the simulating friction and lubrication conditions in sheet metal forming can be divided into tests that simulate processes and tests which simulate tribological conditions. These tests simulate friction conditions in the specific contact areas between the sheet and tools. The coefficient of friction is calculated based on the assumed friction model. Experimental tests of the friction coefficient value only allow the determination of friction in selected areas of the drawpiece, which leads to the need to use multiple friction tests. In addition to many simulation tests used in the industry, numerical methods have been used to study the interactions in the sheet-tool interface [9–11].

Knowledge of the friction is necessary to prepare an appropriate design of the die tool and to forecast the material flow during the forming of drawpieces, especially those with complex shapes. This provides a good justification for work on the proper planning of friction tests using the right simulator. In this paper, there are three typical friction tests: The strip drawing test, bending under tension test, and draw bead test were used to characterise the friction phenomena of the DC04 steel sheet tested under dry friction and lubricated conditions.

2. Material and Methods

2.1. Material

Friction tests were performed on deep-drawing quality (DDQ) steel sheets with a thickness of 1 mm assigned to form complex-shaped drawpieces. The mechanical properties determined in a uniaxial tensile test (as given in Table 1) according to the EN ISO 6892-1: 2016 are the yield stress $R_{p0.2}$, ultimate strength R_m, strain hardening coefficient K, and strain hardening exponent n in the Hollomon equation. Three samples were tested for each orientation of the specimen: Along the rolling direction (0°) and transverse (90°) to the rolling direction (RD).

Table 1. Mechanical properties of the deep-drawing quality (DDQ) steel sheet.

Orientation	$R_{p0.2}$, MPa	R_m, MPa	K, MPa	n
0°	162	310	554	0.217
90°	163	312	530	0.210

2.2. Surface Characterisation

Measurement of the surface roughness parameters was carried out using a Talysurf CCI Lite 3D optical profiler equipped with a Nikon 5x/0.13 lens. The height resolution was 0.01 nm and the sampling interval in two orthogonal directions was 3.2 μm. The values of basic surface roughness parameters are listed in Table 2.

Table 2. Basic surface roughness parameters of the DDQ steel sheet *.

Sa (μm)	Sq (μm)	Sz (μm)	Sp (μm)	Sv (μm)	St (μm)
1.31	1.53	9.41	4.41	6.98	11.39

* *Sa*—Average roughness, *Sq*—Root mean square roughness parameter, *Sz*—zero-point peak-valley surface roughness, *Sp*—Highest peak of the surface, *Sv*—Maximum pit depth, *St*—Total height.

2.3. Methods

2.3.1. Bending under Tension Test

The bending under tension (BUT) test developed by Littlewood and Wallace [12] is attributed to the friction modeling on the edge of the die. The test consists of drawing the strip of metal around a cylindrical counter-sample (Figure 1). Figure 2 shows the configuration of the BUT test.

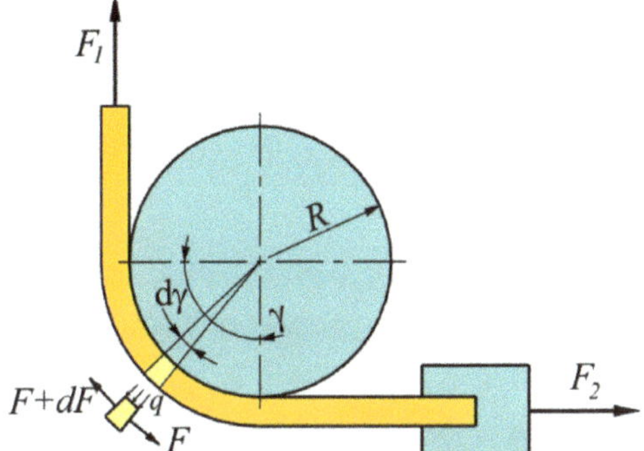

Figure 1. Forces acting on an elementary section of the strip.

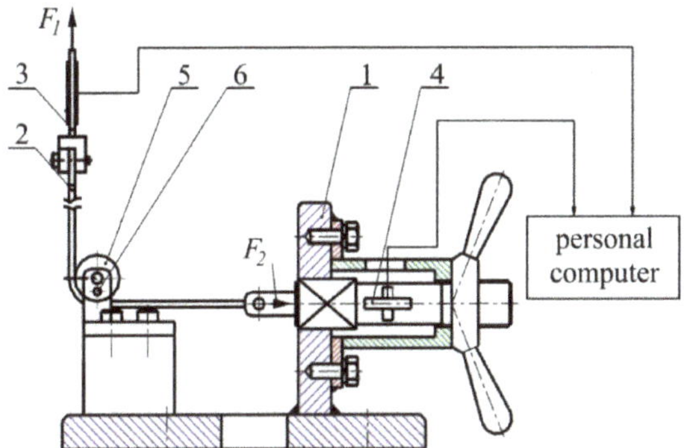

Figure 2. Schematic diagram of the measuring device for bending under tension (BUT): 1—Frame; 2—Specimen; 3—Vertical load cell; 4—Horizontal load cell; 5—Working roller; 6—Blocked pin.

The BUT test allows one to not only determine the value of the coefficient of friction, but also its change during the process of stretching the specimen over the roller. This change may be related to the change of the sheet topography and increased normal pressure as a result of sheet deformation [13] and a change under the contact conditions related to the strain hardening phenomenon. The presence of friction between the roller and the sheet results in $F_1 > F_2$ (Figure 1). The tensile forces $F1$ and F_2 are measured simultaneously during the test. Increasing the upper grip displacement increases the sample deformation until fracture. Assuming that the value of the coefficient of friction is the same for the entire contact surface and the wrap angle γ is constant during the deformation, the coefficient of friction can be determined from the equation:

$$\mu = \frac{2}{\pi} ln\left(\frac{F_1}{F_2}\right) \tag{1}$$

In the present study, specimens for the BUT test were carefully prepared to assure a constant width of 10 mm.

2.3.2. Draw Bead Test

The draw bead test (DBT), according to the concept which was developed by Nine [14], is attributed to friction modelling at the draw bead in sheet metal forming. The curvature of the metallic sheet passing through the draw bead model (Figure 3) is changed several times, the sheet is alternately bent and straightened. The idea behind the method is to provide the ability to separate the deformation resistance of the sheet from the friction. In the DBT, the values of the pulling force and the clamping force are measured when pulling the strip over fixed and rotatable rollers (Figure 3). Drawing the sheet metal over a set of rotatable rolls allows one to minimise the resistance due to friction. The pulling force in this case may be connected with the deformation resistance of the sheet. The system of fixed rollers represents the total resistance of sheet drawing through the draw bead.

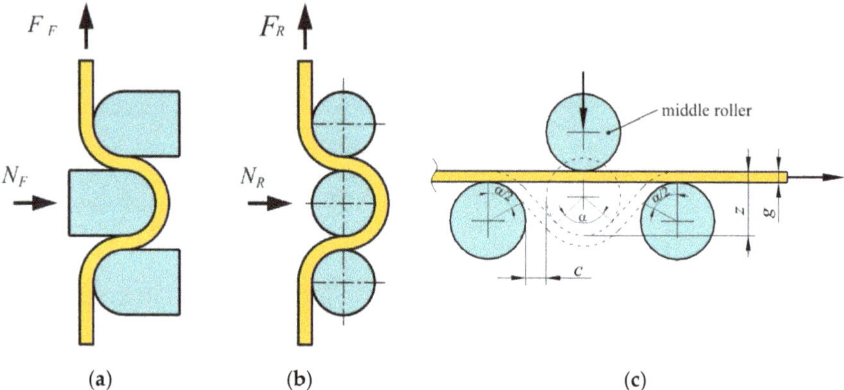

(a)   (b)   (c)

Figure 3. The concept of determining friction: (**a**) Fixed rollers, (**b**) rotatable rollers, (**c**) geometrical parameters of draw bead.

When the wrap angle α (Figure 3c) is not equal to 180° the value of the coefficient of friction is determined from the equation [15]:

$$\mu = \frac{sin\alpha}{2a} \frac{F_F - F_R}{N_F} \tag{2}$$

where α is the quarter contact angle of the actual engagement of the strip over the middle roll.

In general, the simulator (Figure 4) is equipped with four replaceable rollers.

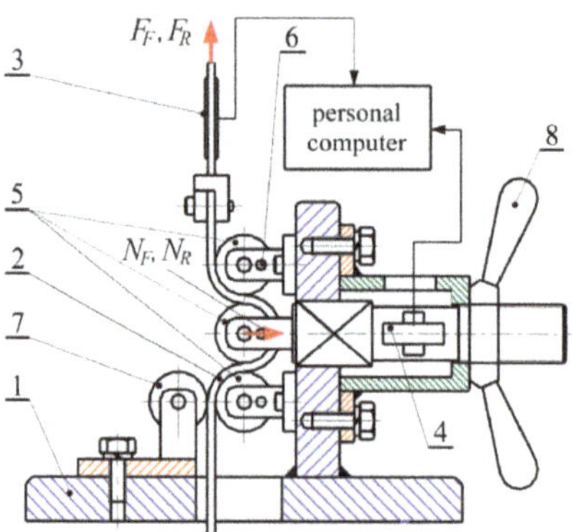

Figure 4. Schematic representation of the measuring device for DBT: 1—Frame; 2—Specimen; 3—Vertical load cell; 4—Horizontal load cell; 5—Working rollers; 6—Blocked pin; 7—Supporting roller; 8—Nut.

During the maximum displacement of the middle roller equal to its diameter, the wrap angle of the middle roller is not equal to 180°, because it is necessary to ensure adequate clearance c (Figure 3c) between the counter-samples to prevent the specimen from blocking. This is especially important when testing sheets with a high susceptibility to galling and low susceptibility to strain hardening. Inadequate clearance c may lead to specimen destruction. Excessive drawing resistance may occur and, consequently, the sample may even break at large wrap angles and when testing mild sheets. According to the investigations of Trzepieciński et al. [3] the clearance c (Figure 3c) is maintained at $c = g + g/2$. In the present study, the samples were prepared as strips having a width of 20 mm and about 400 mm in length, cut along the RD and the transverse direction of the sheet.

2.3.3. Strip Drawing Test

In the sheet forming process, the strip drawing test (SDT) is attributed to modelling the friction phenomenon between the punch and the die wall. This test consists in drawing a strip of metal placed between non-rotating counter-samples, usually flat or with a cylindrical shape. The occurrence of frictional forces on the two contact surfaces is conducive to achieving a greater accuracy of measurement of the coefficient of friction. As in most of the friction tests, a flat specimen several times wider than the thickness is drawn in a plane strain state. Parameters affecting the change in the frictional resistance are the counter-sample pressure force, lubrication conditions, sample drawing speed, surface roughness of the counter-samples, and temperature.

Figure 5 shows the configuration of the SDT. The sheet strip was placed between two fixed cylindrical rollers with an equal radii of 20 mm and roughness $Ra = 0.32$ μm and $Ra = 0.63$ μm. The value of the Ra parameter was measured along the generating line of the roller surfaces. The rollers were made of a cold worked tool steel.

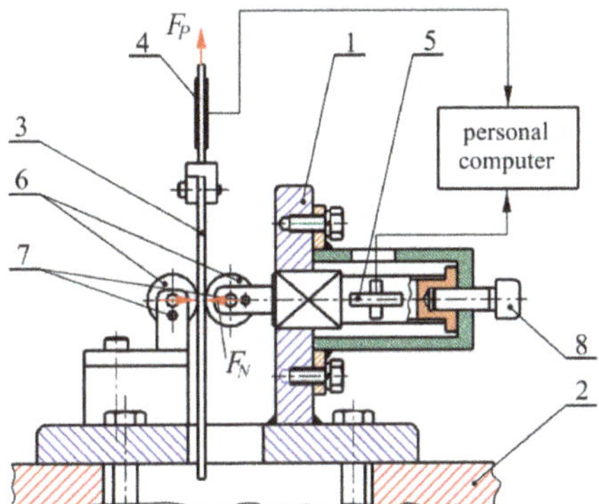

Figure 5. Schematic representation of the measuring device for strip drawing test (SDT): 1—Frame; 2—Support table; 3—Specimen; 4—Vertical load cell; 5—Horizontal load cell; 6—Working rollers; 7—Blocked pins; 8—Set bolt.

The samples were prepared as strips having a width of 20 mm and a length of about 260 mm and were cut along the rolling direction (0°) and transverse to the RD (90°). Values of both forces, the clamping force F_N, and the pulling force F_P, were constantly recorded using the electrical resistance strain gauge technique. The value of the coefficient of friction was evaluated during the test and the average value of the coefficient of friction was determined using Equation (3).

$$\mu = \frac{F_P}{2F_N} \tag{3}$$

The friction tests were carried out for two contact conditions: Dry and lubricated. Dry tests were carried out by degreasing the surface of the specimens using acetone while the machine oil L-AN 46 with a viscosity of 44 mm^2·s^{-1} at 40 °C was used for lubricated conditions. A sliding velocity of 0.001 m·s^{-1} was used, which is relatively high compared to the industrial values [3].

Each friction test has been repeated three times, and the average value of the coefficient of friction has been determined.

3. Results and Discussion

3.1. Bending under Tension Test

To describe how the coefficient of friction changes as the sample elongates, the coefficient of relative elongation is introduced:

$$\varepsilon_r = \frac{l_1 - l_0}{l_0} \cdot 100\% \tag{4}$$

where l_0 is the length of the sheet strip before deformation, and l_1 is the length of the sheet strip after deformation.

In Figure 6, results from the BUT test are presented. The value of the coefficient of friction increases as the relative elongation increases in both dry (Figure 6a) and lubricated (Figure 6b) conditions. The most linear relation between the relative elongation and the value of the coefficient of friction is observed for the rollers with a surface roughness of $Ra = 0.32$ μm and a specimen orientation of $0°$. The increase in friction with the relative elongation is apparently due to the roughening of the strip by plastic deformation.

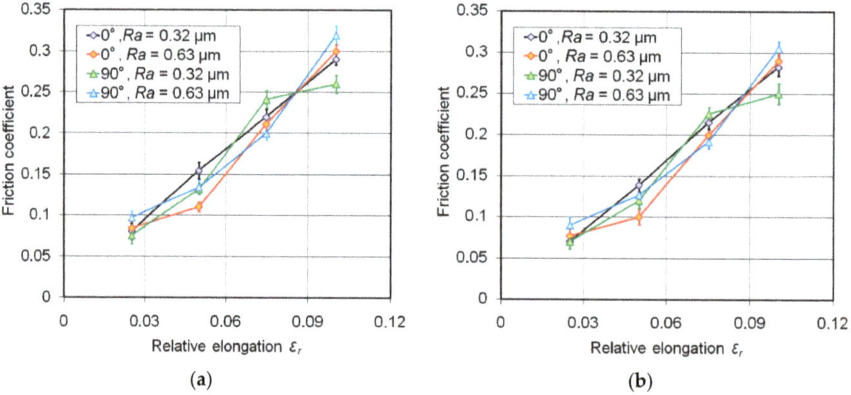

Figure 6. Effect of the relative elongation ε_r on the coefficient of friction for (**a**) dry and (**b**) lubricated conditions.

To examine the effectiveness of sheet lubrication the L_e-index was introduced, defined as follows:

$$L_e = \frac{\mu_d - \mu_o}{\mu_d} \cdot 100\% \tag{5}$$

where μ_d is the coefficient of friction determined under dry conditions and μ_o is the coefficient of friction determined in the presence of a lubricant.

The greatest efficiency of lubrication (ca. 7–13%) is observed for the small values of sheet deformation (Figure 7). Under these conditions, the surface pressures are the smallest. With the increase of pressure, the lubricant reduced the friction to a smaller extent. This may be a result of changes of metal sheet surface topography under the deformation process, which causes the real contact area to increase with the normal pressure. The real contact area depends on, for instance, the roughness parameters of the sheet metal, the inclination to strain hardening of roughness asperities, and the geometry of the contact surface. This makes it difficult to generalise and interpret the results obtained for the variation of the coefficient of friction.

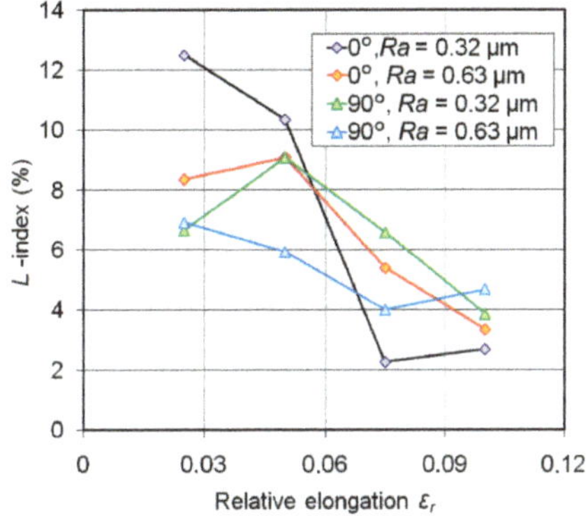

Figure 7. Effect of the relative elongation ε_r on the effectiveness of lubrication.

3.2. Draw Bead Test

The results of the DBT demonstrate that the value of the coefficient of friction tested under dry friction conditions is higher than in the case of the lubricated sheets (Figure 8). Under both friction conditions, the samples tested with the rollers with a surface roughness of $Ra = 0.32$ µm demonstrated a higher value of coefficient of friction compared to the samples tested with the rollers with a surface roughness of $Ra = 0.63$ µm. For all sheets an increase in the draw bead height z (Figure 3c) leads to an increase in the value of the coefficient of friction (Figure 8). The trend to an increase is similar for both roughness of rolls analysed. The bead height z directly influences the wrap angle of the middle roller. It should be noted that, according to the Amontons–Coulomb theory, the friction force does not depend on the contact area. However, the change in the wrap angle of the rollers has an effect on the character of plastic deformation of the specimen and in consequence on the value of pulling and normal forces. Increasing the draw bead height increases the plastic resistance to drawing the sample and increases nominal pressures. Under these conditions, there is an increase in the share of the mechanical interactions of the surface asperities, which increases the total resistance during drawing the sample through the draw bead. The opposite occurs when using rollers with a roughness of 0.32 µm.

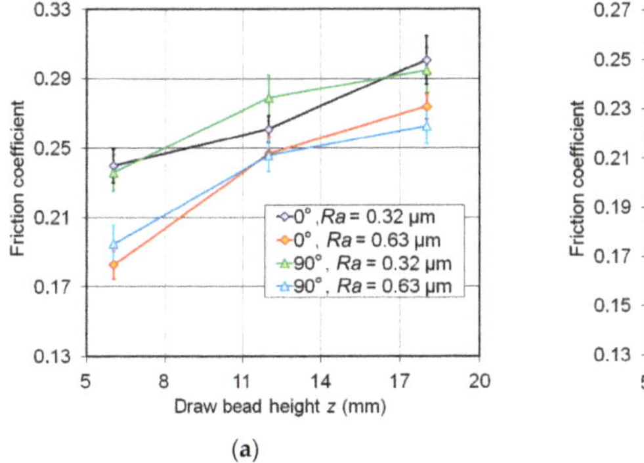
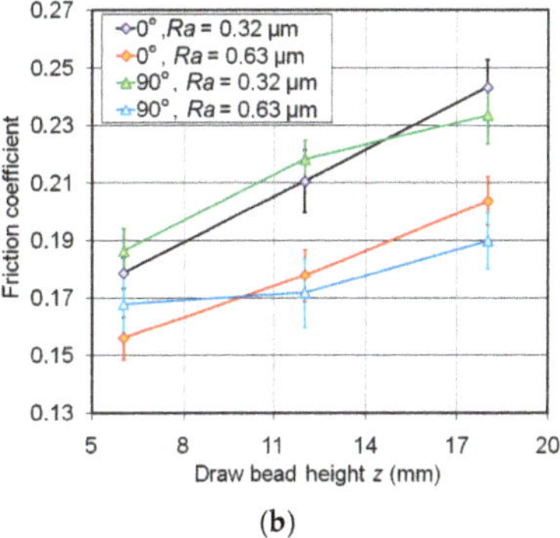

(a) (b)

Figure 8. Effect of the draw bead height z on the coefficient of friction for (**a**) dry and (**b**) lubricated conditions.

In Figure 9, the effect of the draw bead height on the effectiveness of lubrication is presented. An increase in the draw bead height results in a continuous decrease in the lubrication efficiency value. For draw bead heights of 12–18 mm a lubrication effectiveness was stabilised at the level of 19–22%. In the case of friction on rollers with a roughness of 0.63 µm, as the draw bead height z increases, the lubrication efficiency increases, but only up to z = 12 mm (Figure 9). Further increase of the draw bead height results in a decrease of the lubrication efficiency. The value of the coefficient of friction under oil lubrication conditions is reduced by more than 25% compared to dry conditions. In the case of draw bead heights of 12 mm and 18 mm, the L_e-index for the roller with Ra = 0.32 µm is lower than that with Ra = 0.63 µm. The rollers with a lower surface roughness are characterised by a small volume of surface roughness valleys, which trap the lubricant. The increased draw bead height led to an increase of the contact area between the rollers and the specimen surface. Although the roughness asperities of the rollers with Ra = 0.63 µm are more at risk of flattening, the resulting surface roughness valleys, which form reservoirs of the lubricant, are higher than in the case of the roller with a surface roughness of Ra = 0.32 µm.

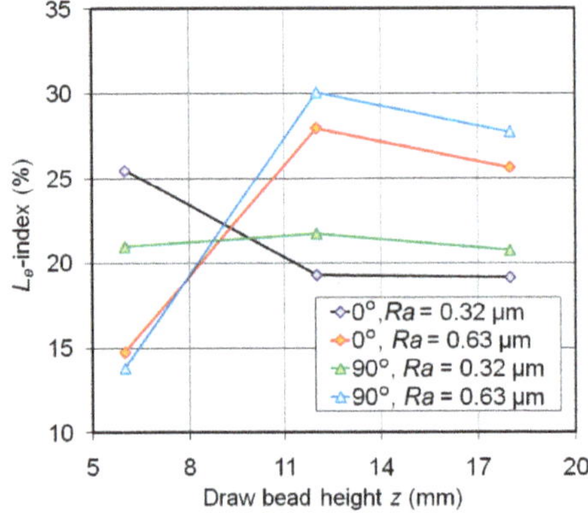

Figure 9. Effect of the draw bead height on the effectiveness of lubrication.

3.3. Strip Drawing Test

The value of the friction coefficient decreases as the contact pressure increases for both dry (Figure 10a) and lubricated conditions (Figure 10b). This trend is in general observed for both orientations. The contact pressure in the strip drawing test may be determined using equation [16]:

$$p_H = \frac{\pi}{4}\sqrt{\frac{F_c/w\cdot E^*}{2\pi R}} \tag{6}$$

where F_c is the clamping force, w is the specimen width, R is the counter-sample radius and E^* is a combined modulus of elasticity:

$$E^* = \frac{2E_1E_2}{E_2\left(1 - v_1^2\right) + E_1\left(1 - v_2^2\right)} \tag{7}$$

where E_1 and E_2 are Young's moduli of the sheet and counter-sample materials, respectively; v_1 and v_2 are Poisson's ratios of the sheet and counter-sample materials, respectively.

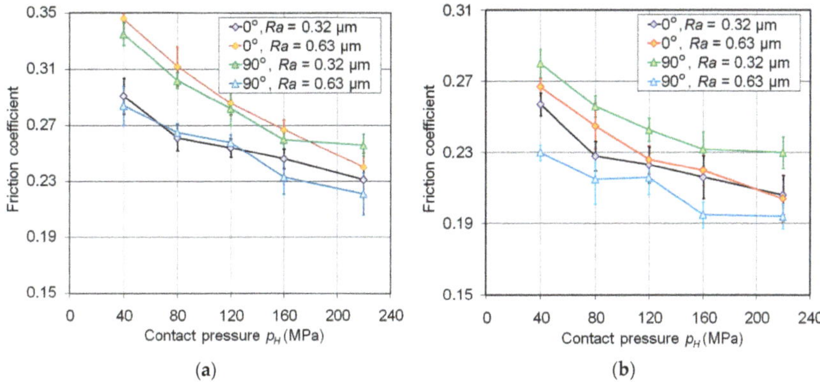

Figure 10. Effect of the contact pressure on the value of the coefficient of friction for (**a**) dry and (**b**) lubricated conditions.

The decreasing of the coefficient of friction with the increased contact pressure may be the result of the dependence of friction on the clamping (normal) force where beyond a certain load the relationship between the friction force and clamping force is nonlinear and the value of the friction coefficient is not constant. In other words, the coefficient of friction changes with a normal load. The nonlinear dependence between the clamping force and the value of the coefficient of friction was previously studied by the author and the results are reported elsewhere [3].

For both contact conditions, when the sheet was tested using a roller with $Ra = 0.32$ μm, the value of the coefficient of friction for samples cut along the RD (0°) was lower than in the case of samples cut perpendicular (90°) to the RD. In the case of the test conducted with the roller with $Ra = 0.63$ μm, the situation is the contrary. This may be attributed to the directional topography. In the range of contact pressures between 40 MPa and 120 MPa, with specimens tested under dry friction conditions, the values of the coefficient of friction for the following conditions: Orientation of 0° and $Ra = 0.32$ μm, and orientation of 90° and $Ra = 0.63$ μm, are similar. Furthermore, the values of the coefficient of friction for the following conditions: Orientation of 0° and $Ra = 0.63$ μm, and orientation of 90° and $Ra = 0.32$ μm, are also similar. The value of the coefficient of friction for lubricated conditions and a specimen orientation of 90° decreases as the surface roughness of the roller increases (Figure 10b). The contrary trend is observed for the specimen orientation of 0°. However, if the contact pressure is higher than 120 MPa, the values of the coefficient of friction for a specimen orientation of 0° and both of the surface roughness (0.32 μm and 0.63 μm) are within the range of the standard deviation.

At low interface pressures and both contact conditions, the decreased roughness develops less asperity contact, so the friction coefficient decreases. The high contact pressures decrease the ability of the lubricant to separate the contacting surfaces, so the friction coefficient stabilises. At higher interface pressures, the friction mainly depends on the surface roughness of the bodies in contact and the specimen orientation. In the case of the high surface roughness and dry friction conditions, the load pressure acts on the asperities, which results in a higher degree of surface flattening and increased friction. In this case the difference between the friction measured on the samples cut along the RD and transverse to the RD may be due to surface flattening. The high surface roughness ensures better lubrication due to the high volume of the valleys in the roughness profile. In the case of the high surface roughness and lubrication conditions, the load pressure acts on the asperities, which results in a higher degree of surface flattening and increased friction.

It is well known that effective lubrication results in low friction, which reduce the loads imposed on the workpiece. The presence of an effective lubricant film also reduces the amount of direct contact of roughness asperities between the tooling and workpiece. This effect appears to be due to the manner in which the lubricant is captured by roughness on the surface of the sheet.

As shown in Figure 11, the L_e-index value has a linear relationship with the value of contact pressure. In the case of the two combinations of rollers and specimen orientations: Orientation of 0° and Ra = 0.63 μm, and orientation of 90° and Ra = 0.32 μm, the value of the L_e-index clearly decreases with an increasing value of contact pressure p_H. For a sample orientation at 0° tested with the roller characterised by Ra = 0.32 μm, on the other hand, a similar effectiveness of lubrication is observed in the whole range of the pressure values, approximately 12%. In the case of the higher surface roughness of the contact bodies, a higher volume of lubricant may be trapped in the surface pits and consequently the effectiveness of lubrication is higher. This is confirmed by the data in Figure 11. The lubrication is more effective for rollers characterised by Ra = 0.63 μm than in the case of rollers with a roughness of Ra = 0.32 μm.

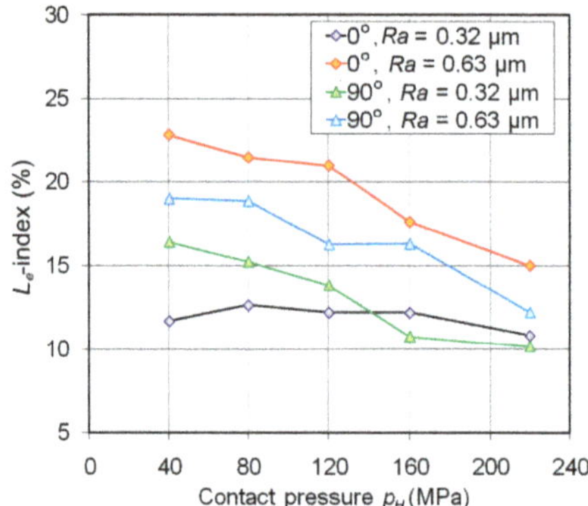

Figure 11. Effect of contact pressure on the value of the L_e-index.

The coefficient of friction is the result of the mutual interaction between the rough surfaces of the tool and the sheet metal to be formed whose topography is constantly changing. In each of the friction tests, there are different kinematics of the mutual movement of the surface in contact and the sheet is subjected to different pressure values. In addition, a very important factor is the difference between the mechanical strength of the tool and the sheet. This determines the contribution of the mechanism of surface roughening in total friction. The mechanical strength of the tool material is always greater than the strength of the material to be formed. In the case of the low tool roughness, there is a reduced effect of flattening the surface asperities of the sheet metal by the roughness asperities of the tool surface. However, under these conditions there is an increased real contact area and the valleys in the topography of the tool surface retain less lubricant. When the surface roughness of the tool is high then it is obvious that the lubricant may be delivered in sufficient quantities to the contact interface. However in this case, the friction is mainly influenced by the ploughing of the sheet surface by the hard asperities of the tool surface.

4. Conclusions

This paper presents and discusses the results of friction tests of deep drawing steel sheets in various friction process conditions. The following conclusions are drawn from the research:

- the specimens tested in the BUT test exhibit a tendency to an increase in the value of the coefficient of friction with the increasing relative elongation; this relationship is observed for both the specimen orientations and contact conditions analysed; the lubricant reduced the coefficient of friction, but the intensity of its action in the test results analysed depends on the specimen orientation and the value of the surface roughness of the counter-samples; for both sample orientations, the greatest efficiency of lubrication is observed for the small values of sheet deformation,

- in the DBT, the most pronounced differences in friction were due to the surface roughness, while the orientation showed to have only a minor influence; moreover, an increase in the draw bead height leads to an increase in the value of the coefficient of friction,
- in the SDT, the value of the coefficient of friction decreases as the contact pressure increases for both dry and lubricated conditions; the effectiveness of the reduction of the coefficient of friction by the lubricant clearly depended on the specimen orientation and surface roughness of the counter-samples.

Differences in the coefficient of friction obtained from various tribological tests indicate a strong dependence of the kinematic conditions of the process on parameters directly related to the characteristics of the friction process.

Funding: This research received no external funding.

Conflicts of Interest: The author declares no conflict of interest.

References

1. Altan, T.; Tekkaya, A.E. *Sheet Metal Forming Processes and Applications*; ASM International: Materials Park, OH, USA, 2012.
2. Nielsen, C.V.; Bay, N. Overview of friction modelling in metal forming processes. *Procedia Eng.* **2017**, *207*, 2257–2262. [CrossRef]
3. Trzepieciński, T.; Bazan, A.; Lemu, H.G. Frictional characteristics of steel sheets used in automotive industry. *Int. J. Automot. Technol.* **2015**, *16*, 849–863. [CrossRef]
4. Masters, L.G.; Williams, D.K.; Roy, R. Friction behaviour in strip draw test of pre-stretched high strength automotive aluminium alloys. *Int. J. Mach. Tools Manuf.* **2013**, *73*, 17–24. [CrossRef]
5. Trzepieciński, T.; Lemu, H.G. Effect of computational parameters on springback prediction by numerical simulation. *Metals* **2017**, *7*, 380. [CrossRef]
6. Trzepieciński, T. 3D elasto-plastic FEM analysis of the sheet drawing of anisotropic steel sheet. *Arch. Civ. Mech. Eng.* **2010**, *10*, 95–106. [CrossRef]
7. Bay, N.; Olsson, D.D.; Andreasen, J.L. Lubricant test methods for sheet metal forming. *Tribol. Int.* **2008**, *41*, 844–853. [CrossRef]
8. Menezes, P.L.; Kumar, K.; Kishore; Kailas, S.V. Influence of friction during forming processes—A study using a numerical simulation technique. *Int. J. Adv. Manuf. Technol.* **2009**, *40*, 1067–1076. [CrossRef]
9. Evin, E.; Tomáš, M.; Výrostek, M. Verification the numerical simulation of the strip drawing test by its physical model. *Acta Mech. Slovaca* **2016**, *20*, 14–21. [CrossRef]
10. Trzepieciński, T.; Lemu, H.G. Proposal for an experimental-numerical method for friction description in sheet metal forming. *Stroj. Vestn. J. Mech. Eng.* **2015**, *61*, 383–391. [CrossRef]
11. Üstünyagiz, E.; Nielsen, C.V.; Christiansen, P.; Martins, P.A.F.; Bay, N. Continuous strip reduction test simulating tribological conditions in ironing. *Procedia Eng.* **2017**, *207*, 2286–2291. [CrossRef]
12. Littlewood, M.; Wallace, J.F. The effect of surface finish and lubrication on the fictional variation involved in the sheet-metal-forming process. *Sheet Met. Ind.* **1964**, *41*, 925–1930.
13. Zhang, S.; Hodgson, P.D.; Duncan, J.L.; Cardew-Hall, M.J.; Kalyanasundaram, S. Effect of membrane stress on surface roughness changes in sheet forming. *Wear* **2002**, *253*, 610–617. [CrossRef]
14. Nine, H.D. Drawbead forces in sheet metal forming. In *Mechanics of Sheet Metal Forming*, 1st ed.; Koistinen, D.P., Wang, N.M., Eds.; Springer: Boston, MA, USA, 1978; pp. 179–211.
15. Nanayakkara, N.K.B.M.P.; Kelly, G.L.; Hodgson, P.D. Determination of the coefficient of friction in partially penetrated draw beads. *Steel Grips* **2004**, *2*, 677–680.
16. Ter Haar, R. Friction in Sheet Metal Forming, the Influence of (Local) Contact Conditions and Deformation. Ph.D. Thesis, Universiteit Twente, Enschede, Holland, 1996.

Review

Recent Developments and Trends in Sheet Metal Forming

Tomasz Trzepieciński

Department of Materials Forming and Processing, Rzeszow University of Technology, al. Powst. Warszawy 8, 35-959 Rzeszów, Poland; tomtrz@prz.edu.pl; Tel.: +48-17-743-2527

Received: 27 April 2020; Accepted: 9 June 2020; Published: 10 June 2020

Abstract: Sheet metal forming (SMF) is one of the most popular technologies for obtaining finished products in almost every sector of industrial production, especially in the aircraft, automotive, food and home appliance industries. Parallel to the development of new forming techniques, numerical and empirical approaches are being developed to improve existing and develop new methods of sheet metal forming. Many innovative numerical algorithms, experimental methods and theoretical contributions have recently been proposed for SMF by researchers and business research centers. These methods are mainly focused on the improvement of the formability of materials, production of complex-shaped parts with good surface quality, speeding up of the production cycle, reduction in the number of operations and the environmental performance of manufacturing. This study is intended to summarize recent development trends in both the numerical and experimental fields of conventional deep-drawing, spinning, flexible-die forming, electromagnetic forming and computer-controlled forming methods like incremental sheet forming. The review is limited to the considerable changes that have occurred in the SMF sector in the last decade, with special attention given to the 2015–2020 period. The progress observed in the last decade in the area of SMF mainly concerns the development nonconventional methods of forming difficult-to-form lightweight materials for automotive and aircraft applications. In evaluating the ecological convenience of SMF processes, the tribological aspects have also become the subject of great attention.

Keywords: electromagnetic forming; finite element method; flexible-die forming; flow-forming; incremental sheet forming; mechanical engineering; metal forming; numerical modeling; plastic working; sheet metal forming; solid granular medium forming; spinning; warm forming

---

1. Introduction

Sheet metal forming (SMF) techniques are widely used in many industries to produce final-shaped components from a workpiece. In an SMF process, a thin piece of metal sheet is stretched into a desired shape by a tool without wrinkling or excessive thinning. In the past decade, methods for forming high-strength material with low plasticity and difficult-to-form metals have been developed for cold, warm and hot forming conditions [1,2].

The mechanical properties of the metallic sheet are an important factor and inadequate consideration of this factor in the design of SMF manufacturing processes causes buckling, excessive thinning, tearing and wrinkling of the components. Other factors that affect the final shape of the components include the geometry of the tool (i.e., punch-to-die clearance, die and punch radii) [3,4], friction conditions (i.e., dry or lubricated contact, lubricant type, contact pressure) [5–9], technological parameters (i.e., forming temperature, forming speed,) [10–12], properties of tool material [13–15] and initial surface topography [16–18]. The existence of friction forces at the workpiece–tool interface determines the nonuniformity of the sheet deformation and quality of the surface of the final part [19]. Considering the prediction of formability, numerical models are widely employed in both industry and research development. Ma and Sugitomo [20] developed a LS-DYNA customized friction subroutine and verified it experimentally. This considered the change of COF with sliding distance, sliding velocity, contact pressure, plastic strain, frictional work and temperature. Ma and Sugimoto [21] suggest that lubrication phenomena can be positively changed during pulse–servo motion. Simulation results using a nonlinear friction model for a pulse servo motion agreed very well with experimental measurement for evaluating forming cracks. A comprehensive review of developments and trends in friction testing for conventional sheet metal forming and incremental sheet forming has been discussed by Trzepiecinski and Lemu [22].

In recent years there has been a dynamic development of two- and three-dimensional numerical modeling of the sheet metal forming (SMF) processes using the finite element method (FEM) [23–25], boundary element method (BEM) [26], finite difference method (FDM) [27], computational fluid dynamics (CFD) [28,29], finite volume method (FVM) [30], neural networks [24], multi grid and mesh free methods [23], crystal plasticity finite element (CPFEM) [2,31,32], discrete element method (DEM) [33,34], extended finite element (XFEM) [35], an arbitrary Lagrangian–Eulerian (ALE) [36,37], cellular automata (CA) [38,39] and fast Fourier transformation (FFT) [40].

During the last decade the above mentioned approaches to the numerical modeling of sheet metal forming operations have been primarily been focused on the development of optimization procedures that could be applied to complex processes, on improving the description of material behavior in both multiphysics models [38–41] and a multiscale modeling framework based on the microscale crystal plasticity theory and on a macroscale element-free methodology for computational simulation of polycrystalline metals [42,43].

Ablat and Quattawi [44] discussed different methods of solution in the simulation of SMF. According to Makinouchi et al. [45] numerical formulations of SMF can be classified into three main categories, which are, static explicit, static implicit and dynamic explicit. Numerical modeling of the forming of new metallic materials, i.e., multiphase steels, requires full-field models [29]. Full-field homogenization refers to the spatially resolved solution of a representative volume element (RVE) by means of FEM [46,47] or spectral methods using a fast Fourier transform approach (FFT) [37]. In this case a microstructure morphology as well as phenomena occurring during forming are precisely simulated at various length scales [2].

To overcome the limitations of FEM in modeling microstructures [48,49], numerical simulation techniques have been developed for incorporating material behavior on different length scales. Computational materials science (CMS) with the emerging concept of digital materials representation (DMR) [50] has been intensively investigated during the last decade. This concept provide a digital model of the microstructure where all important features are represented explicitly and can offer a support to the description of a material's behavior during the forming of new products with special in-use properties [2,51]. In 2020 Han et al. [52] proposed a microstructure-based multiscale modeling of large strain plastic deformation by coupling a full-field crystal plasticity-spectral solver with an implicit finite element solver. The model which was developed takes both dislocation density and phenomenological hardening law into account and is suitable for modeling materials with complex microstructural characteristics (e.g., grain morphology, multiple phases and textures).

During the last decade, the coupling of individual computational methods to provide both multiscale and multiphysics responses has proved to have enormous predictive capabilities. In order to analyze the sheet metal forming processes, combinations of a range of CPFEM and XFEM numerical techniques have been applied. CPFEM can explicitly consider lattice rotation, and thus capture the geometric softening effect which is the main mechanism of shear band formation in strain hardening metals under quasi-static loading [2,28]. The CPFEM formulations for polycrystalline materials could be realized in macro-, meso- and microscales, classified by the size represented by each element in the FEM. The XFEM technique is an inexpensive, powerful, secure and time-saving numerical formulation for the analysis of crack problems [32].

The issues related to the increased predictive capability of numerical simulations with simultaneous reduction in computational time have also gained much attention in the area of the modeling of sheet metal forming [53,54]. To overcome the limitations of classical mesh numerical techniques which require significant computational effort in remeshing steps, meshless or mesh-free methods have been developed. These methods require only nodal data without explicit connectivity between nodes [21,55]. In meshless methods, i.e., SPH, the interpolation accuracy is not significantly affected by the nodal distribution, so interpolation is free of the mesh requirement [56,57].

Numerical modelling is used to predict material flow [58], stress distribution [59,60], deformation [58,61], temperature distribution [62,63], prediction of phase transformations [64,65], springback [66,67], sheet thickness change [68] as well as for determining forming forces [69,70], locations of potential wrinkling and cracking [66,71] and for predicting forming limit diagrams (FLDs) [72,73]. Numerical simulations coupled with advanced 3D-adaptive remeshing procedures and fully coupled damage constitutive equations play an important role in the control and optimization of material flow in SMF. Digital material representation concepts are also created to evaluate the influence of local microstructural features in the form of twins on material behavior [74]. Numerical simulation has become an indispensable tool for the prediction of crack occurrence during sheet forming and fracture behavior under strong impact loading when designing and fabricating automotive parts made of high-strength steels. In the past decade, various approaches to the prediction of the initiation and propagation of forming cracks have been presented and several fracture criteria have been proposed. Ma et al. [75] employed the plastic strain criterion depending on the triaxial nature of stress in the prediction of fracture in stamping parts using the simple tensile test. They also applied the digital image grid method (DIGM) to measure the strain localization behavior and local strain distribution. Based on DIGM, a new method for the identification of the ductile damage limit of steel sheets was proposed with the aid of the historical path of nonlinear local strain and local fracture strain that had been measured. The commonly used Cockroft damage criterion, in which the plastic strain and the maximum principal stress are integrated, is an effective method to predict fracture under various loading conditions [76]. In a later study, Ma et al. [77] combined the measured transient displacement field with the FEM and a measurement-based FEM (M-FEM) was developed for the computation of the distribution of the local stress and strains, and the accumulation of ductile damage in a tensile test piece.

High-strength steel is increasingly finding uses in automotive body parts, whose properties tend to increase shape deviations (springback) after the forming stage. Many countermeasures are commonly used, including the use of lock beads [78,79], die-shape compensation [80,81] and the improvement of forming process design [82,83]. The numerical prediction of elastic deformations of metallic sheets is also vitally important. To meet their needs, the accuracy of springback prediction has been significantly improved over the last decade [84], and the sheet metal forming simulation system JSTAMP has been providing advanced capabilities [85–87]. The compensation capabilities in JSTAMP are powerful methods to compensate the stamping tools, and the compensated CAD surfaces of the stamping tools can be directly exported from JSTAMP to CAM for machining [87]. In a similar manner to the prediction of springback, surface defects must be also controlled in the stamping process in order to make high quality outer panels of auto bodies. Measuring methods based on stoning and light reflection and the three-point curvature method are commonly used to detect surface defects. The cheapest and most convenient method is the stoning method supported by the JSTAMP system which does without any expensive measuring devices except a stone block [88].

This article summarizes recent trends in both the numerical and experimental contributions of SMF developed in the last decade. The main techniques of SMF, i.e., conventional deep-drawing, spinning, shear forming, flow forming, electromagnetic forming (EMF), flexible-die forming, electro-hydraulic forming, shear forming, solid granular medium forming, and incremental sheet forming are addressed.

2. Methods

This systematic review of the latest progress in the area of SMF was prepared following the PRISMA guidelines [89]. To meet the PRISMA checklist, the following assumptions were made:

- To fulfil the goal of this study, international databases were explored including (in alphabetical order: Bielefield Academic Search Engine, DOAJ Directory of Open Access Journals, eLibrary;ru, GoogleScholar, INGENTA, ScienceDirect, Springer, Web of Science, WorldCat, WorldWideScience;
- Year restrictions: the databases were explored for the period of the past decade with special attention to the 2015–2020 period;
- The English language is selected as the main source of review;
- Duplicated articles found in different databases were not considered;
- Papers which did not fit the goal of this study were excluded;
- No search engines were used; papers were reviewed manually;
- References available in the articles found were also considered.

3. Conventional Sheet Metal Forming

3.1. Sheet Microforming

Micromanufacturing technologies have been developed to meet an increased tendency to product miniaturization, and among these, microforming is a promising method of producing microparts. Engel and Eckstein [90] defined microforming as the fabrication of metallic microparts or structures with at least two dimensions in the submillimeter range for microsystems or microelectromechanical systems (MEMS). Due to the size effect, there are many issues including deformation behavior (anisotropy, flow stress, limit strains), tools, machines, deformation defects in micro forming and damage accumulation which could be barriers to the production of micro- and mesoparts (Figure 1) [91,92]. In the case of material properties, grain size is the most important parameter which limits part geometry and limits strains in microforming. In the past decade many investigations have been devoted to studying the size effect on the fracture mechanism [93], plastic deformation [94–97], formability of material in microforming [98], Hall–Petch effect [99], elastic recovery [100,101], fracture behavior [102,103], surface roughness [104,105] and the hardness of parts [105,106]. The sheet microforming process is recognized as one of the most efficient processes for producing microparts from sheet metals.

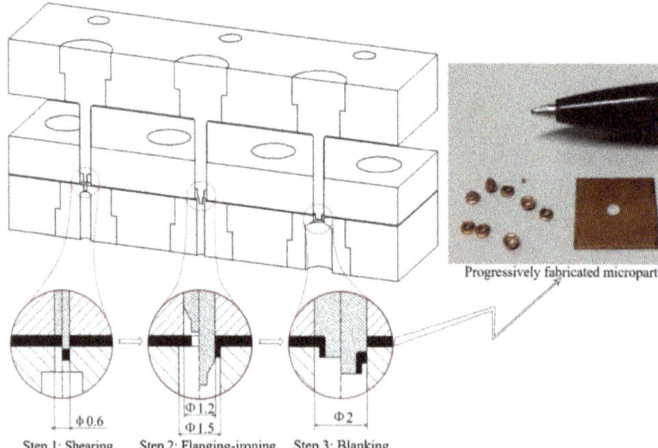

Figure 1. An example of the experimental die layout and finished microparts (reproduced with permission from [107]; copyright © 2018 Elsevier B.V.).

Parallel to conventional sheet microforming, micro-hydromechanical deep drawing [108] is widely investigated. The hydroforming pressure and rate of applying pressure have a significant effect on formability [109]. The increase of hydroforming pressure and stamping force yields a better surface finish. In rubber pad micro forming (RPM), the rubber pad is compressed and then deformed to push the blank into the die cavity. In this method only a single rigid tool is needed, and it avoids the punch cavity misalignment problem. Several researchers employed the RPM to study the effect of tool geometry, interfacial friction and hardness on the deformation process of embossing microchannels [107,110]. The dimensional accuracy of RPM parts is difficult to control due to the large elastic deformation of the pressure carrying medium, viz., the rubber [107]. Compared to conventional SMF processes involving rigid dies, microsheet formation by the rubber-pad forming process has many advantages [111]: (i) the process only uses one rigid die, (ii) there is no problem in precise orientation of the soft punch in relation to the rigid die, (iii) the cost and processing time can be greatly reduced compared to conventional SMF. Many researchers [110,112,113] have successfully employed rubber-pad-forming (Figure 2) to fabricate high-performance proton exchange membrane fuel cells.

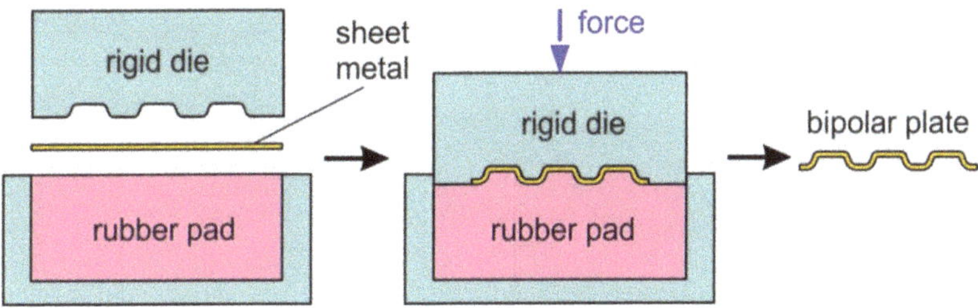

Figure 2. Diagram of the rubber-pad-forming process.

Laser shock microforming is an alternative method of microforming and is conceived as a non-thermal method of laser forming for thin metal sheets using the shock wave induced by laser irradiation (Figure 3a) [114–116]. High velocity forming processes have many potential advantages including higher resistance to wrinkling and higher forming limits. The pressure distribution is a crucial issue in high velocity forming processes. When the forming pressure is not uniform, puckers will be generated by uneven material flow velocity [117]. The plastic deformation induced by the shock wave and the direct plasma pressure applied on the material generate a residual stress distribution in the material finally leading to its bending (Figure 3b). The final bending of the specimens can be controlled over a relatively wide range by a stable quasi-proportional relationship to the number of pulses applied, and water confinement for the plasma leads to the ability to increase the pressure around 10 times and the final deformation has a significant increase.

In the case of thin metal sheets, the uneven pressure will promote the onset of localized necking instability which will lead to fracture. Therefore, Shen et al. [118] proposed a mechanism for a rubber-induced smoothing effect on the confined laser shock in order to smooth the laser shock wave (Figure 4). It has been suggested that the smoothing effect is mainly due to the radial expansion of the plasma cloud on the rubber surface.

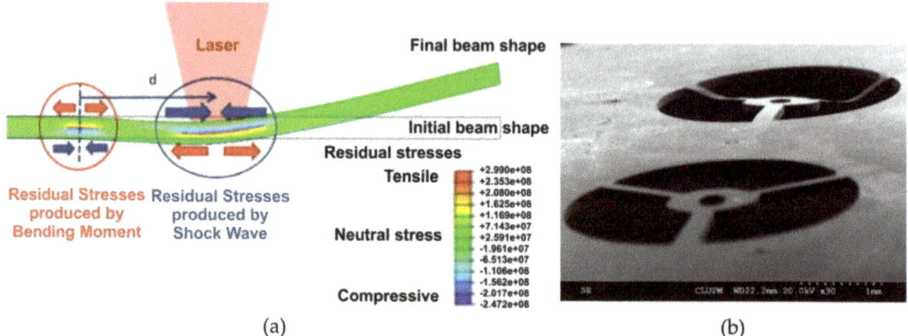

(a) (b)

Figure 3. (**a**) schematic diagram of deformation and stresses for a single-pinned strip and (**b**) scanning electron microscopy micrograph of a three-bridge actuator (reproduced with permission from [114]; copyright © 2011 Elsevier B.V.).

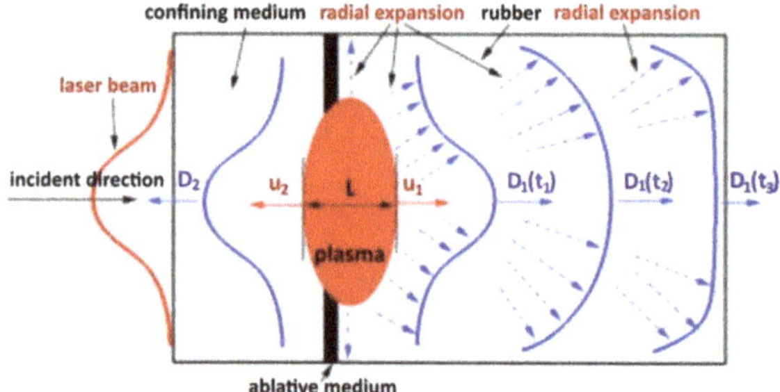

Figure 4. Mechanism of rubber-induced smoothing effect (reproduced with permission from [118]; copyright © 2015 Elsevier B.V.).

New methods and processes for a mastered mass production of micro parts which are smaller than 1 mm have been described in the SPB 747 report of the Collaborative Research Center of the University of Bremen (Germany) on "Micro Cold Forming". This report is focused on the description of micro forming processes [119], process design [120], tooling [121], and quality control and the characterization [122] of high-precision micro parts.

3.2. Warm/Hot Forming

Warm forming is usually carried out at $0.35T_m < T < 0.55T_m$, where T_m is the melting temperature of the metal [123]. It is important to distinguish warm forming from hot forming. Hot forming is carried out at temperatures $T > 0.55T_m$. In hot forming the temperatures exceed the melting temperature of the material and this allows simultaneous recrystallization, which controls the refinement of grain size [124,125]. In this way warm forming improves the formability of a material by lowering the yield stress. In comparison to hot forming processes, warm forming requires higher forces for deformation because of the greater material flow stress [126–128].

The quenching operation in the hot stamping process has a significant influence on the phase transition and mechanical properties of the hot-stamping steel. A proper quenching technique is quite important to control the microstructure and properties of an ultra-high-strength hot-stamping steel [129]. Kayan and Kaftanoglu [130] proposed non-isothermal deep drawing which is applied to DP600 HSLA and IF steels in elevated temperature conditions. They found that the process increases the LDR and there is no significant change in the microstructure of the material due to warm forming.

The application strategy that was developed can solve the problems encountered in applications at ambient conditions, such as dimensional instabilities due to springback and high residual stresses.

The formability of the material increases with an increase in the forming temperature and is also affected by the microstructure, which changes according to the temperature at which the material is deformed [131–133]. In the case of austenitic stainless steel, martensite formation is not only affected by temperature, but also influenced by the rate at which the material is deformed [125].

Warm forming is used for stainless steels [134], aluminum alloys [135], magnesium alloys [136] and dual-phase steels [137] to reduce springback and the forming forces [42]. The springback phenomenon is one of the main problems associated with the assurance of the desired shape and dimensional accuracy of the formed part after its removal from the stamping tool [138,139].

Compared with conventional SMF processes, forming carried out at an elevated temperature requires an initial increase of the blank temperature before the forming stage [140–142]. Two main strategies are used for the heating process. The whole blank is heated in the furnace to receive a uniform temperature in the whole blank or the workpiece is heated locally using different techniques [143]. The whole blank heating strategy can be accomplished externally in an oven or internally through conduction from a heated tool. The strategy commonly used in industry is uniform heating of the blank [144]. The effect of the heating method and temperature on the formability of metals was widely studied by many authors [125,145,146].

A robotized assembly stand with induction heater was proposed to minimize the loss of heat into the environment at the stage of transferring the sheet from the heater to the stamping die (Figure 5) when warm forming a 17–4PH stainless steel turbine engine strut [123] (Figure 6a). The effect of overheating of the blank on the dimensional deviation of the drawpiece due to the transfer of the blank from the heater to the stamping die using an industrial robot is also considered. The temperature of the tools during the experiments was stabilized by cooling channels located in the stamping die (Figure 6b).

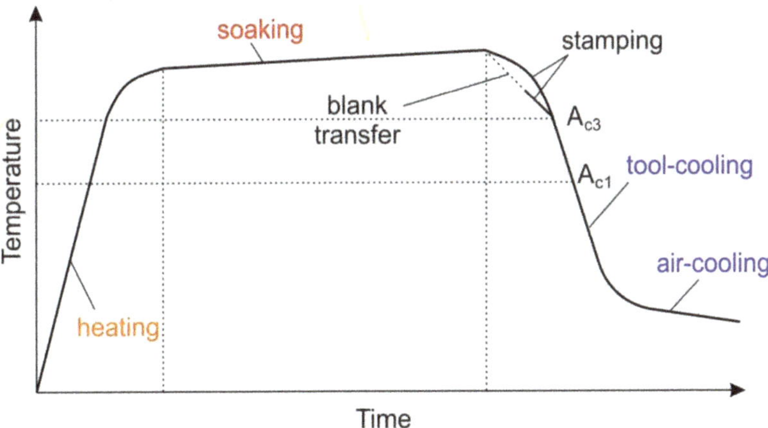

Figure 5. Temperature change versus time in the warm stamping process.

During forming of a sheet heated to a temperature up to 600 °C, a very large springback of the drawpiece material is observed. The forming temperature that guaranteed receiving the component with the required shape deviations appropriate to the forming temperature was about 680 °C. This led, however, to the blank overheating and reaching a temperature of 820 °C.

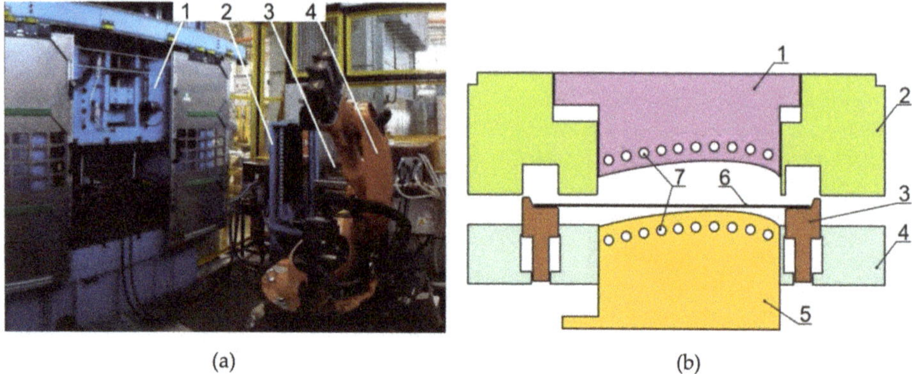

Figure 6. (**a**) Industrial stand for warm SMF (1–press, 2–workpiece store, 3–induction heater, 4–robot arm) and (**b**) scheme of stamping die (1–die, 2–upper blank holder, 3–positioner/knock-out, 4–lower blank holder, 5–punch, 6–workpiece, 7–cooling channels) (reproduced with permission from [123]; copyright © 2018 Elsevier B.V.).

In order to increase the collision-safe performance and reduce weight, the use of ultra-high-strength steel (UHSS), multiphase transformation-induced plasticity [TRIP] steels and advanced high-strength steels (AHSS) has gained attention in the automobile industry [147,148]. Besides the AHSSs, dual-phase (DP) steels consisting of martensite and ferrite phases are characterized by a proper balance between formability and high strength, and thus there is a growing interest in this type of sheet metal in the automotive industry. Due to the high-strength properties of DP steels, forming at elevated temperatures is required [149]. The thermal-mechanical behavior of DP590 sheet metal examined with uniaxial tensile tests as well as biaxial tensile tests at elevated warm-forming temperatures (20–190 °C) can be found in [149]. A dynamic strain aging effect was observed via thermal-uniaxial tensile tests and the Swift hardening model with temperature-dependent parameters. Inflow resistance of the flange surface of the UHSS parts increased due to rapid cooling caused by contact with the tool, and as a result formability in hot stamping deteriorated in comparison with cold forming conditions when the forming speed was low [150].

Warm forming at temperatures from 200 °C to 600 °C has several advantages compared with cold forming at room temperature: better shape accuracy, better stretch flange formability and lower press load. The results of the stretch formability of uncoated high-strength steel sheets and galvannealed HSS sheets in warm forming assessed by spherical stretch forming tests at temperatures from room temperature to 600 °C can be found in [151]. It was concluded that the COF has a significant influence on formability in the low temperature range of 200–400 °C. Springback of HSSs may be much reduced by a lower forming speed. These phenomena directly correspond to the material's viscoplastic behavior at elevated temperature [152].

Nowadays, aluminum alloys are considered desirable for the automotive and aerospace industries because of their superior corrosion resistance, excellent high-strength to weight ratio, recyclability and high weldability [126]. The application of forming at elevated temperatures to age-hardening aluminum alloys requires knowledge of the interactions between temperature, degree of deformation, precipitation kinetics and exposure time [153,154]. The precipitation that can happen during the warm drawing process influences the drawing force and the maximum drawing force depending on the moment the precipitation takes place. A comprehensive review of theoretical and numerical methods (Figure 7) used for prediction of the formability of lightweight materials for sheet metal forming applications can be found in [155]. The increase in plastic strain during the two-stage warm drawing process resulted in a gradual increment in cup wall strength as substantiated with the microhardness results.

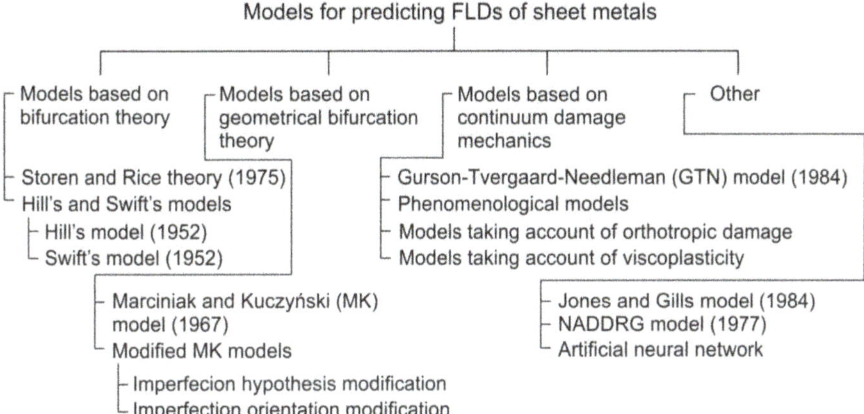

Figure 7. Numerical and theoretical models used for the prediction of formability.

Warm forming is an effective solution to improving formability for both heat-treatable and non-heat-treatable aluminum alloys. The investigations on this topic in the last decade are mainly focused on the development of effective numerical codes for thermomechanical FE-based analyses [156,157], design of warm forming tools [123,141], modeling of the mechanical behavior of materials at elevated temperatures [139,158–160] and the development of alternative warm forming methods [161].

Magnesium, with a better recyclability and higher strength-to-density ratio, has become an alternative to aluminum and steel, especially in the automotive industry which is looking for ways to limit the emission of greenhouse gases and meet the demand by customers for fuel efficient vehicles [154,162]. The room temperature formability of magnesium alloys strongly depends on deformation twinning and texture [163]. Although magnesium alloy sheets have low ductility at room temperature due to the small number of slip systems, it is well-known that formability at temperatures around 200 °C can be drastically improved [164]. As the temperature increases, prismatic and pyramidal slip systems can be activated, thus improving the formability [165]. As shown by Maksoud et al. [166], as the forming temperature of magnesium alloy increased from 300 °C to 400 °C, the grain size quickly increased, and performance decreased.

The formability of sheet metal not only depends on the material's properties, but also on the friction on the tooling/workpiece interface [167,168]. According to the study by Abu-Farha et al. [169], the fissure orientation in the specimens tested using the elevated-temperature pneumatic stretching test apparatus (Figure 8) was also affected by the rolling direction and relative size of major strain and minor strain.

In recent years, many researchers have looked at the process of deep-drawing laminated sheets [125]. However, most of the investigations carried out on laminated sheets were conducted during the deep drawing process at room temperature [170,171]. A comprehensive investigation on the warm deep-drawing process (Figure 9) in AA1050/St304 and AA5052/St304 laminated sheets at three different temperatures and various grain sizes was carried out by Afshin and Kadkhodayan [126]. It was found that the growth of grain size led to an increasing coefficient of friction (COF) which had negative effects on the material formability. Raising the blank holder force had a dominant impact on formability, which led to the warm deep-drawing process being performed with a smaller load requirement at higher temperatures.

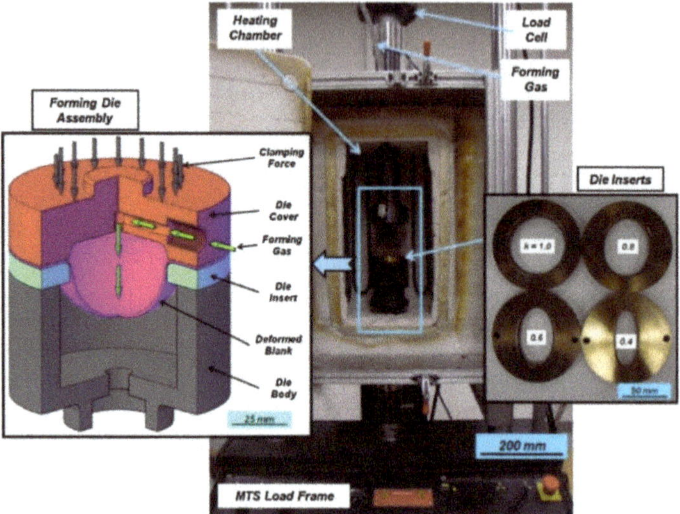

Figure 8. View of the elevated-temperature pneumatic stretching test apparatus, showing an enlarged schematic diagram of the set up including the forming die assembly (left), furnace with forming assembly (central) and die inserts (right) (reproduced with permission from [169]; copyright © 2012 Elsevier B.V.).

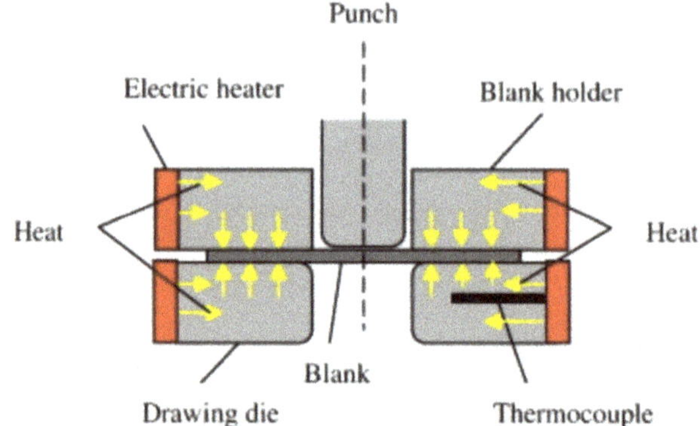

Figure 9. Schematic diagram of the equipment for deep-drawing (reproduced with permission from [126]; copyright © 2015 Elsevier, Ltd.).

The main problems occurring during SMF processes at elevated temperatures are wear and high friction due to high adhesion of the workpiece material to the tool surface and surface initiated fatigue [172,173]. An understanding about the occurrence and prevention of a galling mechanism [172,174] is investigated by using special tribotesters [172,175] or typical tribometers [176] destined for wear analysis.

Friction in warm and hot forming conditions is in general higher that in the case of cold forming conditions. The tribology of warm and hot metal forming is very much of concern because there are several influential variables such as materials, surface coatings, lubricants, contact pressure and temperature [177,178]. It usually requires special heavy-duty lubricants with fillers. In warm and hot metal forming, the graphite-based "black" lubricants have been abandoned by many industries now using "white" lubricants on either a polymer base, carvone base or liquid glass base [162]. A review of related effects of oxidation, surface coating, lubrication and a tribometer in metal forming at elevated temperatures can be found in these papers [125,179,180].

4. Incremental Sheet Forming

4.1. Background

In incremental sheet forming (ISF), called "dieless forming", a sheet metal blank is simply clamped with the help of a blank holder and a tool with a hemispherical end incrementally forms the sheet into the desired shape. The basis for the development of the incremental forming method is the patent developed by Leszak [181] in 1967 before it was technically feasible. Initially the methodology of single-point incremental forming (SPIF) methods was based on conventional spinning, allowing one to obtain axisymmetric parts. Widespread use of numerically controlled machines and robots allowed the introduction of ISF into industrial practice. Some researchers have developed special purpose machines to carry out incremental sheet forming [182–186]. Although ISF has a very reasonable economic performance in small-lot production and is fine for manufacturing those elements which cannot be obtained using conventional sheet forming methods, several attempts have been made to use the forming technique in mass production [187–189]. There are many factors which influence the applicability of ISF and the accuracy of the parts thus formed [190–194]: tool path strategy, mechanical properties of the sheet metal (material anisotropy, ability to strain harden, elastic properties), the technological parameters (i.e., tool diameter, tool path strategy, depth value between two tool passages, tool rotational speed, friction conditions) and design parameters (geometry of the product, sheet thickness). One surface of the drawpiece in SPIF revealed small linear grooves as a result of the interaction of the tool tip with the workpiece (Figure 10). The surface finish of the second side of the drawpiece is the result of small-scale roughness induced by large surface strains which leads to an orange peel phenomenon (Figure 11) [195].

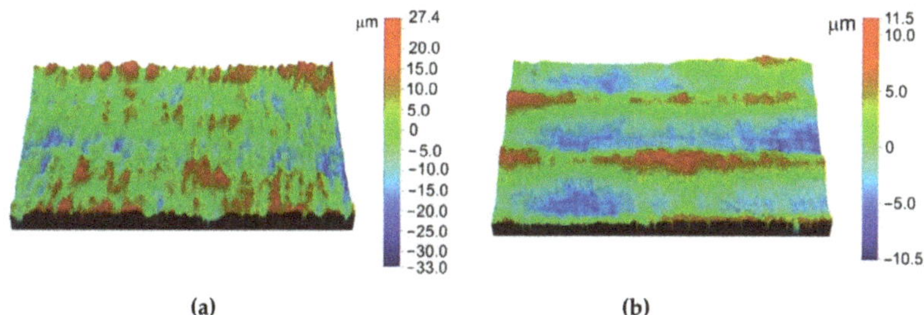

(a)   (b)

Figure 10. Surface topography of the outer (**a**) and inner (**b**) surface of a conical drawpiece with a slope angle of 71°.

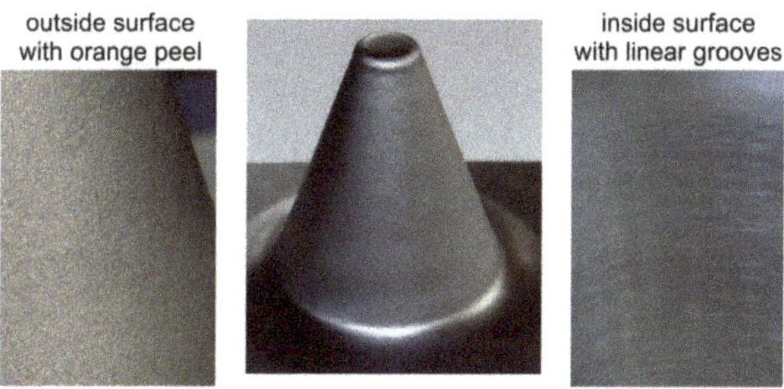

Figure 11. View of the inner and outer surface of a conical drawpiece.

ISF is mainly carried out in the four different ways shown in Figure 12, which are also considered as process principles. Figure 12a represents SPIF in which a rotating tool moves over the clamped edge of the workpiece and produces the desired shape. The SPIF type is a two-point incremental form (TPIF), which can be produced using a partial die (Figure 12b) [196] or a specific die (Figure 12c) [197], which is also is called positive incremental forming (PIF), while all the other methods are negative incremental forming (NIF) [198]. In the TPIF methods (Figure 12d), there is an additional movement of the fixing grip of the plate, which produces effects on the improvement of the accuracy of the drawpieces [196,199].

Parallel investigations are also conducted on the use of a freely rotating [200] or non-rotatable tool [201]. The linear motion of the tool is usually in the range of 300–3000 mm·min^{-1}. The rotational speed of the spindle can reach up to $2 \cdot 10^3$ rpm [201], however in most SPIF methods the rotational speed of a spindle with a rounded tip is in the range 200 to 800 rpm.

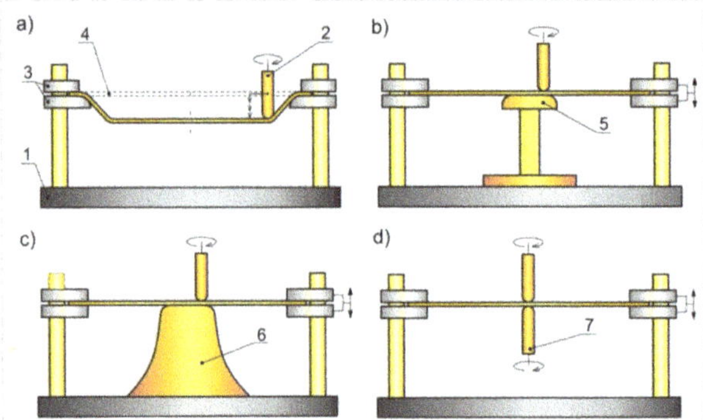

Figure 12. Incremental forming processes: (**a**) single-point incremental forming (**b**) two-point incremental forming with a partial die, (**c**) TPIF with specific die, (**d**) TPIF with counter tool: 1–base, 2–forming spindle, 3–fixing grip, 4–workpiece (initial position), 5–partial matrix, 6–specific matrix, 7–auxiliary spindle.

4.2. Incremental Sheet Forming with Metallic Tool

Most variants of ISF that have been investigated in recent years are intended to fully improve the formability of sheet materials. Xu et al. [183] studied electrically assisted incremental sheet forming with a combination of an electricity-assisted method with double sided incremental forming (E-DSIF) (Figure 13) and the newly designed slave tool force control device used to ensure stable tool–sheet contact (Figure 14). E-DSIF reduced the springback of finished parts during the unclamping and trimming stages.

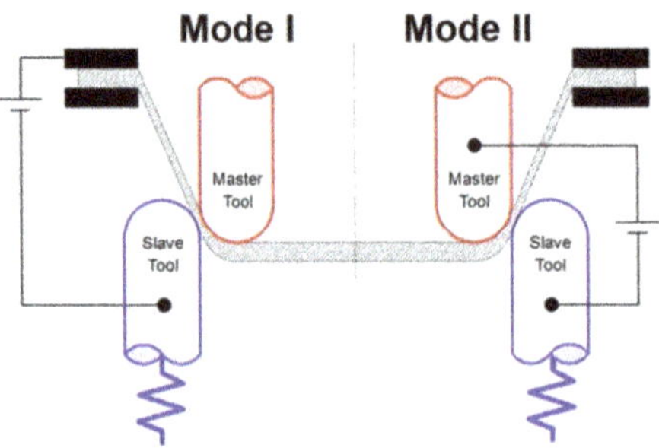

Figure 13. Schematic diagram of the circuit connection in E-DSIF (reproduced with permission from [183]; copyright © 2016 Elsevier, Ltd.).

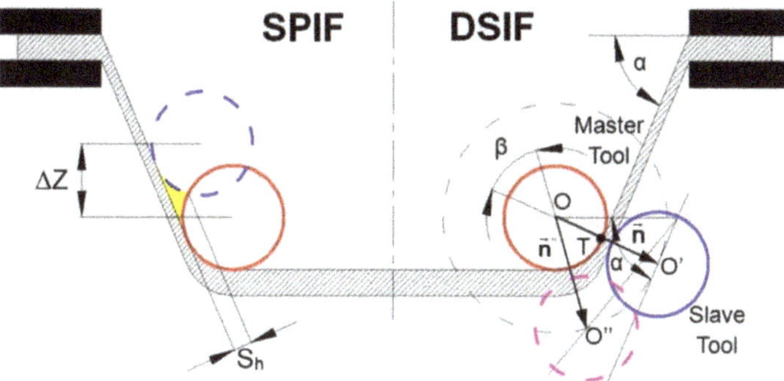

Figure 14. Illustration of the SPIF and DSIF toolpath strategies (reproduced with permission from [183]; copyright © 2016 Elsevier, Ltd.).

Al-Obaidi et al. [186] designed a fixture for hot single-point incremental forming of glass-fiber-reinforced polymer (GFRP) supported by hot air. The GFRP sheet was sandwiched between combinations of two polytetrafluoroethylene (PTFE; commonly known as Teflon) layers and metal sheets (Figure 15). The Teflon layer was used to reduce the flow of the melted matrix polymer out of the woven fiber. The aim of the work was to develop a way to shorten the production process for medical implants which will dramatically reduce the cost of their manufacture. In general, the relationship between the depth of the formed part and the heat initiated was found to make the overall workpiece temperature homogeneous. However, as was concluded, the influence of the material and process properties needs broad-ranging investigation to eliminate the defects and to improve the quality of the formed parts.

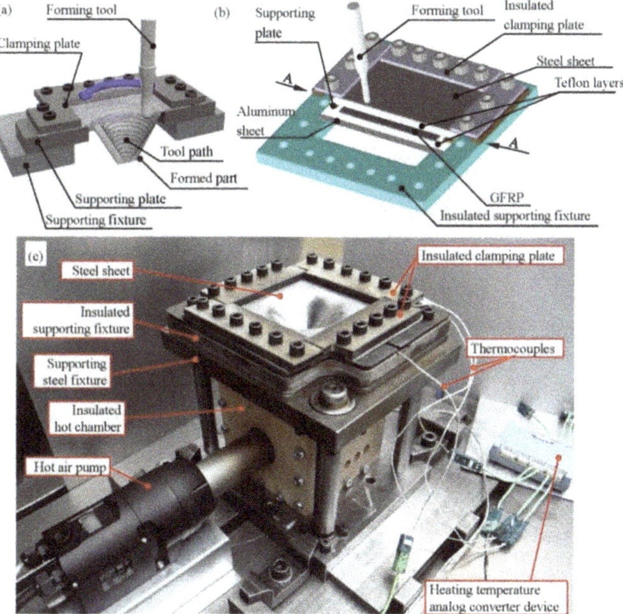

Figure 15. (a) principle of SPIF, workpiece combination, (b) clamping fixture and (c) setup assembly (reproduced with permission from [186]; copyright © 2019 The Society of Manufacturing Engineers. Published by Elsevier, Ltd.).

In recent years, SPIF has been combined with heating techniques to heat up and form difficult-to-form metals into complex shapes by using a laser [202–206], a halogen lamp [207], oil [208], induction heat [209,210] and friction generated by incremental tool application [211] or hot air blowers [212–214]. The apparatus developed by Obaidi et al. [209] and shown in Figure 16 permits local heating of the advanced high-strength steel sheet to elevated temperatures of more than 850 °C, achieving a reduction in the forming forces and residual stress, and improving formability. Some of the authors used the high rotation speed of the forming tool in order to generate friction heat, which can be utilized to improve the formability of the workpiece material [211,215–223]. The formability of hard-to-deform alloys can also be enhanced by direct or indirect electric heat [183,224–235]. Palumbo and Brandizzi [236] used combined heat (electric band + stir rotation friction) to SPIF the deformation of titanium alloy in which static electric heating was employed to pre-heat the sheets. Friction heating was used to further increase the temperature. A detailed state-of-the-art review of the development of heat-assisted incremental sheet forming has been discussed by Liu [237].

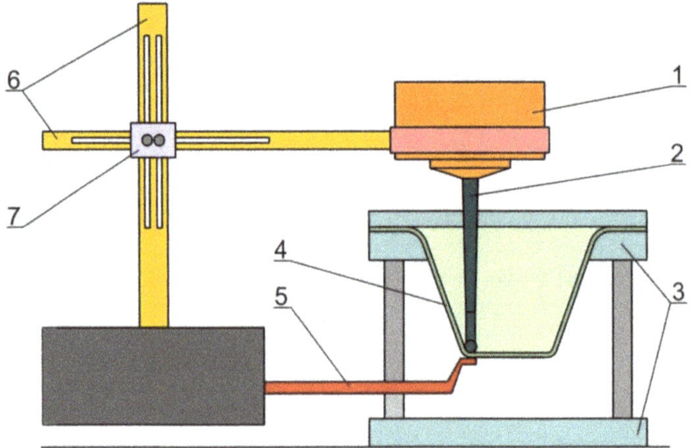

Figure 16. Apparatus for induction heat-assisted SPIF: 1–spindle, 2–tool, 3–clamping fixture, 4–sheet metal, 5–inductor, 6–sliding fixtures, 7–sliding jig.

Excessively high temperature generated at the tool/workpiece interface affects the quality of the sheet surface after forming and the durability of the tool tip [238–240]. The appearance of frictional heat and its effect on the accuracy of the formed part and the formability of the workpiece material has been explored by many researchers in relation to the alloys [200,241,242]. The role of friction in the formability of polymeric materials during the ISF process has been widely discussed by several authors [243–247], who draw conclusions on the important role of friction due to the increase in the temperature of the sheet metal. Much experimental work and many numerical simulations have been conducted to support the designs of the tooltip profile [248,249], coatings of the sheets and tool tips [249,250], tool path strategy and forming parameters [251–253] and residual stresses [254,255] in ISF. It is believed [256] that the tooltip/sheet coatings capable of thermal segregation in the tool–sheet interaction have a huge impact on mitigating the temperature-induced problems in ISF.

A principle of the hybrid combination of stretch forming and SPIF is presented in Figure 17. Stretch forming will not yield the geometry of the final part. Such features as grooves, pockets and corrugations that are too small to be formed by stretch forming are formed using asymmetric incremental sheet forming [257]. In order to minimize thinning and improve the thickness distribution along the formed part, Tandon and Sharma [258] proposed a hybrid incremental stretch drawing process which is designed to combine incremental sheet metal forming with a deep drawing process. Initially the workpiece is deformed by a large sized tool and then an ISF tool formed the final shape of the part. It is found that the proposed process is able to reduce thinning by as much as approx. 300%, considering the same forming depth for the ISF process. Similar investigations on hybrid ISF conducted by Tandon and Sharma [258] and Jagtab and Umar [259] were also studied by Araghi et al. [260] and Lu et al. [261].

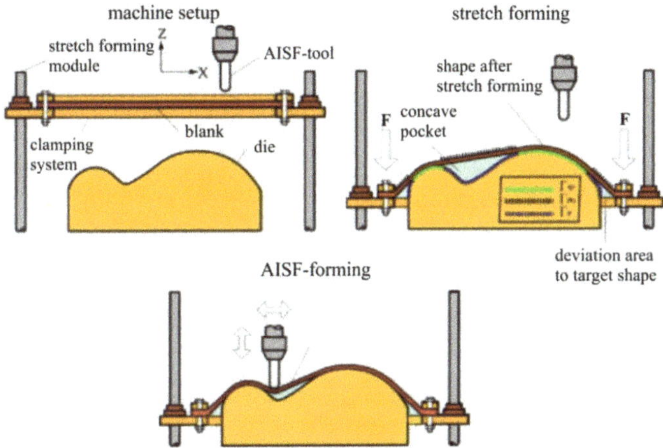

Figure 17. Principle of the combination of processes (reproduced with permission from [257]; copyright © 2009 CIRP. Published by Elsevier, Ltd.).

Supporting the sheet in ISF leads to localized deformation under the tool which reduces the flexibility of the ISF process. Therefore, Allwood et al. [262] proposed the use of partially cut-out blanks to create a weakness in the sheet at the perimeter of the required part and to localize the sheet deformation to the region that will be covered by the tool (Figure 18). It was concluded that the use of partially cut-out blanks does not give a useful benefit in ISF. Compared to ISF with no backing plate, partially cut-out blanks develop localized deformation earlier.

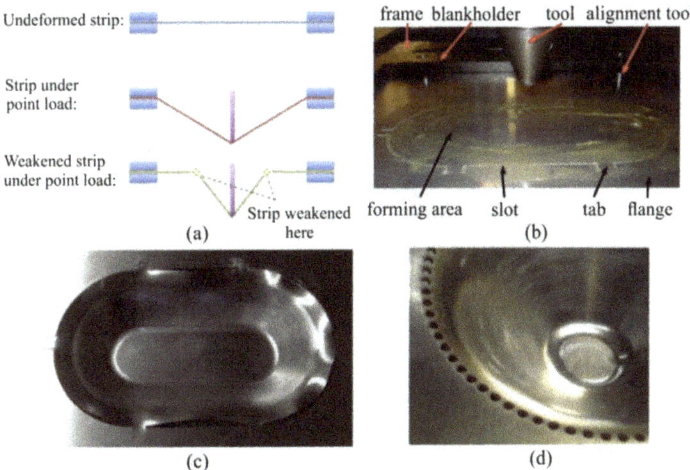

Figure 18. Partially cut-out blanks: (**a**) the concept intended to localize deformation, (**b**) partially cut-out blank before forming and (**c**) after forming, (**d**) after successfully forming (reproduced with permission from [262]; copyright © 2010 Elsevier B.V.).

5. Flexible-Die Forming

5.1. Background

Flexible-die forming on sheet metal employs a pressure–transfer medium as a female die (or punch) and a female punch (or die) [263]. The pressure–transfer medium can be polyurethane, oil, viscous medium, rubber or compressed air. Flexible-die forming technology has undergone rapid development as one of the mainstream forming technologies for thin walled parts. In addition to viscous pressure forming (VPF) and hydroforming by deep drawing (HDD), these technologies are developing rapidly, and aspects of experiments, theoretical analyses, and numerical computation are provided at least once per week. Flexible forming processes, like stamping, deep drawing or bending, are widely used in the aircraft industry to deform difficult-to-form materials by a conventional deep drawing process [264–266].

5.2. Rubber-Pad Forming

Rubber-pad forming (RPF) employs a rubber pad on the form block. The rubber pad acts as a hydraulic fluid in exerting a uniformly distributed pressure on a workpiece surface as it is pressed around the form block. With some advantages like ease of operation, reduced springback and high surface quality, rubber-pad forming is widely used in many real-life industrial situations [41,267]. Flexible forming processes, like stamping, deep drawing or bending, are widely used in the aircraft industry to deform difficult-to-form materials by a conventional deep drawing process. RPF has been used to remove surface scratches of the sheet during forming. Several researchers studied the RPF process. Belhassen et al. [268] introduced FEM to analyze the RPF process in AA6061-T4 sheet metal. An elastic-plastic constitutive model with a J_2 yield criterion and mixed nonlinear isotropic/kinematic hardening coupled with Lemaitre's ductile damage has been adopted during forming.

Irthiea et al. [269] report the results of FE simulation and experimental research on micro deep drawing processes of 304 stainless steel sheets using a flexible die. Two novel approaches were considered with regard to the positive and negative initial gap between an adjustment ring and a workpiece and a blank holder (Figure 19). Initial gaps affect the final cup profiles, in particular at the shoulder corner radius (Figure 20). The numerical predictions conducted in Abaqus/Standard software reveal the capability of the proposed technique to produce micro metallic cups with high quality and a large aspect ratio.

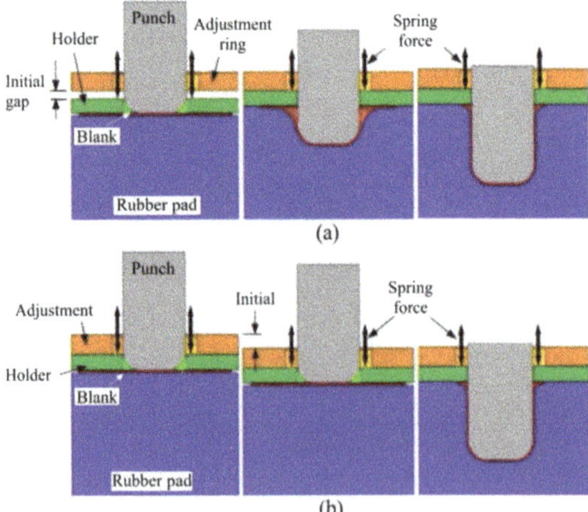

Figure 19. Flexible forming technology with (**a**) positive and (**b**) negative initial gap (reproduced with permission from [269]; copyright © 2014 Elsevier, Ltd.).

The results showed that reducing the initial gap causes a reduction in the initial thickness of the final products. Fabrication of the blank holder with a smooth surface and a rigid punch with a rough surface would improve the formability of the workpiece material. The proposed technique has desirable advantages of significantly lower overall cost and high quality in the fabrication of micro metallic cups.

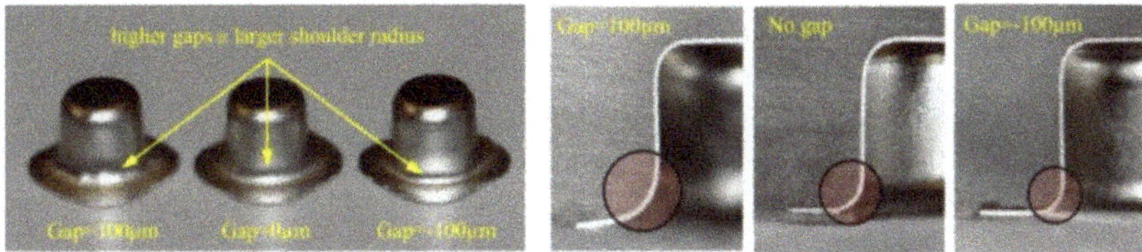

Figure 20. Cups formed with different initial gaps (reproduced with permission from [269]; copyright © 2014 Elsevier, Ltd.).

Trzepiecinski et al. [270] analyze the forming process of a fan engine bearing housing made of 17–4PH stainless steel sheet. Due to the high ratio of yield stress to ultimate tensile strength and large amount of springback of the formed material, the forming process has been divided into two stages in order to ensure suitable shape and dimensional accuracy of the drawpiece: forming of the drawpiece using a rubber punch (Figure 21a) and calibration of the drawpiece at elevated temperature. A 3D finite element-based coupled thermomechanical model was built using the commercial FE-package eta/DynaForm. The distribution of the drawpiece shape error (Figure 21b) obtained by the GOM ATOS system and thickening measurements confirmed that the experimental method thus developed is suitable for the manufacture of bearing housings for aircraft fan engines. As predicted by the eta/Dynaform program, the component is free from defects (Figure 21c). The formation of wrinkles was observed on the free surfaces of the part, but this part of the drawpiece is cut off after the forming process is finished.

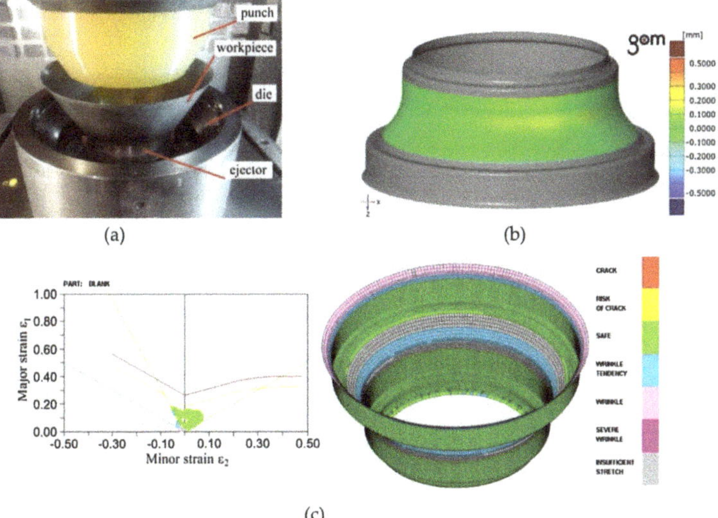

Figure 21. Rubber-based forming: (**a**) tool setup, (**b**) The distribution of the shape error obtained using the GOM ATOS instrument, (**c**) The major and minor strains and FLD (left) and the final shape of the bearing housing (right) after the first stage of forming (reproduced with permission from [270]; copyright © 2017 The Society of Manufacturing Engineers. Published by Elsevier, Ltd.).

The rubber forming process (RFP) has also been successfully applied to straight flanging. Flanging is used to deform the edge of the part to create a mating surface to increase the stiffness of the part. The influence of process parameters (time and pressure) on the RFP was studied by Chen et al. [271]. The key phenomenon which limits the ability to obtain a part with the desired geometry is springback. It was found that springback of the straight angle increased with a decrease in blank thickness. An increase of forming pressure and time of rubber forming had little effect on the springback when the blank coincided with the die face. It should be pointed out that the springback of flanging in the rubber forming process was smaller than that of stamping and can be eliminated with a ratio of the bending radius r (die radius) to the sheet thickness t: $r/t < 2$.

Rubber forming technology is also a method that can feasibly be used for micropatterning of thin metallic bipolar plates [272]. In this process, the workpiece is placed on the rubber pad and then the blank and the rubber pad are pressed simultaneously by the die. With this technology, the force from the upper die and the repulsive force resulting from elastic deformation of the rubber pad are transmitted to the entire surface of the workpiece. Jin et al. [272] analyzed the effects of the hardness and thickness of the rubber pad, physical properties of the materials, forming speed and pressure of the punch on the channel depth. They found that (i) the lower the hardness of the rubber pad, the more beneficial it was for the deformation of the rubber and the channel is deeper (ii) the thicker the rubber pad, the deeper was the channel in the bipolar plate due to it not losing the load imposed by the punch toward the outside. Most researchers studied the forming of a straight channel based on the 2D models.

Liu et al. [273] studied the deformation mechanisms during the forming of a channel in metallic bipolar plates (Figures 22 and 23) using the two styles of deformation (concave and convex). Experimental methods and numerical simulations showed that the thickness distribution of the parts manufactured by the concave approach is less uniform than that of the convex shape. The ratio of channel height h to rib width w (h/w) is a determining parameter in obtaining the desired geometry of the part. For the concave deformation style, the maximum thickness reduction also increases as the value of h/w increases. By contrast, for the convex style, the maximum thickness reduction decreases with increasing h/w.

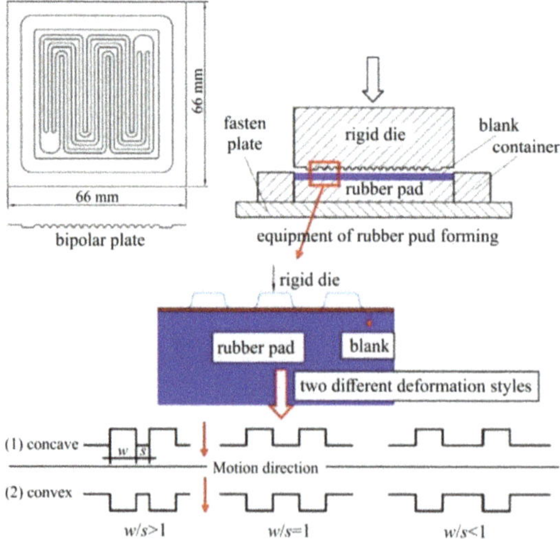

Figure 22. Fabrication of a bipolar plate by rubber-pad forming (reproduced with permission from [273]; copyright © 2010 Elsevier B.V.).

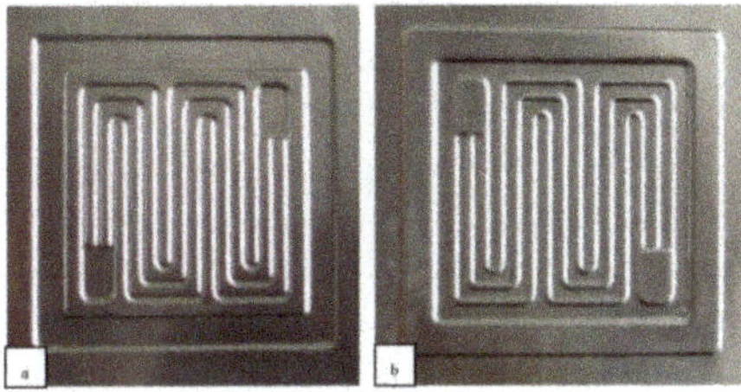

Figure 23. Bipolar plate fabricated by rubber-pad forming: (**a**) front of the bipolar plate and (**b**) back of the bipolar plate (reproduced with permission from [273]; copyright © 2010 Elsevier B.V.).

5.3. Multipoint Die Forming

A multipoint forming (MPF) die uses reconfigurable matrices of pins, which can move normally to the die base to create the die surface. A reconfigurable tool consists of a large number of adjustable pins, the form of the ends of which define the specified shape of the part [274]. Over the years, several designs of reconfigurable tooling have been developed [275–277]. The MPF process is very flexible permitting rapid changes in tool configuration to be accommodated without affecting tooling costs [278]. The concept of the use of a discrete-pin die for forming sheet metal was first introduced about 40 years ago [279].

Many investigations have been conducted in recent years to develop the multipoint forming technique. Most of these are focused on avoiding defects without taking into account the effects of the forming process on the quality of the finished parts. Paunoiu et al. [280] developed numerical simulation models using dynamic FE-based Dynaform software and investigated the effect of tool geometry and the use of an elastic cushion on deformation behavior. The elastic cushion had a positive effect on the part surface and a negative effect on shape accuracy. The elastic cushion had a negative effect on shape accuracy and a positive effect on the part surface.

To reduce tool costs to a significantly lower value compared to that for conventional reconfigurable MPF tools, the multipoint sandwich forming (MPSF) was developed. Pins are used only in the bottom die and the space between them is large (Figure 24), so that fewer of them are required to span a given area and their height is adjusted manually using their threaded shanks [274]. The upper die is composed of urethane. To provide a continuous surface for the bottom die, a steel sheet is deformed between the appropriately positioned pins and the urethane upper die [274].

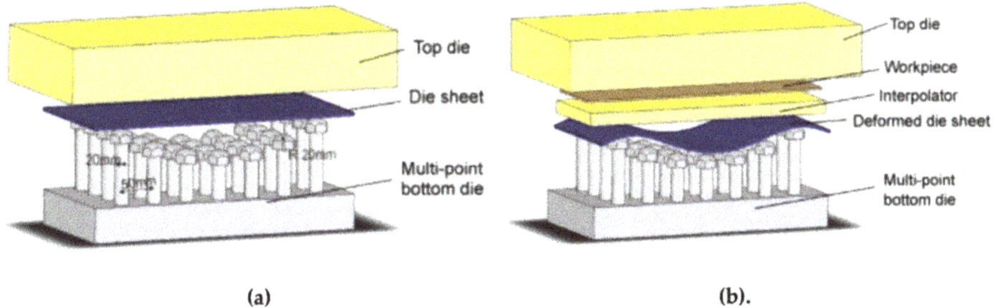

Figure 24. Multipoint sandwich forming; set up for the deformation of (**a**) die sheet and (**b**) workpiece (reproduced with permission from [274]; copyright © 2008 Elsevier, Ltd.).

Multifunctional light–weight sandwich panels, due to their profitable load-bearing characteristics, have been widely used in a variety of applications such as ship structures, aircraft, high-speed trains and the automotive industries [281,282].

Sandwich panels are mostly limited to the flat-panel type however in many engineering applications it is necessary to use three-dimensional panels rather than flat sheets [170,283,284]. This creates the need to work on effective methods for forming honeycomb-type structures. Cai et al. [285] applied an MPF tool (Figure 25a) to form double curved sandwich panels with egg-box-like (EBL) cores. Among various types of topological configurations of sandwich cores, EBL-cored panels are ultra-lightweight structures. Experiments using MPF- and FE-based analyses reveal that the formability of a sandwich panel with EBL cores (Figure 25b) in the plastic forming process is mainly dominated by tree mechanisms: face sheet plastic dimpling, the failure modes of face sheet plastic wrinkling and core cell fracture. The analytical formulations presented allow one to facilitate the prediction and successful prevention of failure modes in the plastic forming process of sandwich panels (Figure 26).

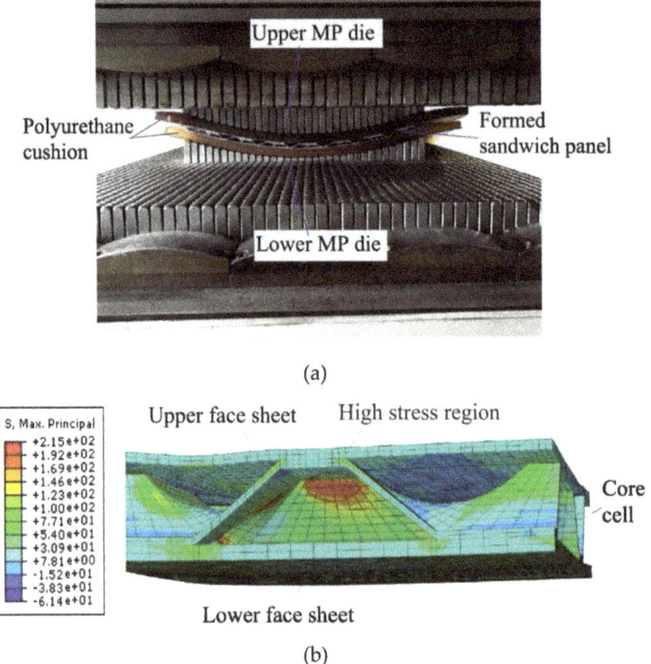

Figure 25. (**a**) Multipoint forming (MPF) of egg-box-like (EBL)-cored sandwich panel and (**b**) distribution of maximum principal stress on the EBL core cell of the panel (reproduced with permission from [285]; copyright © 2018 Elsevier, Ltd.).

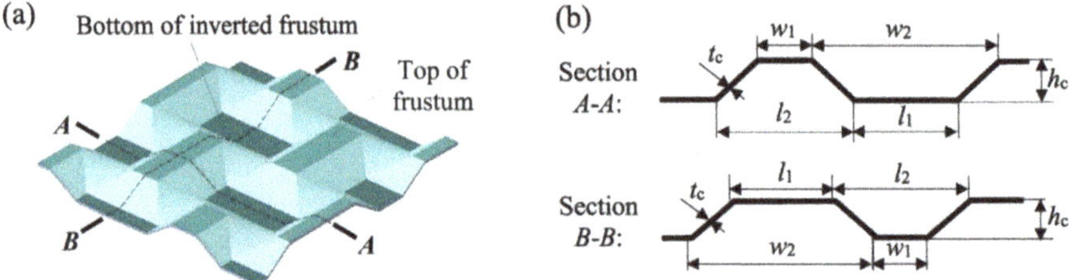

Figure 26. (**a**) sketch of the EBL core and (**b**) cross-sectional geometry of the EBL core (reproduced with permission from [285]; copyright © 2018 Elsevier, Ltd.).

The cylindrical bending process of bidirectional trapezoidal core sandwich structures has been very limited in recent decades. In 2018 Liang et al. [286] applied multipoint bend-forming (MPBF) to aluminum alloy welded metal sandwich panels with a new type of bidirectional trapezoidal core (Figure 26). The main forming defects of the sandwich panels and the deformation characteristics were analyzed numerically. The dimple and straight plane effects are found to be the most frequent forming defects in the bend-forming process. In addition, the main factors that affect forming defects are bending radius and face sheet thickness.

5.4. Solid Granular Medium Forming

Solid granular medium forming (SGMF) adopts spherical non–metal or metal particles with a diameter in the range of 0.05–2.00 mm as filling medium [287]. The particle medium can overcome loading induced by liquid media under elevated temperature conditions. Furthermore, friction produced by the particles adhering to the sheet surface results in a uniform wall thickness of the formed parts. The advantage of SGMF over the remaining flexible-die forming techniques is good surface quality, high dimensional precision, simple implementation, high production flexibility and a better fit of the die [287]. SGMF has found particular application in the manufacturing of workpieces with rigorous requirements with regards to geometric accuracy and surface finish. This technology can simplify complex sheet metal production and provides production with good flexibility, fine structures and less elasticity. Although LSDF requires high pressure from the liquid–pressure system, a sophisticated liquid pressure system design and practical cavity sealing plans, it provides for the production of fine structures with good flexibility [263].

Chen et al. [288] implemented the press hardening process (Figure 27) for tubular components by using SGMF, which have much higher stiffness when compared to sheet metal parts. A multitype granular medium (quartz sand, sintered zirconium silicate and cerium-stabilized zirconium dioxide) reduced the interfacial friction force and showed better formability. The effects of interfacial friction between the particles of the granular medium and the formed sheet were represented by tube length, type of granular medium and the COF.

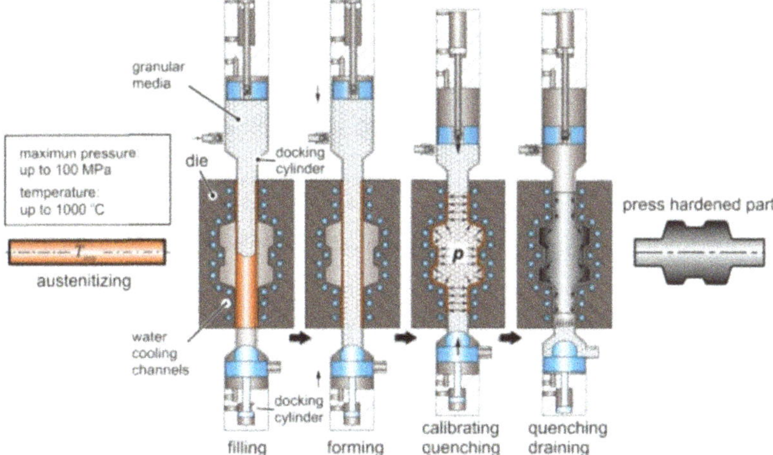

Figure 27. Granular medium-based tube press hardening (reproduced with permission from [288]; copyright © 2015 Elsevier B.V.).

The advantages of coupling FEM and DEM solutions are as follows: applicability to non-axisymmetric geometries of components, automatic detection and solution of the contact phenomena between the granular medium and the workpiece, and applicability to quasi-static processes which contain complex friction contact conditions and large deformations. Dong et al. [289] carried out FE-based numerical simulation of SGMF of back pressure deep drawing and concluded that the forming of thin-walled rotary parts with complex cross-sections is enormously influenced by the process parameters: the back pressure load and the blank holder gap. Dong et al. [51] applied extended Drucker–Prager linear FEM-based numerical simulation of SGMF of the AA6061 extruded tube (Figure 28). The effects on the forming performance of the workpiece and the approach to the control of blank holding were analyzed through this model. The results of the investigations confirmed the ability of the constitutive equation to be reflected in the establishment of the material model of the granule medium.

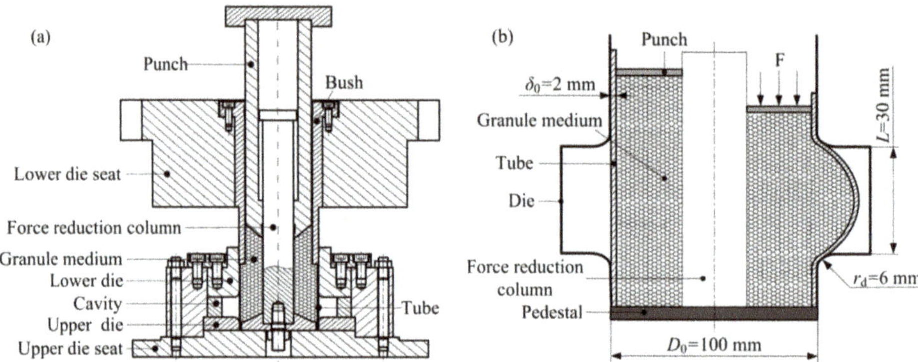

Figure 28. (**a**) forming mold design and (**b**) finite element method (FEM)-based numerical model (reproduced with permission from [51]; copyright © 2017 Elsevier B.V.).

6. Electromagnetic Sheet Forming

Electromagnetic forming (EMF) is a kind of powerful and high-speed forming technique where the strain rate of the sheet metal is of the order of approximately $10^3 \cdot s^{-1}$ [287,290–294] and the deformation velocity can reach up to 300 m·s^{-1} [295]. EMF uses a Lorentz's force occurring in a pulse magnetic field generated by circuits conducting high-oscillation electric current [296]. Thin metallic sheet can be accelerated to high velocity in a period of a millisecond [297]. During EMF, a large electric current pulse passes through a conductive coil by discharging a capacitor bank. The electric current produces a transient magnetic field around the coil that induces eddy currents in a metal workpiece.

EMF has many advantages that make it an attractive alternative to conventional SMF methods:

- It is an environmentally friendly process; no lubricants are needed;
- There is no mechanical contact with the work piece;
- The formability limit is increased during electromagnetic forming due to high deformation velocity, the forming limit of the aluminum alloys can be improved by 10%–14% [298] or even 2–3-fold [299], compared to the quasi-static loading conditions;
- Controllability and repeatability of the formed parts are ensured;
- The greatest advantage is a significant increase in workpiece ductility over conventional SMF methods;
- Parts formed by EMF possess the merits of low springback [300], uniform strain distribution and good surface quality.

In the last decade, EMF has been studied experimentally and numerically by many researchers. The nonuniform deformation behavior results in a reduction in the shape and dimensional accuracy of the final part. To precisely control the material behavior in EMF and obtain parts without defects, various techniques were employed, such as selecting optimized process parameters [301,302], electromagnetic calibration [300,302,303], applying two-step forming [304] and using tailored forming coils [305,306], prediction of formability and failure [307]. The current methods for the evaluation of limit strains at the onset of necking and fracture have recently been discussed in [308].

Cui et al. [309] proposed electromagnetic incremental forming (EMIF) technology in which it is feasible to produce a large part with a small working coil and small discharge energy, which enhances the flexibility of EMF. The principle of EMIF is shown in Figure 29a. The magnetic force is used to launch the sheet at very high speeds and to obtain the desired shape. Figure 29b shows that the working coil moves to a specific position and the metallic sheet deforms in many cycles of charge and discharge. Finally, the local deformation accumulates into the final deformation as shown in Figure 29c. Two different strategies are proposed to predict the electromagnetic incremental sheet and tube forming process. For the first method, the coil can move directly to a special position and remesh technology is used to consider the effect of the workpiece deformation and the movement of the coil on magnetic analysis. In the case of the second method, technology like the "birth–death element" is used to indirectly describe the movement of the coil. It was found that the second method cannot be used for the EMF process if an overlap region exists in two adjacent discharge regions.

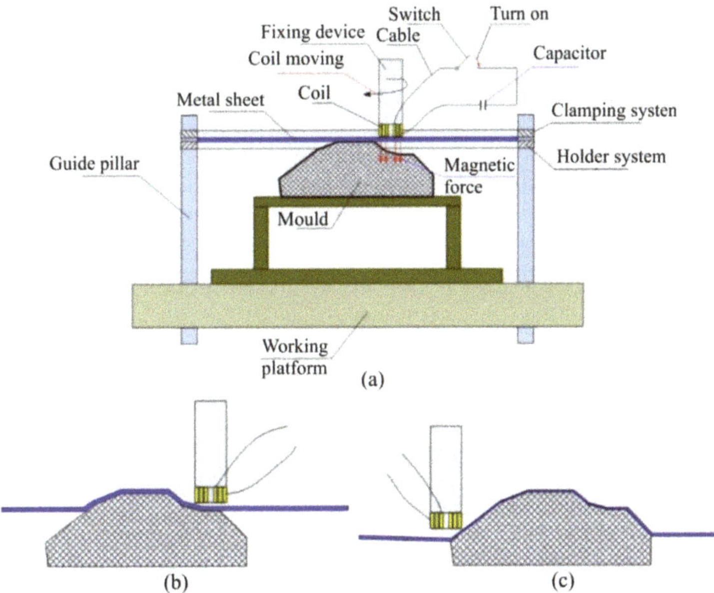

Figure 29. (**a**) schematic diagram of the forming tool, (**b**) local deformation and (**c**) termination of deformation (reproduced with permission from [309]; copyright © 2014 Elsevier B.V.).

Lai et al. [310] proposed a forming facility for large sheet metal forming which is lightweight compared to other conventional tools. The EMF facility they developed consists of four major sub-systems: a forming coil system, a die system, an inertia confinement system and a pulsed electromagnetic blank holding force system (Figure 30). The pulsed electromagnetic blank holding force system is a newly developed approach that utilizes a pulsed attractive Lorentz force as a blank holder force. The resulting measurements of sheet thickness and deviation from the desired shape are very promising, providing forming of the large sheet metal components. Tan et al. [311] introduced EMF to form a panel with stiffened grid ribs. The forming rules of the grid–rib panels during the EMIF process were revealed by analyzing the evolution of the stress, strain and forming depth. It was found that the forming depth is mainly attributed to the forces exerted on the web, although electromagnetic force is applied on both the ribs and the web.

Because of the large degree of stretching and bending deformation in hole-flanging regions [312] and the relatively low formability of aluminum alloys at room temperature [313], the workpiece tends to tear or crack along the circumference at the flanged edge formed by conventional stamping [314,315].

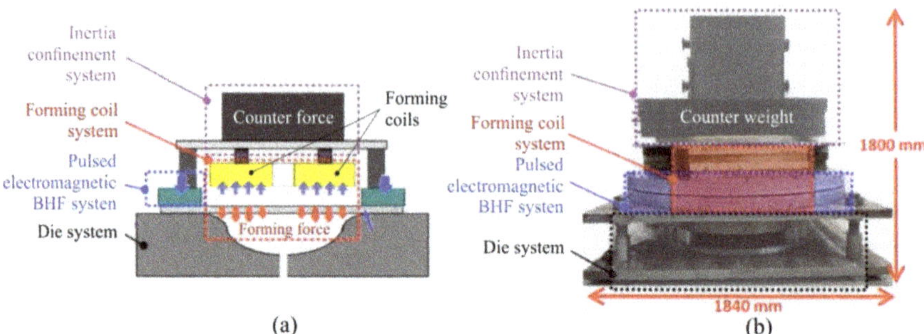

Figure 30. Electromagnetic forming: (**a**) schematic diagram and (**b**) experimental setup (reproduced with permission from [310]; copyright © 2017 Elsevier, Ltd.).

To overcome this problem, Su et al. [303] proposed a new two-step electromagnetic forming process which combines EMF with electromagnetic calibration for local features of large-size sheet metal parts. During this process, the workpiece is first electromagnetically formed by a flat spiral coil and then electromagnetically calibrated by a helical coil with a similar shape to the final profile of the workpiece (Figure 31). The experimental results show that there are critical discharge voltages for both EMF and electromagnetic calibration which lead to the minimum die-fitting gap. It was also found that the proposed two-stage forming process improved the stress distribution in the workpiece and reduced the bending moment, which is responsible for the minimization of springback and shape and dimensional errors.

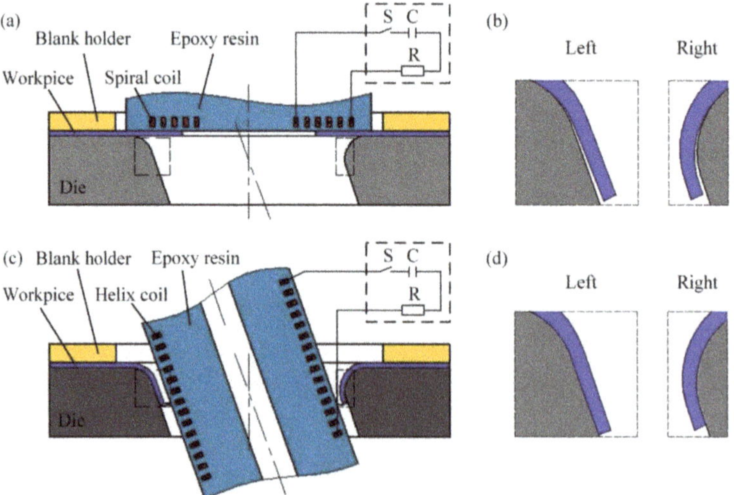

Figure 31. Schematic diagram of a two-step electromagnetic forming (EMF): (**a**) electromagnetic forming, (**b**) workpiece after electromagnetic forming, (**c**) electromagnetic calibration, (**d**) final part after EM calibration (reproduced with permission from [303]; copyright © 2018 Elsevier, Ltd.).

Similarly, Su et al. [316] studied the uneven deformation behavior of a 2219 aluminum alloy workpiece formed by electromagnetic flanging. The authors established a numerical model (Figure 32) in LS-DYNA 8.0 software to study the effect of different axial angles between the flanging direction and the normal direction of the sheet (Figure 33). They concluded that uneven deformation behavior is essentially due to the uneven deformation requirement. The electro-magnetic force and the area of the deformation region have a significant influence on the uneven deformation behavior. Furthermore, the die constraint indirectly affects the uneven deformation behavior of the sheet metal by changing the evolution of the deformation region. The high-speed electromagnetic impaction behavior according to the Gurson–Tvergaard–Needleman (GTN) model parameters obtained is studied numerically by Feng et al. [317]. The analysis of the void volume fraction and plastic strain distributions are analyzed during the process of high-speed EMF, which indicates the validity of using the GTN damage model to describe or predict the fracture.

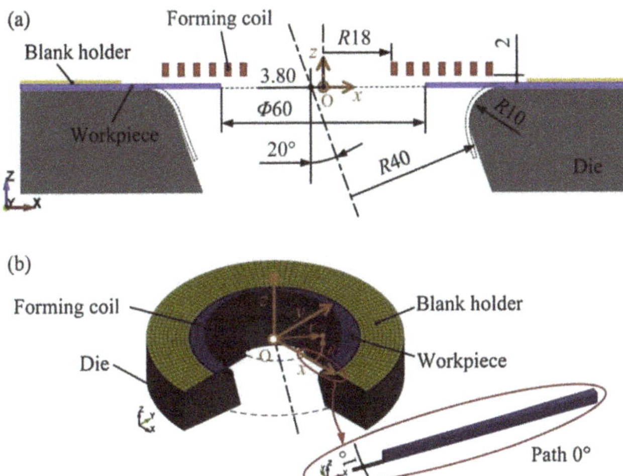

Figure 32. Numerical model of EMF for the hole-flanging process: (**a**) geometry and dimensions, (**b**) FE mesh (reproduced with permission from [316]; copyright © 2020 The Society of Manufacturing Engineers. Published by Elsevier, Ltd.

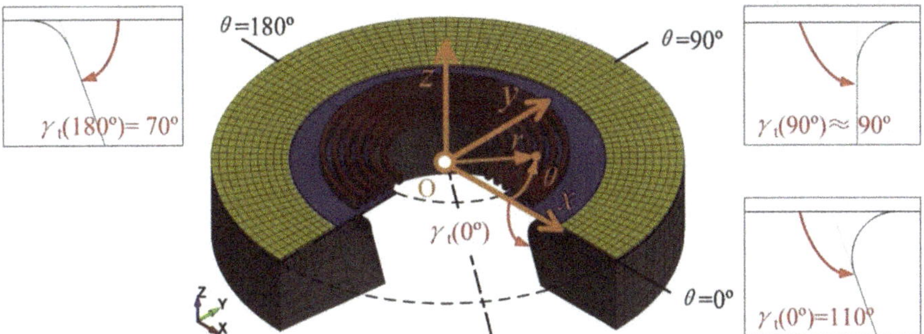

Figure 33. Flanging angles along typical parts (reproduced with permission from [316]; copyright © 2020 The Society of Manufacturing Engineers. Published by Elsevier, Ltd.).

Radial Lorentz force augmented deep drawing was developed by Lai et al. [293] to enhance the material flow of the flange in an electromagnetic deep drawing (EMDD) process. This combines an axial Lorentz force on the unsupported region of the sheet metal and an additional radial inward

Lorentz force at the periphery of the sheet metal. The Lorentz forces are flexibly controlled using a dual-coil electromagnetic forming system (Figure 34). In conventional EMF, there is only one driving coil. Experimental and high-speed multiphysics coupled dynamic numerical contributions on circular aluminum 1060-H24 sheet workpieces permit the draw–in of the flange to linearly increase with the discharge voltage of the axial Lorentz force. Increasing the radial Lorentz driving force leads to an exponential increase of the draw–in of the flange.

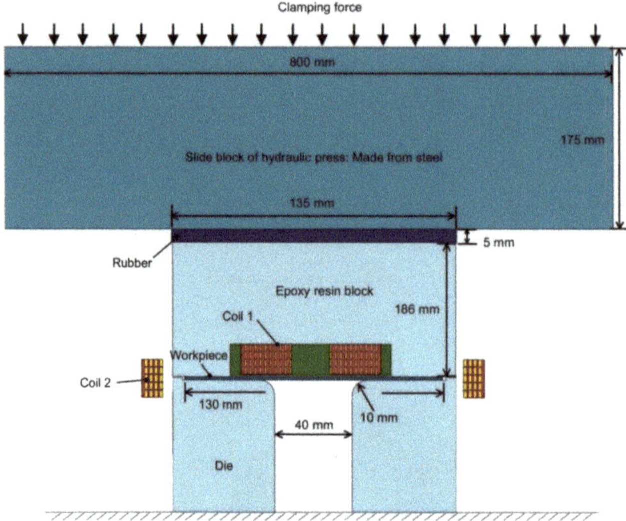

Figure 34. Experimental setup of the electromagnetic deep drawing (EMDD) process (reproduced with permission from [293]; copyright © 2017 Elsevier B.V.).

7. Electrohydraulic Forming

Advanced high-strength steels and aluminum alloys are among the materials that are currently seeing increased potential application in efforts at weight reduction. A problem common to the use of these materials is their limited formability in SMF operations compared to carbon steels. Electrohydraulic forming (EHF) is able to overcome the limitations of conventional SMF methods [290,318]. In EHF a high pressure, high temperature plasma channel is created between the tips of the electrodes during a high voltage discharge. The electrical energy is stored in a bank of capacitors. The shockwave in the liquid initiated by the expansion of the plasma channel then propagates towards the blank at high speed, and the mass and momentum of the water in the shock wave forms the sheet metal blank into the die [319]. A typical configuration of EHF shown in Figure 35 includes the electrodes, discharge chamber, die, pulse generator (which consists of a high-voltage/high-current discharge switch D), high-voltage low-inductive bank of capacitors C and a charging/amplifying/rectifying circuit T-R-A.

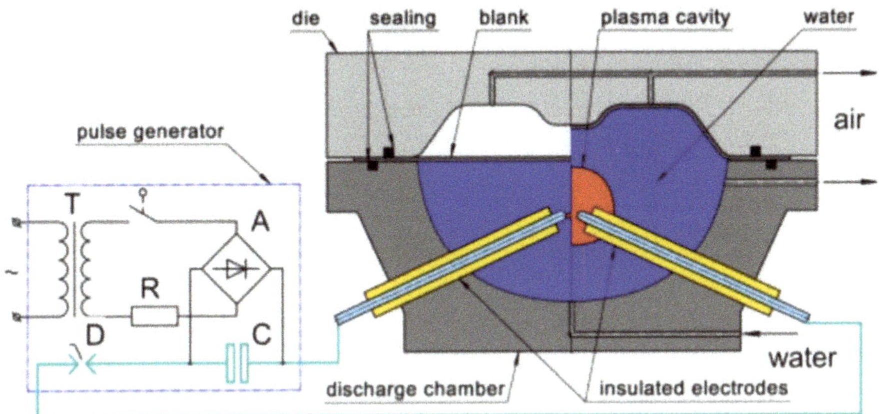

Figure 35. Experimental setup of the EMDD process (reproduced with permission from [319]; copyright © 2015 Elsevier B.V.).

Most practical applications of EHF in the last century were related to the aerospace industry, where, typically, rather small parts lying within the overall size range of 300–400 mm or less were stamped [320]. EHF allows a reduction in capital investment since EHF requires one-sided dies rather than two-sided ones and a reduction in manufacturing costs in the low volume applications of the aerospace industry.

With popularization of the production and applications of high-strength steel sheets, EHF is being taken into consideration as a potential technology for automotive applications [320]. Substantial development of the EHF process includes a durable electrode system, a durable sealing system, an efficient water–air management system and a numerical modeling technique, as reported in [321]. According to Golovashchenko et al. [322], a fundamental advantage of EHF is that it is capable of filling rather complex sheet metal forming shapes in one operation by providing a series of discharges which can fully form the workpiece into the cavity of the stamping die and calibrate springback without opening the chamber and removing water from the tool. The relative improvement in plane strain formability of dual-phase steel in EHF conditions is between 63% and 190%, compared to the quasi-static limiting dome height test [320]. The high-velocity impact of the sheet against the die in EHF leads to the suppression of void nucleation and a growth in dual phase steels, significantly delaying the onset of failure [323]. Two different strategies of EHF can be distinguished differing in the formability of the sheet material: electrohydraulic free-forming conditions, where the strain rates are rather moderate, and die-forming conditions, where the strain rates are much higher and through-thickness compressive and shear stresses have a significant effect [324].

The formability of the blank material can be significantly improved in the process of warm electrohydraulic forming (WEHF) [325]. The workpiece in hybrid WEHF can be warmed before placing it on the chamber or using an induction coil. According to Figure 36, in order to prevent cooling the blank due to contact between the blank and water, an air gap has been created between the workpiece and water. During the forming of 1000-series aluminum alloy using WEHF at elevated temperatures, a 23.6% increase in failure strain is observed at WEHF when compared to EHF (Figure 37) [325].

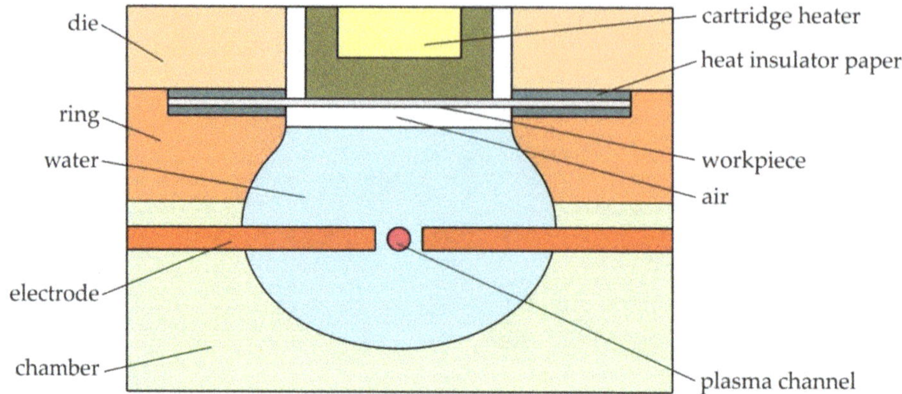

Figure 36. Schematic diagram of the cross-section of warm electrohydraulic forming (WEHF).

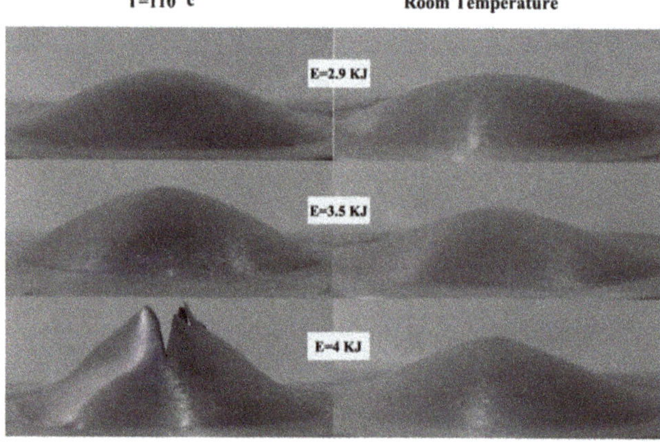

Figure 37. Comparison of the final shapes of specimens formed at 110 °C and 27 °C using different energies E (reproduced with permission from [325]; copyright © 2017 the authors. Published by Elsevier, Ltd.).

8. Spinning and Shear Spinning

8.1. Background

Spinning is the shaping of a rotating disc or drawpiece by applying local pressure using a spinning tool (Figure 38). A forming tool, in the form of a mandrel or a roll, can roll or slide over the surface of the sheet. A characteristic feature of the spinning process is that the thickness of the sheet metal, from which the element is formed, changes within only a very small range. The deformed disc or blank gradually adopts the shape of the spinning block, which is usually made of metal. In the case of spinning components with complex shapes, partial or uniform spinning blocks are used. The spinning tool and workpiece performs a rotational motion. Spinning ability is measured by the limiting spinning ratio, which is the ratio of the maximum original blank diameter that can be successfully spun forming a cup, in a single pass, to the mandrel diameter [326–330].

The classification of traditional spinning processes has mainly been developed according to the relative position between the roller and the blank, the deformation characteristics of the blank material, the temperature of the blank, as well as spinning with or without a mandrel [331–333]. The classification of the traditional spinning processes is presented in Figure 39.

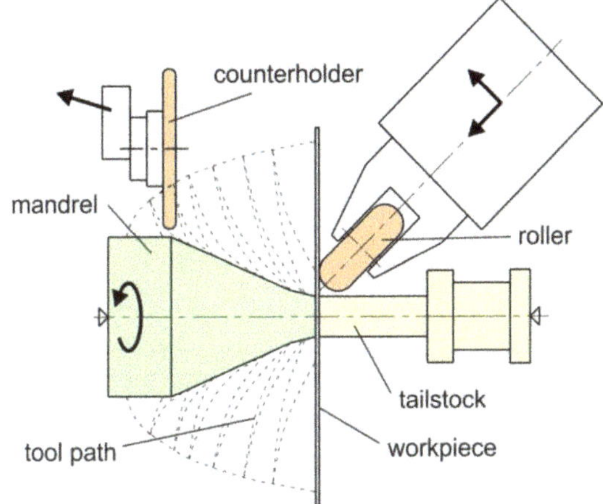

Figure 38. Schematic illustration of conventional sheet metal spinning.

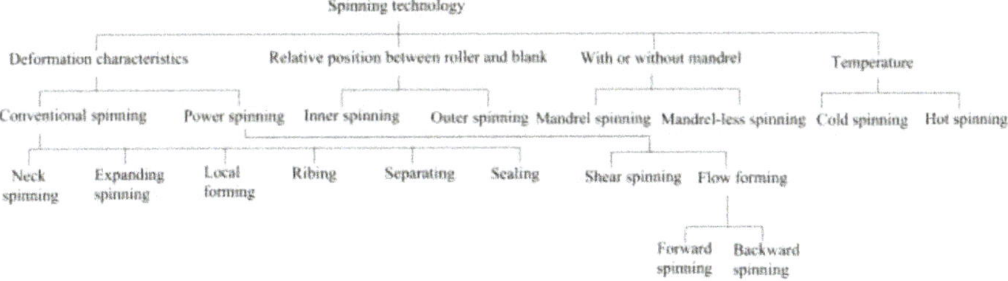

Figure 39. Classification of the conventional spinning process (reproduced with permission from [330]; copyright © 2014 Elsevier, Ltd.).

Shear spinning is the process combining conventional spinning with simultaneous intensive thinning of the wall. Shear spinning is applied to discs or drawpieces from which cylindrical, conical or curvilinear shapes with a thick bottom and thin walls are obtained. Unit pressures in shear spinning processes reach 3000 MPa. The process parameters, i.e., circumferential speed, reaching up to $5~\mathrm{m\cdot s^{-1}}$ and a feed rate in the range of 0.01–0.25 mm per revolution, are assumed. The finished component formed by shear spinning is characterized by a very smooth surface and increased mechanical properties due to the work hardening phenomenon.

Flow forming, also known as tube spinning, is one of the techniques closely allied to shear forming [334,335]. In this process (Figure 40), the sheet metal is displaced axially along a mandrel, while the internal diameter remains constant. Flow forming is usually employed to produce cylindrical components [332,336]. There are many varieties of flow forming methods with regard to tool design. Most modern flow forming machines employ two or three rollers and their design and structural strength is more complex than that of spinning and shear forming machines. Using spinning methods, it is possible to produce axisymmetric and asymmetric components with complex shapes, a task which is difficult or sometimes even impossible by conventional stamping methods. Spinning methods are used in particular for difficult-to-form alloys [334,337].

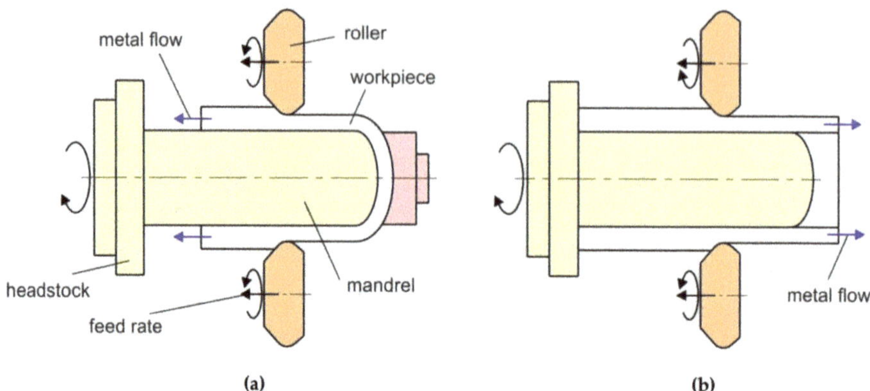

Figure 40. (**a**) forward and (**b**) backward flow forming.

8.2. Conventional Axisymmetrical Spinning

The most challenging aspect in this process is the low formability of the workpiece material due to the formation of wrinkling in the free flange [326]. Several attempts have been made to study the spinning process, particularly the thickness, geometry and profiles of the final component, failure modes and material formability [338–341]. The most important parameters affecting wrinkling are, large mandrel diameter, small modulus of plastic buckling and small thickness of the original blank [342–345]. In addition, several studies [346–348] have shown that spinning of thicker sheet metals will fail due to the formation of wrinkling in the unsupported part of the blank at higher roller feeds [327,347,349]. The tendency for wrinkling to occur increases significantly with higher roller angles and higher roller feeds. In the case of the deep spinning process, the formation of wrinkling can be limited by a roller aided by a constant clearance blank holder (Figure 41). Experimental investigations when forming commercially pure aluminum Al99.5 sheets confirm the ability of the proposed tool to suppress wrinkling in order to improve spinning formability [326]. Based on overcoming fracture failure and wrinkling, many attempts have also investigated using conical, flat conical or D-shaped rollers [327] with various working conditions to enhance spinning formability.

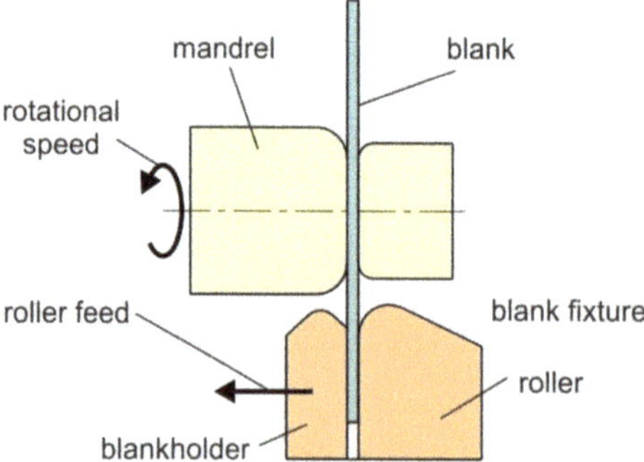

Figure 41. Schematic diagram of deep spinning.

Using the spinning process to produce axisymmetric parts produces a thinner cup wall with smaller roller nose radii, smaller roller feeds and higher roller angles. The inner profile of the final spun cup is larger than the mandrel profile due to springback [350]. The springback phenomenon is more visible with smaller roller nose radii, higher conical roller angles and higher roller feeds. Automation of the spinning process requires techniques to predict workpiece failure based on the path of the tool.

Analytical methods to predict workpiece failure are of practical use for online correction of the tool path. Therefore, Russo et al. [351] proposed a method in which a haptic device is connected to a computer numerical control (CNC) spinning machine; this device allows a human operator to control the working roller manually while feeling the force applied to the workpiece. Haptic devices provide users with force feedback and have found applications in gaming and robotic surgery [332,352]. A control system consisting of force, position and workpiece shape sensors allows the collection of information on the tool path followed by the operator and on its effect on the mechanics of the forming process.

8.3. Asymmetric Spinning

The challenge of extending the conventional spinning processes to other areas of application has given rise to investigation of the fabrication of non-axisymmetric parts. Four approaches to this have been identified; using a radially offset mandrel, using spring-controlled rollers, using a feedback control system and using a radially offset roller (Figure 42).

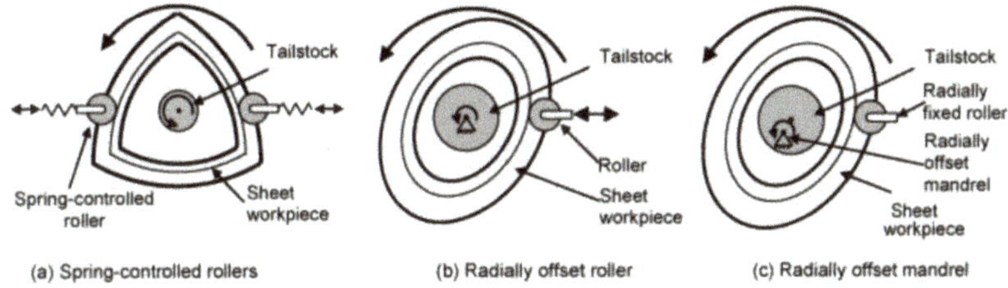

Figure 42. Schematic diagram of asymmetric spinning: (**a**) spring-controlled rollers, (**b**) radially offset roller, (**c**) radially offset mandrell (reproduced with permission from [333]; copyright © 2010 Elsevier B.V.).

To overcome the limitations of ordinary asymmetric spinning, Shimizu [339] developed a machine (Figure 43) for the synchronous spinning of asymmetric truncated cone-shaped components. In this case, the mandrel feed, roller feed and mandrel motion were synchronized by pulse control. The successful application of the proposed method for the formation of a truncated pyramid-shaped product with a sidewall and an asymmetric truncated, elliptical, cone-shaped product confirmed the flexibility of this method.

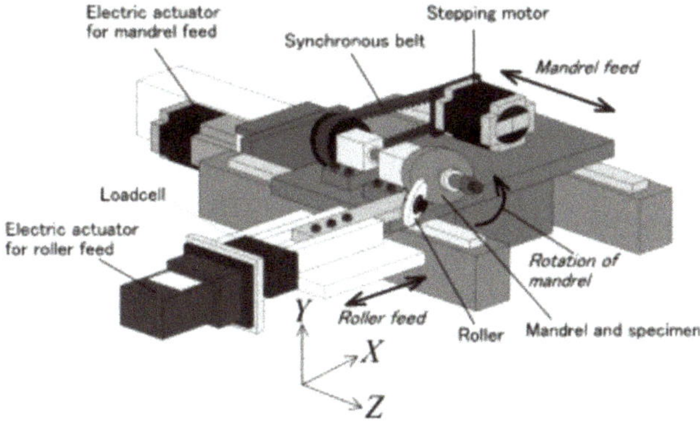

Figure 43. Schematic diagram of asymmetric spinning (reproduced with permission from [339]; copyright © 2010 Elsevier B.V.).

Methods to design the multipass tool paths and blanks required for the spinning of both asymmetric and axisymmetric components without a mandrel were investigated by Russo et al. [353]. They applied the flexible spinning method (Figure 44) developed by Music and Allwood [354]. The results show that increasing the degree of asymmetry of the target part only weakly influences the forming weight achievable in mandrel-free spinning. This method is seen as having great potential to fabricate multiple geometries flexibly and to reduce the costs of prototyping considerably.

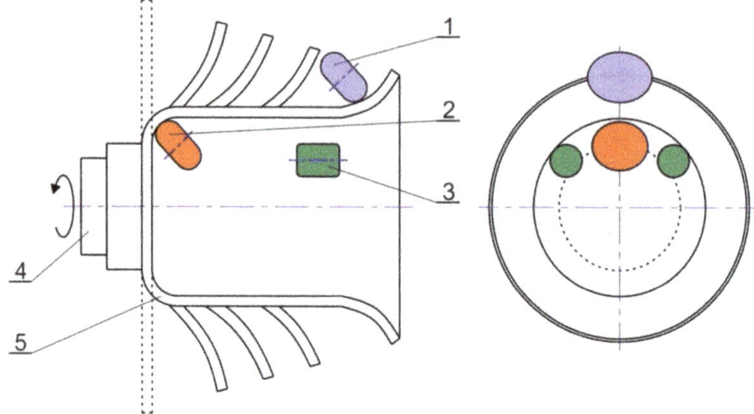

Figure 44. Schematic diagram of asymmetric spinning: 1–working roller, 2–blending roller, 3–support roller, 4–tailstock, 5–workpiece.

Sugita and Arai [342] were the first to perform multipass spinning on a part with asymmetric vertical walls in a rectangular box (Figure 45a). They designed a synchronous multipass spinning machine (Figure 45b) which could apply to both axisymmetric and asymmetric cups. Their setup employed a mandrel. They found that a lower height could be achieved in an asymmetric component than in an axisymmetric one. Arai [355] applied a hybrid force/position control system to allow the roller to track an asymmetric mandrel. This approach has been successfully applied to form asymmetric drawpieces in shear spinning.

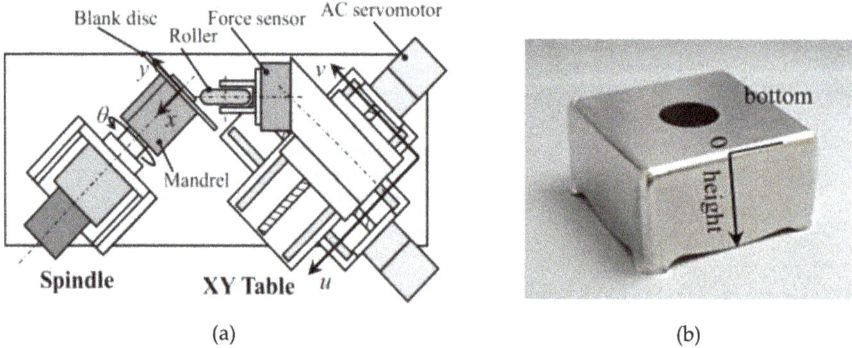

Figure 45. (**a**) Synchronous multipass spinning machine and (**b**) an example of the rectangular box shape formed using rotational pass (reproduced with permission from [342]; copyright © 2015 Elsevier B.V.).

8.4. Heat-Assisted Spinning

In conventional spinning processes, the forming limit of the workpiece material is restricted due to the work hardening effect [356]. An effective approach to improve the forming limits is the use of heat treatment which can be achieved by intermediate heating, burner systems or the use of lasers (Figure 46) [333,357]. These approaches worsen the process flexibility by increasing the costs associated with energy and maintenance. The main benefits of laser-assisted metal spinning operations are: (i) improved reproducibility when compared to gas burners, (ii) improved formability of challenging materials such as nickel-based alloys [334,358,359], magnesium alloys [360,361] and titanium alloys [362–364], (iii) reduced forming forces and (iv) improved component quality due to locally limited heating. In their study, Nguyen-Tran et al. [365] summarize the previously reported electroplastic behavior of various metals or metal alloys and recent electrically assisted manufacturing processes.

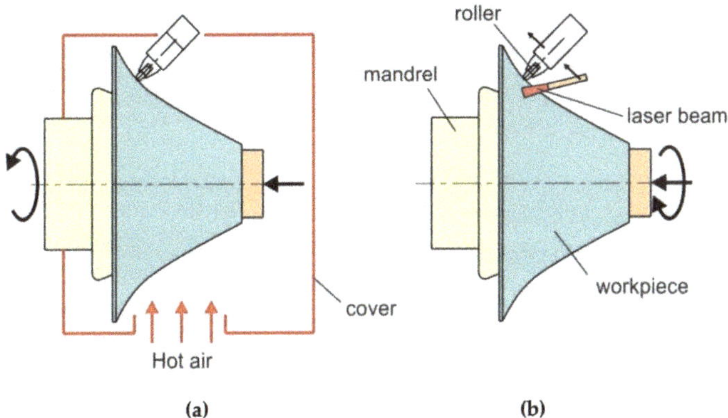

Figure 46. Hot spinning with (**a**) hot air and (**b**) laser.

Another approach is to use the friction-spinning process (Figure 47). As a consequence of a localized warm-forming operation, multifunctional components can be manufactured with locally varying mechanical properties that satisfy the demands of lightweight design. Figure 48 shows the forming strategies in conventional and heat assisted spinning. The conventional spinning process is characterized by the use of additional equipment and complex tool path geometries. If an additional reduction in wall thickness of the drawpiece side wall is required, it is necessary to use a multipass strategy. Furthermore, it is possible to achieve a defined adjustment of the hardness distribution by varying the parameters. The possibility of extending the forming limits of conventional spinning through in-process heat generation is confirmed by Lossen and Homberg [356].

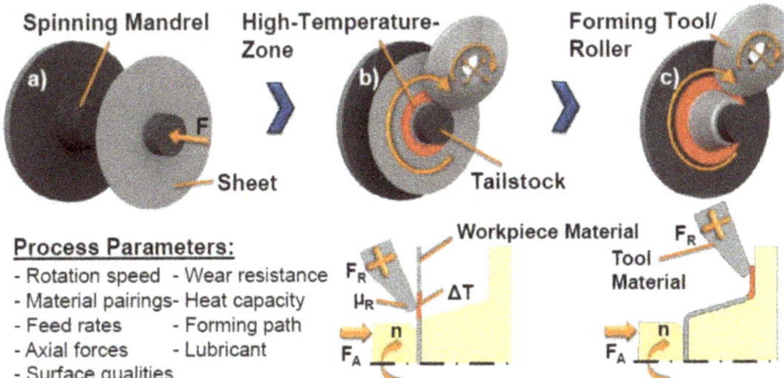

Figure 47. Friction-assisted spinning (reproduced with permission from [356]; copyright © 2014 the author(s). Published by Elsevier, Ltd.).

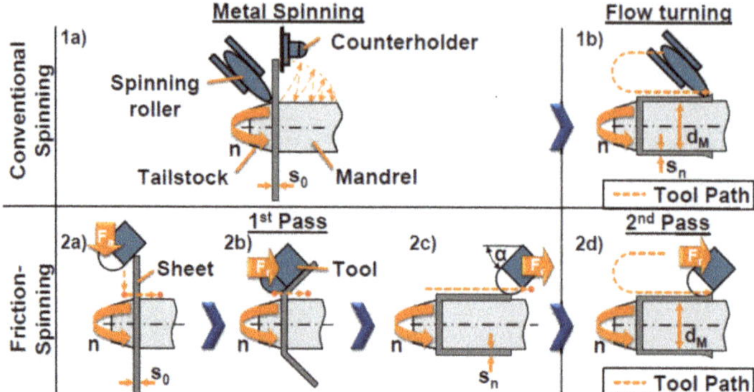

Figure 48. Comparison of the principles of cup forming by friction-assisted spinning and conventional spinning with one-pass and two-pass forming strategies (reproduced with permission from [356]; copyright © 2014 the author(s). Published by Elsevier, Ltd.).

Homberg et al. [366] used frictional heat between the workpiece and friction tool/roller (Figure 49) to increase workpiece temperature. In the process a workpiece is set in rotation in a conventional spinning machine while a pressure or friction plate works in an axial direction on the material to be deformed. The combination of spinning with friction-induced heat makes it possible to achieve a larger deformation to deform tailor-made components which are functionally graded and to produce a workpiece with more complex geometry [367]. Therefore, the use of frictional heat may provide a potential heating method which can be widely applied in the hot spinning process.

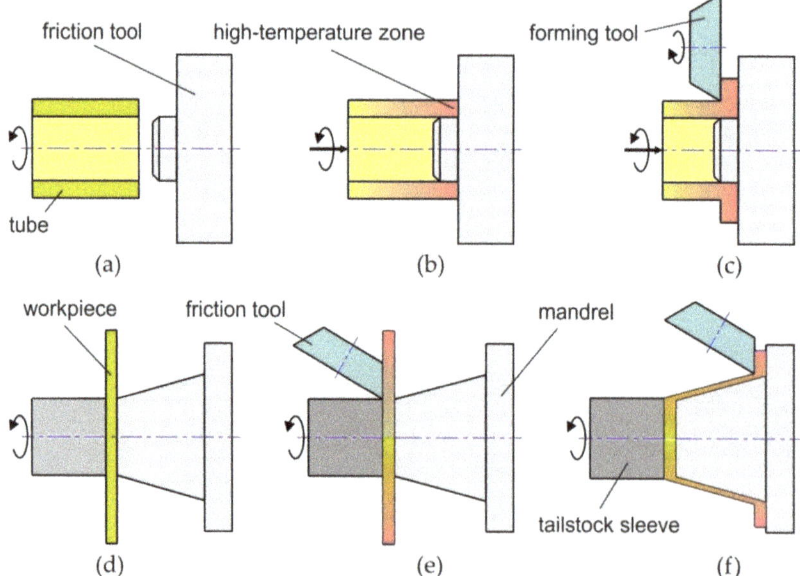

Figure 49. (a–f) Sequential stages of friction spinning.

In recent years, the demands on lightweight construction with respect to its functional qualities and potential to save resources by reducing weight have increased significantly. Tailor-made components with functionally graded properties manufactured by a novel so-called friction-spinning process meet these demands very well. This new process combines elements from metal spinning and friction welding to a new thermomechanical process for the manufacture of complex hollow parts made of tubes, profiles or sheet metals.

8.5. Shear Spinning and Flow Forming

In shear spinning forming, a sheet blank is formed by a roller into an axisymmetric part with a desired shape and thickness distribution. The thickness is deliberately reduced to obtain the desired distribution. Due to the localized deformation of the material, shear spinning has major advantages, such as high utilization of the material, simple tooling and low forming force. The work hardening phenomenon improves the mechanical properties of spun components. Hence, this often eliminates the requirement for any additional heat treatment to be carried out on finished parts [368,369]. Shear spinning is widely used to manufacture components in the aerospace, weapon, aviation and automotive industries [330,332,333]. Materials that are difficult to deform at room temperature, such as titanium alloys [370,371], nickel-based alloys [357,372] and stainless steel [373], are commonly deformed using the shear spinning process.

Dieless shear spinning (DLSS), as a kind of conventional shear spinning, can form many types of product using only an elastic multipurpose base instead of a mandrel [124,374]. At first, a blank sheet is laid over an elastomer base and is rotated on a lathe (Figure 50). Then a bulged circle is formed on it like a ring doughnut when the hemisphere tip of the bar tool is pressed on it. The component is completed when its wall height is increased by repetition of piling up the traveled steps [124]. DLSS is considered an innovative method since it is possible to form components with a vertical wall. DLSS is found to be an effective way of forming the skin of the air intake lip of airplane jet engine nacelles [375].

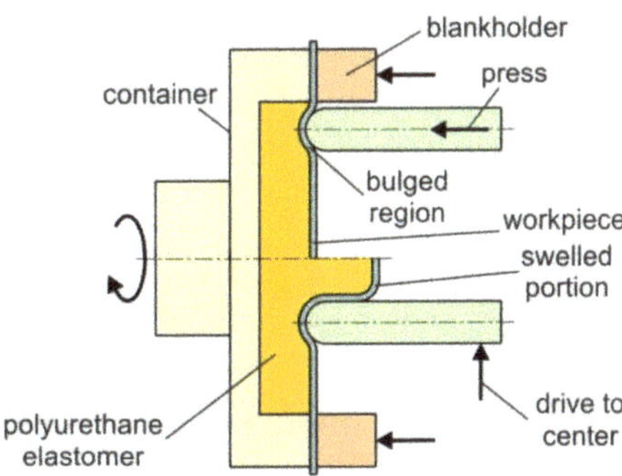

Figure 50. Dieless shear spinning.

In double-sheet dieless shear spinning, two aluminum disk blanks arranged concentrically are formed into a truncated cone. The cover thicknesses of the spun workpieces are smaller than the value that conforms to the sine law using the same blank thickness. The wall thicknesses of the base spun workpieces are greater than the sine law value under the same circumstances. The theoretical mechanics analysis of the double-sheet shear spinning process is carried out through mechanics and finite element analysis. Jia et al. [376] analyzed the mechanics of double-sheet dieless shear spinning (Figure 51). In a double-sheet DLSS the roller path coincides with that of traditional shear spinning. It is predicted that the wall thickness distribution can be different from the sine law while retaining the same shape of the lower workpiece with the single-sheet working condition.

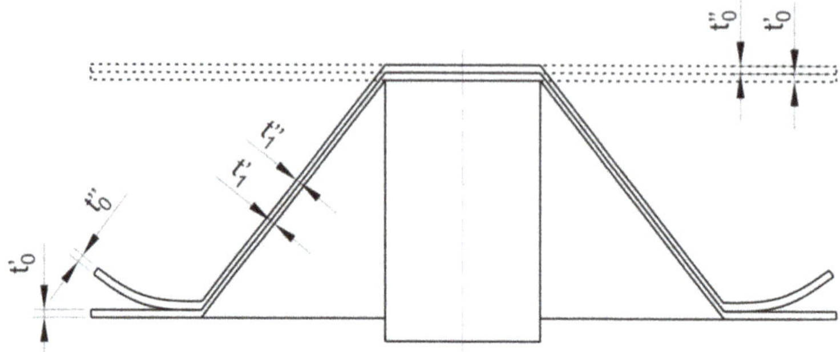

Figure 51. Schematic diagram of double-sheet shear spinning.

Han et al. [370] proposed innovative flame-heating integration into a spinning machine to apply new processing strategies for flame-assisted dieless shear and multipass metal spinning. The results of forming pure titanium TA2 and its alloy TC4 plates show that the forming accuracy is acceptable when adopting the system under two forming temperatures, 450 °C and 850 °C. In double-sheet dieless shear spinning, two aluminum disk blanks arranged concentrically are formed for a truncated cone. The cover thicknesses of the spun workpieces are smaller than the value that conforms to the sine law using the same blank thickness. The wall thicknesses of base spun workpieces are greater than the sine law value under the same circumstances.

Flow forming (Figure 52) is one of the incremental bulk-metal-forming processes, used for the manufacture of tubular parts [377]. Flow-forming was used to process a wide range of materials [378,379]: Inconel, Hastelloy, titanium, maraging steels, stainless steels and precipitated hardened stainless steels [167]. Multipass tube spinning at elevated temperature is generally applied when processing titanium parts due to its advantages of simple tooling, low forming force and high material utilization [332,377,380,381]. Later research is focused more on the analysis of different kinematic and dynamic characteristics of the flow forming process and assurances of geometric accuracy of the formed components. Sivanandini et al. [378] have shown potential applications of the flow forming method in automotive, defense and aerospace applications.

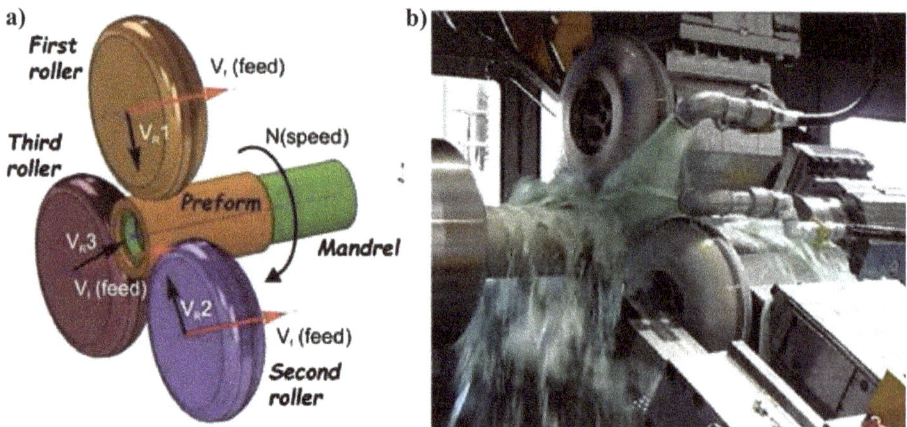

Figure 52. (**a**) schematic diagram of flow forming and (**b**) the flow forming machine (reproduced with permission from [377]; copyright © 2018 Elsevier B.V.).

9. Conclusions

This study gives an overview of the main topics concerning the development of sheet metal forming methods including spinning and shear spinning, flow forming, incremental sheet forming including water jet sheet forming, flexible-die forming, multipoint die forming, solid granular medium forming, electromagnetic and electrohydraulic forming. The progress observed in the last decade in the area of SMF concerns mainly the development of nonconventional methods of forming difficult-to-form lightweight materials for automotive and aircraft applications. The most important trends in the development of modern machines include the improvement of their design to increase their performance with significant production flexibility, reducing production costs and the development of structures adapted to unconventional methods of plastic forming. To state the major achievements in the previous sections, the following conclusions are highlighted:

- Friction and lubrication are key factors in sheet metal forming which decide product quality and productivity as well as the environmental performance of manufacturing. The main problems observed in SMF concern processes carried out at elevated temperatures where high levels of friction and wear occur due to high adhesion between the tool surface and the workpiece and surface fatigue is then initiated. National and international regulations have meant that the use of environmentally friendly lubricants has become increasing important. At the same time friction conditions have to be minimized in order to reduce loads. An example of environmentally friendly technology is hydroforming, which eliminates metallic contact of the forming tool with the workpiece. Thus, friction forces and tool wear are eliminated. In evaluating the ecological convenience of SMF processes the tribological aspects will receive great attention in the near future.

- In new vehicle development, weight reduction is one of the most important driving forces. The main trend involved in producing lightweight automotive structures with low cost manufacturing and reduced vehicle weight is the use of advanced high-strength steels and aluminum alloys with very high strength and good formability. However, some manufacturing problems related to residual stresses and springback must be overcome. Nonconventional forming techniques (i.e., electromagnetic or electrohydraulic forming, solid granular medium forming) with high deformation rates cover these requirements. Hot forming utilizing the advantageous effect of higher temperatures on the improvement of formability should be also mentioned. The advantage of hydroforming is that it shapes the product in one operation, which speeds up the production cycle and eliminates the need for interoperational storage.

- The utilization of conventional forming is more time-consuming and costly for processes recently applied in the production of parts in small batch production. Consequently, there is a need to disseminate an alternative process to reduce the manufacturing costs and time while forming individual parts, like medical implants. Processes which show benefits in this respect are methods of incremental sheet forming which do not require the manufacture of dies for operation and have the ability to shape elements on a conventional CNC milling machine. The problems related to material springback may be effectively reduced by rapid change of the forming strategy and tool path.

- The increasing complexity of components formed by SMF techniques as well as the continuous extension of the process window have made it necessary to develop new methods to efficiently analyze the forming strategies. Current numerical analysis software offers the capability of designing the tooling and process parameters in a virtual environment. Application of general purpose FEM, DEM, ALE and CFD codes may be regarded as one of the possible routes in the simulation of sheet metal forming. The use of simulation programs requires a thorough knowledge of continuum mechanics and FE programming to develop a suitable numerical analysis for a forming process. In the future, most experimental and numerical research work must be focused on the development of a macroscopic constitutive model based on a physical mechanism, grain-twinning interactions and a crystal plasticity model considering the grain–grain interaction and new constitutive models to establish the material behavior of new multifunctional materials.

- The goal of the designers of modern machines and instrumentation for SMF is to ensure that the product can be manufactured in one work cycle and on one machine. The development of machine tool design for plastic working is focused on improving the quality of the product and increasing productivity while maintaining the economic aspects of the manufacturing process. Progress in the implementation of automatic control systems and product manipulation during and after machining is clearly visible in the area of plastic forming processes. Despite the constant tendency to reduce the size of the production series related to the short-term launching of new series of vehicles, it should be hoped that the demand of the automotive and aviation industries for innovative technologies will translate into a greater interest in metal forming technologies. In this manufacturing area, all the considerations given in the previous section should be analyzed taking into account the sustainable break-even point. The development of innovative SMF processes will depend upon the improvement of materials and require adaptive control systems and integrated design of the forming process equipment and tooling.

Funding: This research received no external funding.

Conflicts of Interest: The author declares no conflict of interest.

References

1. Hagenah, H.; Schulte, R.; Vogel, M.; Hermann, J.; Scharrer, H.; Lechner, M.; Merklein, M. 4.0 in metal forming - questions and challenges. *Procedia CIRP* **2019**, *79*, 649–654. [CrossRef]
2. Gronostajski, Z.; Pater, Z.; Madej, L.; Gontarz, A.; Lisiecki, L.; Lukaszek-Solek, A.; Luksza, J.; Mróz, S.; Muskalski, Z.; Muzykiewicz, W.; et al. Recent development trends in metal forming. *Arch. Civ. Mech. Eng.* **2019**, *19*, 898–941. [CrossRef]
3. Han, S.S. The influence of tool geometry on friction behavior in sheet metal forming. *J. Mater. Process. Technol.* **1997**, *63*, 129–133. [CrossRef]
4. Wang, P.Y.; Wang, Z.J.; Xiang, N.; Li, Z.X. Investigation on changing loading path in sheet metal forming by applying a property-adjustable flexible-die. *J. Manuf. Process.* **2020**, *53*, 364–375. [CrossRef]
5. Shisode, M.P.; Hazrati, J.; Mishra, T.; de Rooij, M.; van den Boogaard, T. Modeling mixed lubrication friction for sheet metal forming applications. *Procedia Manuf.* **2020**, *47*, 586–590. [CrossRef]
6. Evin, E.; Tomáš, M. Verification of models implemented in the simulation software. *Mater. Sci. Forum* **2020**, *994*, 223–231. [CrossRef]
7. Löfgren, H.B. A first order friction model for lubricated sheet metal forming. *Theor. Appl. Mech. Lett.* **2018**, *8*, 57–61. [CrossRef]
8. Figueiredo, L.; Ramalho, A.; Oliveira, M.C.; Menezes, L.F. Experimental study of friction in sheet metal forming. *Wear* **2011**, *271*, 1651–1657. [CrossRef]
9. Wang, W.; Zhao, Y.; Wang, Z.; Hua, M.; Wei, X. A study on variable friction model in sheet metal forming with advanced high strength steels. *Tribol. Int.* **2016**, *93*, 17–28. [CrossRef]
10. Wang, Z.; Gu, R.; Chen, S.; Wang, W.; Wei, X. Effect of upper-die temperature on the formability of AZ31B magnesium alloy sheet in stamping. *J. Mater. Process. Technol.* **2018**, *257*, 180–190. [CrossRef]
11. Kowalik, M.; Trzepiecinski, T. Examination of the influence of pressing parameters on strength and geometry of joint between aluminum plate and sheet metal. *Arch. Civ. Mech. Eng.* **2012**, *12*, 292–298. [CrossRef]
12. Seshacharyulu, K.; Bandhavi, C.; Naik, B.B.; Rao, S.S.; Singh, S.K. Understanding friction in sheet metal forming-A review. *Mater. Today Proc.* **2018**, *5*, 18238–18244. [CrossRef]
13. Jaworski, J.; Trzepieciński, T. Research on durability of turning tools made of low-alloy high-speed steels. *Kov. Mater.-Met. Mater.* **2016**, *54*, 17–25. [CrossRef]
14. Kirkhorn, L.; Bushlya, V.; Andersson, M.; Ståhl, J.E. The influence of tool steel microstructure on friction in sheet metal forming. *Wear* **2013**, *302*, 1268–1278. [CrossRef]

15. Sulaiman, M.H.; Farahana, R.N.; Bienk, K.; Nielsen, C.V.; Bay, N. Effects of DLC/TiAlN-coated die on friction and wear in sheet-metal forming under dry and oil-lubricated conditions: Experimental and numerical studies. *Wear*, 2019; 438–439, 203040.

16. Wang, C.; Ma, R.; Zhao, J.; Zhao, J. Calculation method and experimental study of coulomb friction coefficient in sheet metal forming. *J. Manuf. Process.* **2017**, *27*, 126–137. [CrossRef]

17. Flegler, F.; Neuhäuser, S.; Groche, P. Influence of sheet metal texture on the adhesive wear and friction behaviour of EN AW-5083 aluminum under dry and starved lubrication. *Tribol. Int.* **2020**, *141*, 105956. [CrossRef]

18. Makhkamov, A.; Wagre, D.; Baptista, A.M.; Santos, A.D.; Malheiro, L. Tribology testing to friction determination in sheet metal forming processes. *Ciência Tecnol. dos Mater.* **2017**, *29*, e249–e253. [CrossRef]

19. Sigvant, M.; Pilthammar, J.; Hol, J.; Wiebenga, J.H.; Chezan, T.; Cerleer, B.; van den Boogard, T. Friction in sheet metal forming: Influence of surface roughness and strain rate on sheet metal forming simulaion results. *Procedia Manuf.* **2019**, *29*, 512–519. [CrossRef]

20. Ma, N.; Sugitomo, N. Development and application of non-linear friction models for metal forming simulation. *AIP Conf. Proc.* **2011**, *1383*, 382–389.

21. Ma, N.; Sugimoto, N. Nonlinear friction model for servo press simulation. *AIP Conf. Proc.* **2013**, *1567*, 918.

22. Trzepiecinski, T.; Lemu, H.G. Recent developments and trends in the friction testing for conventional sheet metal forming and incremental sheet forming. *Metals* **2020**, *10*, 47. [CrossRef]

23. Yuan, B.; Wang, Z. A multi-deformable bodies solution method coupling finite element with meshless method in sheet metal flexible-die forming. *Procedia Eng.* **2017**, *207*, 1641–1646. [CrossRef]

24. Jamli, M.R.; Farid, N.M. The sustainability of neural network applications within finite element analysis in sheet metal forming: A review. *Measurement* **2019**, *138*, 446–460. [CrossRef]

25. Ma, N.; Zhu, T.; Ogawa, R.; Liu, Y.; Harada, Y. Development of 3D Thick Shell and its Application to Sheet Metal Forming Simulation. In Proceedings of the 10th International Conference and Workshop on Numerical Simulation of 3D Sheet Metal Forming Processes Numisheet, Bristol, UK, 4–9 September 2016; pp. 5–6.

26. Topčagić, Z.; Križaj, D.; Bulić, E. Application of a current sheet in BEM analysis for numerical calculation of torque in the magnetostatic field. *IEEE Trans. Magn.* **2020**, *56*, 1–9. [CrossRef]

27. Siddiqui, M.A.; Correia, J.P.M.; Ahzi, S.; Belouettar, S. A numerical model to simulate electromagnetic sheet metal forming process. *Int. J. Mater. Form.* **2008**, *1*, 1387–1390. [CrossRef]

28. Kumar, R.U. Role of CFD in sheet metal forming. *Int. J. Tech. Innov. Mod. Eng. Sci.* **2018**, *4*, 458–462.

29. Zakaria, A.; Ibrahim, M.S.N.; Dezfouli, M.M.S. CFD evaluation of hot stamping die cooling system. *Int. J. Eng. Technol.* **2018**, *7*, 68–71. [CrossRef]

30. Saad, M.; Akhtar, S.; Srivastava, M.; Chaurasia, J. Role of simulation in metal forming processes. *Mater. Today Proc.* **2018**, *5*, 19576–19585. [CrossRef]

31. Wu, Y.; Shen, Y.; Chen, K.; Yu, Y.; He, G.; Wu, P. Multi-scale crystal plasticity finite element method (CPFEM) simulations for shear band development in aluminum alloys. *J. Alloy. Compd.* **2017**, *711*, 495–505. [CrossRef]

32. Mellbin, Y.; Hallberg, H.; Ristinmaa, M. A combined plasticity and graph-based vertex model of dynamic recrystallization at large deformations. *Model. Simul. Mater. Sci. Eng.* **2015**, *23*, 045011. [CrossRef]

33. da Silva, G.S.; Kosteski, L.E.; Iturrioz, I. Analysis of the failure process by using the Lattice Discrete Element Method in the Abaqus environment. *Theoret. Appl. Fract. Mech.* **2020**, *107*, 102563. [CrossRef]

34. Tallinen, T.; Åström, J.A.; Timonen, J. Discrete element simulations of crumpling of thin sheets. *Comput. Phys. Commun.* **2009**, *180*, 512–516. [CrossRef]

35. Xue, F.; Li, F.; Li, J.; He, M.; Yuan, Z.; Wang, R. Numerical modeling crack propagation of sheet metal forming based on stress state parameters using XFEM method. *Comput. Mater. Sci.* **2013**, *69*, 311–326. [CrossRef]

36. Gasiorek, D.; Baranowski, P.; Malachowski, J.; Mazurkiewicz, L.; Wiercigroch, M. Modelling of guillotine cutting of multi-layered aluminum sheets. *J. Manuf. Process.* **2018**, *34*, 374–388. [CrossRef]

37. Crutzen, Y.; Boman, R.; Papeleux, L.; Ponthot, J.P. Lagrangian and arbitrary Lagrangian Eulerian simulations of complex roll-forming processes. *Comptes Rendus Mécanique* **2016**, *344*, 251–266. [CrossRef]

38. Łach, Ł.; Nowak, J.; Svyetlichnyy, D. The evolution of the microstructure in AISI 304L stainless steel during the flat rolling – modeling by frontal cellular automata and verification. *J. Mater. Process. Technol.* **2018**, *255*, 488–499. [CrossRef]

39. Yang, H.; Wu, C.; Li, H.; Fan, X.G. Review on cellular automata simulations of microstructure evolution during metal forming process: Grain coarsening, recrystallization and phase transformation. *Sci. China Technol. Sci.* **2011**, *54*, 2107–2118. [CrossRef]

40. Kochmann, J.; Wulfinghoff, S.; Reese, S.; Mianroodi, J.R.; Svendsen, B. Two-scale FE–FFT- and phase-field-based computational modeling of bulk microstructural evolution and macroscopic material behavior. *Comput. Methods Appl. Mech. Eng.* **2016**, *305*, 89–110. [CrossRef]

41. Hongyu, W.; Fei, T.; Zhen, W.; Pengchao, Z.; Juncai, S.; Shijun, J. Simulation research about rubber pad forming of corner channel with convex or concave mould. *J. Manuf. Process.* **2019**, *40*, 94–104. [CrossRef]

42. Li, T.; Garg, R.; Galvin, J.; Pannala, S. Open-source MFIX-DEM software for gas-solids flows: Part II-Validation studies. *Powder Technol.* **2012**, *220*, 138–150. [CrossRef]

43. Neuwirth, J.; Antonyuk, S.; Heinrich, S.; Jacob, M. CFD-DEM study and direct measurement of the granular flow in a rotor granulator. *Chem. Eng. Sci.* **2013**, *86*, 151–163. [CrossRef]

44. Martínez-Valle, Á.; Martínez-Jiménez, J.M.; Goes, P.; Faes, K.; De Waele, W. Multiphysics fully-coupled modelling of the electromagnetic compression of steel tubes. *Adv. Mater. Res.* **2011**, *214*, 31–39. [CrossRef]

45. Liu, X.; Sun, W.K.; Liew, K.M. Multiscale modeling of crystal plastic deformation of polycrystalline titanium at high temperatures. *Comput. Methods Appl. Mech. Eng.* **2018**, *340*, 932–955. [CrossRef]

46. Tolipov, A.A.; Elghawail, A.; Shushing, S.; Pham, D.; Essa, K. Experimental research and numerical optimisation of multi-point sheet metal forming implementation using a solid elastic cushion system. *IOP Conf. Ser. J. Phys.* **2017**, *896*, 012120. [CrossRef]

47. Ablat, M.A.; Quattawi, A. Numerical simulation of sheet metal forming: A review. *Int. J. Adv. Manuf. Technol.* **2017**, *89*, 1235–1250. [CrossRef]

48. Makinouchi, A.; Teodosiu, C.; Nakagawa, T. Advance in FEM simulation and its related technologies in sheet metal forming. *CIRP Ann. -Manuf. Technol.* **1998**, *47*, 641–649. [CrossRef]

49. Reis, F.J.P.; Pires, F.M.A. An adaptive sub-incremental strategy for the solution of homogenization-based multi-scale problems. *Comput. Methods Appl. Mech. Eng.* **2013**, *257*, 164–182. [CrossRef]

50. Coenen, E.W.C.; Kouznetsova, V.G.; Bosco, E.; Geers, M.G.D. A multi-scale approach to bridge microscale damage and macroscale failure: A nested computational homogenization-localization Framework. *Int. J. Fract.* **2012**, *178*, 157–178. [CrossRef]

51. Dong, G.J.; Bi, J.; Du, B.; Zhao, C.C. Research on AA6061 tubular components prepared by combined technology of heat treatment and internal high pressure forming. *J. Mater. Process. Technol.* **2017**, *242*, 126–138. [CrossRef]

52. Trzepieciński, T.; Lemu, H.G. Effect of computational parameters on springback prediction by numerical simulation. *Metals* **2017**, *7*, 380. [CrossRef]

53. Madej, L. Digital/virtual microstructures in application to metals engineering—A review. *Arch. Civ. Mech. Eng.* **2017**, *17*, 839–854. [CrossRef]

54. Perzyński, K.; Wrożyna, A.; Kuziak, R.; Legwand, A.; Madej, L. Development and validation of multi scale failure model for dual phase steel. *Finite Elem. Anal. Des.* **2017**, *124*, 7–21. [CrossRef]

55. Han, F.; Roters, F.; Raabe, D. Microstructure-based multiscale modeling of large strain plastic deformation by coupling a full-field crystal plasticity-spectral solver with an implicit finite element solver. *Int. J. Plast.* **2020**, *125*, 97–117. [CrossRef]

56. Gerasimov, D.; Gartvig, A. Parallel computing of metal forming simulation in QForm software. *Comput. Methods Mater. Sci.* **2016**, *16*, 139–142.

57. Alfaro, I.; Yvonnet, J.; Cueto, E.; Chinesta, F.; Doblare, M. Meshless methods with application to metal forming. *Comput. Methods Appl. Mech. Eng.* **2006**, *195*, 6661–6675. [CrossRef]

58. Liewald, M.; Riedmüller, K.R. A new one-phase material model for the numerical prediction of critical material flow conditions in thixoforging processes. *CIRP Ann.* **2019**, *68*, 293–296. [CrossRef]

59. Van Houtte, P.; Gawad, J.; Eyckens, P.; Van Bael, B.; Samaey, G.; Roose, D. Multi-scale modelling of the development of heterogeneous distributions of stress, strain, deformation texture and anisotropy in sheet metal forming. *Procedia IUTAM* **2012**, *3*, 67–75. [CrossRef]

60. Barik, S.K.; Narayanan, R.G.; Sahoo, N. Forming response of AA5052–H32 sheet deformed using a shock tube. *Trans. Nonferrous Met. Soc. China* **2020**, *30*, 603–618. [CrossRef]

61. Raju, C.; Haloi, N.; Narayanan, C.S. Strain distribution and failure mode in single point incremental forming (SPIF) of multiple commercially pure aluminum sheets. *J. Manuf. Process.* **2017**, *30*, 328–335. [CrossRef]

62. Nakata, T.; Xu, C.; Ohashi, H.; Yoshida, Y.; Yoshida, K.; Kamado, S. New Mg–Al based alloy sheet with good room-temperature stretch formability and tensile properties. *Scr. Mater.* **2020**, *180*, 16–22. [CrossRef]

63. Badrish, A.; Morchhale, A.; Kotkunde, N.; Singh, S.K. Influence of material modeling on warm forming behavior of nickel based super alloy. *Int. J. Mater. Form.* **2020**. [CrossRef]

64. Liu, X.; Li, H.; Zhan, M. A review on the modeling and simulations of solid-state diffusional phase transformations in metals and alloys. *Manuf. Rev.* **2018**, *5*, 10. [CrossRef]

65. Rodríguez-Martínez, J.A.; Rusinek, A.; Pesci, R.; Zaera, R. Experimental and numerical analysis of the martensitic transformation in AISI 304 steel sheets subjected to perforation by conical and hemispherical projectiles. *Int. J. Solids Struct.* **2013**, *50*, 339–351. [CrossRef]

66. Neto, D.M.; Oliveira, M.C.; Santos, A.D.; Alves, J.L.; Menezes, L.F. Influence of boundary conditions on the prediction of springback and wrinkling in sheet metal forming. *Int. J. Mech. Sci.* **2017**, *122*, 244–254. [CrossRef]

67. Han, F.; Mo, J.H.; Qi, H.W.; Long, R.F.; Cui, X.H.; Li, Z.W. Springback prediction for incremental sheet forming based on FEM-PSONN technology. *Trans. Nonferrous Met. Soc. China* **2013**, *23*, 1061–1071. [CrossRef]

68. Basak, S.; Panda, S.K.; Lee, M.G. Formability and fracture in deep drawing sheet metals: Extended studies for pre-strained anisotropic thin sheets. *Int. J. Mech. Sci.* **2020**, *170*, 105346. [CrossRef]

69. Bansal, A.; Lingam, R.; Yadav, S.K.; Reddy, N.V. Prediction of forming forces in single point incremental forming. *J. Manuf. Process.* **2017**, *28*, 486–493. [CrossRef]

70. Abdelkader, W.B.; Bahloul, R.; Arfa, H. Numerical investigation of the influence of some parameters in SPIF process on the forming forces and thickness distributions of a bimetallic sheet CP-titanium/low-carbon steel compared to an individual layer. *Procedia Manuf.* **2020**, *47*, 1319–1327. [CrossRef]

71. Hu, Q.; Zhang, F.; Li, X.; Chen, J. Overview on the prediction models for sheet metal forming failure: Necking and ductile fracture. *Acta Mech. Solida Sin.* **2018**, *31*, 259–289. [CrossRef]

72. Amaral, R.; Santos, A.D.; de Sá José, C.; Miranda, S. Formability prediction for AHSS materials using damage models. *J. Phys. Conf. Ser.* **2017**, *843*, 012018. [CrossRef]

73. Oliveira, M.C.; Fernandes, J.V. Modelling and simulation of sheet metal forming processes. *Metals* **2019**, *9*, 1356. [CrossRef]

74. Madej, L.; Ambrozinski, M.; Kwiecien, M.; Gronostajski, Z.; Pietrzyk, M. Digital material representation concept applied to investigation of local inhomogeneities during manufacturing of magnesium components for automotive applications. *Int. J. Mater. Res.* **2017**, *108*, 3–11. [CrossRef]

75. Ma, N.; Takada, K.; Sato, K. Measurement of local strain path and identification of ductile damage limit based on simple tensile test. *Procedia Eng.* **2014**, *81*, 1402–1407. [CrossRef]

76. Takuda, H.; Hama, T.; Nishida, K.; Yoshida, T.; Nitta, J. Prediction of forming limit in stretch flanging by finite element simulation combined with ductile fracture criterion. *Comp. Meth. Mater. Sci.* **2009**, *9*, 137–142.

77. Ma, N.; Sato, K.; Takada, K. Analysis of local fracture strain and damage limit of advanced high strength steels using measured displacement fields and FEM. *Comput. Mater. Contin.* **2015**, *46*, 195–219.

78. Siswanto, W.A.; Anggono, A.D.; Omar, B.; Jusoff, K. An alternate method to springback compensation for sheet metal forming. *Sci. World. J.* **2014**, *2014*, 301271. [CrossRef] [PubMed]

79. Fekete, J.R.; Hall, J.N.; Meuleman, D.J.; Rupp, M. Progress in implementation of advanced high-strength steels into vehicle structures. *Iron Steel Technol.* **2008**, *5*, 55–64.

80. Lingbeek, R.A.; Huetink, H.; Ohnimus, S.; Weiher, J. Iterative Springback Compensation of NUMISHEET Benchmark #1. In Proceedings of the 6th International Conference and Workshop on Numerical Simulation of 3D Sheet Metal Forming Processes, NUMISHEET 2005, Detroit, MI, USA, 15–19 August 2005; pp. 1–6.

81. Ren, H.; Xie, J.; Liao, S.; Leem, D.; Cao, J. In-situ springback compensation in incremental sheet forming. *CIRP Ann.* **2019**, *68*, 317–320. [CrossRef]

82. Li, Y.; Liang, Z.; Zhang, Z.; Zou, T.; Shi, L. An analytical model for rapid prediction and compensation of springback for chain-die forming of an AHSS U-channel. *Int. J. Mech. Sci.* **2019**, *159*, 195–212. [CrossRef]

83. Zheng, K.; Politis, D.J.; Wang, L.; Lin, J. A review on forming techniques for manufacturing lightweight complex—Shaped aluminium panel components. *Int. J. Light. Mater. Manuf.* **2018**, *1*, 55–80. [CrossRef]

84. Alghtani, A.; Brooks, P.C.; Barton, D.C.; Toropov, V.V. Springback Analysis and Optimization in Sheet Metal Forming. In Proceedings of the 9th European LS-DYNA Conference, Menchester, UK, 3–4 June 2013; pp. 1–11.

85. Ma, N.; Sugitomo, N. Dynamic explicit FEM and simulation on sheet metal forming. *J. Jpn. Soc. Technol. Plast.* **2006**, *47*, 29–34. [CrossRef]

86. Kato, Y.; Umezu, Y.; Watanabe, Y. Recent Developments in JSTAMP/NV for the Best Stamping Simulation Environment. In Proceedings of the 11th International LS-DYNA Users Conference, Dearborn, MI, USA, 6–8 June 2010.

87. Shindo, T.; Sugimoto, N.; Ma, N. Springback simulation and compensation for high strength parts using JSTAMP. *AIP Conf. Proc.* **2011**, *1383*, 1086–1091.

88. Ma, N.; Zhu, X. Computational stoning method for surface defect detection. *AIP Conf. Proc.* **2013**, *1567*, 728–731.

89. Transparent Reporting of Systematic Reviews and Meta-Analyses. Available online: http://prisma-statement.org/ (accessed on 7 April 2020).

90. Engel, U.; Eckstein, R. Microforming - from basic research to its realization. *J. Mater. Process. Technol.* **2002**, *125*, 35–44. [CrossRef]

91. Fu, M.W.; Chan, W.L. A review on the state-of-the-art microforming technologies. *Int. J. Adv. Manuf. Technol.* **2013**, *67*, 2411–2437. [CrossRef]

92. Zheng, J.Y.; Yang, H.P.; Fu, M.W.; Ng, C. Study on size effect affected progressive microforming of conical flanged parts directly using sheet metals. *J. Mater. Process. Technol.* **2019**, *272*, 72–86. [CrossRef]

93. Furushima, T.; Tsunezaki, H.; Manabe, K.I.; Alexsandrov, S. Ductile fracture and free surface roughening behaviors of pure copper foils for micro/meso-scale forming. *Int. J. Mach. Tools Manuf.* **2014**, *76*, 34–48. [CrossRef]

94. Mao, M.Y.; Peng, L.F.; Fu, M.W.; Lai, X.M. Co-effect of microstructure and surface constraints on plastic deformation in micro- and mesoscaled forming process. *Int. J. Adv. Manuf. Technol.* **2018**, *98*, 1861–1886. [CrossRef]

95. Cheng, C.; Wan, M.; Meng, B.; Zhao, Z.; Han, W.P. Size effect on the yield behavior of metal foil under multiaxial stress states: Experimental investigation and modelling. *Int. J. Mech. Sci.* **2019**, *151*, 760–771. [CrossRef]

96. Wang, Z.; Li, S.; Wang, X.; Cui, R.; Zhang, W. Modeling of surface layer and strain gradient hardening effects on micro-bending of non-oriented silicon steel sheet. *Mater. Sci. Eng. A* **2018**, *711*, 498–507. [CrossRef]

97. Cheng, T.C.; Lee, R.S. The influence of grain size and strain rate effects on formability of aluminium alloy sheet at high-speed forming. *J. Mater. Process. Technol.* **2018**, *253*, 134–159. [CrossRef]

98. Lin, B.T.; Huang, K.M.; Kuo, C.C.; Wang, W.T. Improvement of deep drawability by using punch surfaces with microridges. *J. Mater. Process. Technol.* **2015**, *225*, 275–285. [CrossRef]

99. Wang, X.; Xu, J.; Jiang, Z.; Zhu, W.L.; Shan, D.; Guo, B.; Cao, J. Size effects on flow stress behavior during electrically-assisted micro-tension in a magnesium alloy AZ31. *Mater. Sci. Eng. A* **2016**, *659*, 215–224. [CrossRef]

100. Xu, Z.; Peng, L.; Bao, E. Size effect affected springback in micro/meso scale bending process: Experiments and numerical modeling. *J. Mater. Process. Technol.* **2018**, *252*, 407–420. [CrossRef]

101. Wang, J.L.; Fu, M.W.; Ran, J.Q. Analysis of the size effect on springback behavior in micro-scaled U-bending process of sheet metals. *Adv. Eng. Mater.* **2014**, *16*, 421–432. [CrossRef]

102. Fu, M.W.; Chan, W.L. Geometry and grain size effects on the fracture behavior of sheet metal in micro-scale plastic deformation. *Mater. Des.* **2011**, *32*, 4738–4746. [CrossRef]

103. Ran, J.Q.; Fu, M.W.; Chan, W.L. The influence of size effect on the ductile fracture in micro-scaled plastic deformation. *Int. J. Plast.* **2013**, *41*, 65–81. [CrossRef]

104. Chan, W.L.; Fu, M.W. Experimental and simulation based study on micro-scaled sheet metal deformation behavior in microembossing process. *Mater. Sci. Eng. A* **2012**, *556*, 60–67. [CrossRef]

105. Chan, W.L.; Fu, M.W. Experimental studies of plastic deformation behaviors in microheading process. *J. Mater. Process. Technol.* **2012**, *212*, 1501–1512. [CrossRef]

106. Rosochowski, A.; Presz, W.; Olejnik, L.; Richert, M. Micro-extrusion of ultra-fine grained aluminium. *Int. J. Adv. Manuf. Technol.* **2007**, *33*, 137–146. [CrossRef]

107. Meng, B.; Fu, M.W.; Shi, S.Q. Deformation characteristic and geometrical size effect in continuous manufacturing of cylindrical and variable-thickness flanged microparts. *J. Mater. Process. Technol.* **2018**, *252*, 546–558. [CrossRef]

108. Sato, H.; Manabe, K.I.; Wei, D.; Jiang, Z.; Alexandrov, S. Tribological behavior in micro–sheet hydroforming. *Tribol. Int.* **2016**, *97*, 302–312. [CrossRef]

109. Mahabunphachai, S.; Cora, O.N.; Koç, M. Effect of manufacturing processes on formability and surface topography of protonexchange membrane fuel cell metallic bipolar plates. *J. Power. Sources* **2010**, *195*, 5269–5277. [CrossRef]

110. Liu, Y.; Hua, L. Fabrication of metallic bipolar plate for protonexchange membrane fuel cells by rubber pad forming. *J. Power Sources* **2010**, *195*, 3529–3535. [CrossRef]

111. Peng, L.; Hu, X.; Mei, D.; Ni, J. Investigation of micro/meso sheet soft punch stamping process – simulation and experiments. *Mater. Des.* **2009**, *30*, 783–790. [CrossRef]

112. Lim, S.S.; Kim, Y.T.; Kang, C.G. Fabrication of aluminum 1050 micro-channel proton exchange membrane fuel cell bipolar plate using rubber-pad-forming process. *Int. J. Adv. Manuf. Technol.* **2013**, *65*, 231–238. [CrossRef]

113. Elyasi, M.; Khatir, F.A.; Hosseinzadeh, M. Manufacturing metallic bipolar plate fuel cells through rubber pad forming process. *Int. J. Adv. Manuf. Technol.* **2017**, *89*, 3257–3269. [CrossRef]

114. Morales, M.; Porro, J.A.; García-Ballesteros, J.J.; Molpeceres, C.; Ocaña, J.L. Effect of plasma confinement on laser shock microforming of thin metal sheets. *Appl. Surf. Sci.* **2011**, *257*, 5408–5412. [CrossRef]

115. Ocaña, J.L.; Morales, M.; García-Ballesteros, J.J.; Porro, J.A.; García, O.; Molpeceres, C. Laser shock microforming of thin metal sheets. *Appl. Surf. Sci.* **2009**, *255*, 5633–5636. [CrossRef]

116. Shen, Z.; Gu, C.; Liu, H.; Wang, X. An experimental study of overlapping laser shock micro-adjustment using a pulsed Nd:YAG laser. *Opt. Laser Technol.* **2013**, *54*, 110–119. [CrossRef]

117. Risch, D.; Beerwald, C.; Brosius, A.; Kleiner, M. On the significance of the die design for electromagnetic sheet metal forming. In Proceedings of the 1st International Conference on High Speed Forming, Dortmund, Germany, 31 March–1 April 2004; pp. 191–200.

118. Shen, Z.; Wang, X.; Liu, H.; Wang, Y.; Wang, C. Rubber-induced uniform laser shock wave pressure for thin metal sheets microforming. *Appl. Surf. Sci.* **2015**, *327*, 307–312. [CrossRef]

119. Kuhfuss, B.; Schattmann, C.; Jahn, M.; Schmidt, A.; Vollertsen, F.; Moumi, E.; Schenck, C.; Herrmann, M.; Ishkina, S.; Rathmann, L.; et al. Micro forming processes. In *Cold Micro Metal Forming*, 1st ed.; Vollertsen, F., Friedrich, S., Kuhfuß, B., Maaß, P., Thomy, C., Zoch, H.W., Eds.; Springer: Cham, Switzerland, 2020; pp. 27–94.

120. Thomy, C.; Wilhelmi, P.; Onken, A.K.; Schenck, C.; Kuhfuss, B.; Tracht, K.; Rippel, D.; Lütjen, M.; Freitag, M. Process design. In *Cold Micro Metal Forming*, 1st ed.; Vollertsen, F., Friedrich, S., Kuhfuß, B., Maaß, P., Thomy, C., Zoch, H.W., Eds.; Springer: Cham, Switzerland, 2020; pp. 95–132.

121. Vollertsen, F.; Seven, J.; Messaoudi, H.; Mikulewitsch, M.; Fischer, A.; Goch, G.; Mehrafsun, S.; Riemer, O.; Maaß, P.; Böhmermann, F.; et al. Tooling. In *Cold Micro Metal Forming*, 1st ed.; Vollertsen, F., Friedrich, S., Kuhfuß, B., Maaß, P., Thomy, C., Zoch, H.W., Eds.; Springer: Cham, Switzerland, 2020; pp. 133–251.

122. Maaß, P.; Piotrowska-Kurczewski, I.; Agour, M.; von Freyberg, A.; Staar, B.; Falldorf, C.; Fischer, A.; Lütjen, M.; Freitag, M.; Goch, G.; et al. Quality control and characterization. In *Cold Micro Metal Forming*, 1st ed.; Vollertsen, F., Friedrich, S., Kuhfuß, B., Maaß, P., Thomy, C., Zoch, H.W., Eds.; Springer: Cham, Switzerland, 2020; pp. 253–310.

123. Trzepieciński, T.; Pieja, T.; Malinowski, T.; Smusz, R.; Motyka, M. Investigation of 17-4PH steel microstructure and conditions of elevated temperature forming of turbine engine strut. *J. Mater. Process. Technol.* **2018**, *252*, 191–200. [CrossRef]

124. Takuhiro, S.; Suzuki, N.; Takeuchi, O. Culinder forming by die-less shear spinning with sheet thickness controlling of its wall. *Procedia Manuf.* **2018**, *15*, 1232–1238. [CrossRef]

125. Lade, J.; Banoth, B.N.; Gupta, A.K.; Singh, S.K. Metallurgical studies of austenitic stainless steel 304 under warm deep drawing. *J. Iron Steel Res. Int.* **2014**, *21*, 1147–1151. [CrossRef]

126. Afshin, E.; Kadkhodayan, M. An experimental investigation into the warm deep-drawing process on laminated sheets under various grain sizes. *Mater. Des.* **2015**, *87*, 25–35. [CrossRef]

127. Panicker, S.S.; Panda, S.K. Investigations into improvement in formability of AA5754 and AA6082 sheets at elevated temperatures. *J. Mater. Eng. Perform.* **2019**, *28*, 2967–2982. [CrossRef]

128. Zhang, Z.; Xu, Y.; Yuan, S. Analysis of thickness variation of reverse deep drawing of preformed 5A06 aluminum alloy cup under different temperatures. *Int. J. Adv. Manuf. Technol.* **2016**, *86*, 521–529. [CrossRef]

129. Liu, S.; Long, M.; Ai, S.; Zhao, Y.; Chen, D.; Feng, Y.; Duan, H.; Ma, M. Evolution of phase transition and mechanical properties of ultra-high strength hot-stamped steel during quenching process. *Metals* **2020**, *10*, 138. [CrossRef]

130. Kayhan, E.; Kaftanoglu, B. Experimental investigation of non-isothermal deep drawing of DP600 steel. *Int. J. Adv. Manuf. Technol.* **2018**, *99*, 695–706. [CrossRef]

131. Singh, K.; Mahesh, K.; Gupta, A.K. Prediction of mechanical properties of extra deep drawn steel in blue brittle region using artificial neural Network. *Mater. Des.* **2010**, *31*, 2288–2295. [CrossRef]

132. Krishna, P.G.; Venugopal, L.; Prabhu, T.R. Experimental investigation of failures in redrawing of Ti-6Al-4V alloy at warm conditions. *Mater. Today Proc.* **2017**, *4*, 8478–8485. [CrossRef]

133. Panicker, S.S.; Prasad, K.S.; Sawale, G.; Hazra, S.; Shollock, B.; Panda, S.K. Warm redrawing of AA6082 sheets and investigations into the effect of aging heat treatment on cup wall strength. *Mater. Sci. Eng. A* **2019**, *768*, 138445. [CrossRef]

134. Hussaini, S.M.; Krishna, G.; Gupta, A.K.; Singh, S.K. Development of experimental and theoretical forming limit diagrams for warm forming of austenitic stainless steel 316. *J. Manuf. Process.* **2015**, *18*, 151–158. [CrossRef]

135. Keum, Y.T.; Han, B.Y. Springback of FCC sheet in warm forming. *J. Ceram. Process. Res.* **2002**, *3*, 159–165.

136. Chen, F.K.; Huang, T.B. Formability of stamping magnesium-alloy AZ31 sheets. *J. Mater. Process. Technol.* **2003**, *142*, 643–647. [CrossRef]

137. Oliveira, M.C.; Alves, J.L.; Chaparro, B.M.; Menezes, L.F. Study on the influence of work-hardening modeling in springback prediction. *Int. J. Plast.* **2007**, *23*, 516–543. [CrossRef]

138. Billur, E. Warm Hydroforming Characteristics of Stainless Steel Sheet Metals. Master's Thesis, Virginia University, Richmond, VA, USA, 2018.

139. Laurent, H.; Coër, J.; Manach, P.Y.; Oliveira, M.C.; Menezes, L.F. Experimental and numerical studies on the warm deep drawing of an Al–Mg Allom. *Int. J. Mech. Sci.* **2015**, *93*, 59–72. [CrossRef]

140. Harrison, N.R.; Friedman, P.A. Warm forming die design, Part II: Parting surface temperature response characterization of a novel thermal finite element modeling code. *J. Manuf. Process.* **2014**, *16*, 312–319. [CrossRef]

141. Harrison, N.R.; Friedman, P.A.; Pan, J. Warm forming die design, Part III: Design and validation of a warm forming die. *J. Manuf. Process.* **2015**, *20*, 356–366. [CrossRef]

142. Davis, J.R. *ASM Specialty Handbook: Heat-Resistant Materials*; ASM International: Materials Park, OH, USA, 1997; pp. 3–66.

143. Martins, J.M.P.; Alves, J.L.; Neto, D.M.; Oliveira, M.C.; Menezes, L.F. Numerical analysis of different heating systems for warm sheet metal forming. *Int. J. Adv. Manuf. Technol.* **2016**, *83*, 897–909. [CrossRef]

144. Hasanuzzaman, M.; Rahim, N.A.; Hosenuzzaman, M.; Saidur, R.; Mahbubul, I.M.; Rashid, M.M. Energy savings in the combustion based process heating in industrial sector. *Renew. Sustain. Energy Rev.* **2012**, *16*, 4527–4536. [CrossRef]

145. Harrison, N.R.; Ilinich, A.; Friedman, P.; Singh, J.; Verma, R. Optimization of high-volume warm forming for lightweight sheet. *SAE Tech. Pap.* **2013**, 2013-01-1170. [CrossRef]

146. Bong, H.J.; Barlat, F.; Ahn, D.C.; Kim, H.Y.; Lee, M.G. Formability of austenitic and ferritic stainless steels at warm forming temperature. *Int. J. Mech. Sci.* **2013**, *75*, 94–109. [CrossRef]

147. Gronostajski, Z.; Niechajowicz, A.; Kuziak, R.; Krawczyk, J.; Polak, S. The effect of the strain rate on the stress-strain curve and microstructure of AHSS. *J. Mater. Process. Technol.* **2017**, *242*, 246–259. [CrossRef]

148. Wiewiórowska, S.; Muskalski, Z. The application of low and medium carbon steel with multiphase TRIP structure in drawing industry. *Procedia Manuf.* **2015**, *2*, 181–185. [CrossRef]

149. Cai, Z.; Wan, M.; Liu, Z.; Wu, X.; Ma, B.; Cheng, C. Thermal-mechanical behaviors of dual-phase steel sheet under warm-forming conditions. *Int. J. Mech. Sci.* **2017**, *126*, 79–94. [CrossRef]

150. Maeno, T.; Mori, K.; Nagai, T. Improvement in formability by control of temperature in hot stamping of ultra-high strength steel parts. *CIRP Ann.* **2014**, *63*, 301–304. [CrossRef]

151. Tokita, Y.; Nakagaito, T.; Tamai, Y.; Urabe, T. Stretch formability of high strength steel sheets in warm forming. *J. Mater. Process. Technol.* **2017**, *246*, 77–84. [CrossRef]

152. Saito, N.; Fukahori, M.; Hisano, D.; Hamasaki, H.; Yoshida, F. Effects of temperature, forming speed and stress relaxation on springback in warm forming of high strength steel sheet. *Procedia Eng.* **2017**, *207*, 2394–2398. [CrossRef]

153. Kumar, M.; Sotirov, N.; Chimani, C.M. Investigations on warm forming of AW-7020-T6 alloy sheet. *J. Mater. Process. Technol.* **2014**, *214*, 1769–1776. [CrossRef]

154. Grimes, R.; Janik, V. Automotive applications for magnesium. In *Encyclopedia of Automotive Engineering*; Crolla, D., Foster, D.E., Kobayashi, T., Vaughan, N., Eds.; John Wiley & Sons, Inc.: Hoboken, NJ, USA, 2014; pp. 3101–3123.

155. Zhang, R.; Shao, Z.; Lin, J. A review on modelling techniques for formability prediction of sheet metal forming. *Int. J. Light. Mater. Manuf.* **2018**, *1*, 115–125. [CrossRef]

156. Martins, J.M.P.; Neto, D.M.; Alves, J.L.; Oliveira, M.C.; Laurent, H.; Andrade-Campos, A.; Menezes, L.F. A new staggered algorithm for thermomechanical coupled problems. *Int. J. Solids Struct.* **2017**, *122–123*, 42–58. [CrossRef]

157. Harrison, N.R.; Rubek, V.; Friedman, P.A. Warm forming die design, part I: Experimental validation of a novel thermal finite element modeling code. *J. Manuf. Process.* **2013**, *15*, 263–272. [CrossRef]

158. Abovyan, T.; Kridli, G.T.; Friedman, P.A.; Ayoub, G. Formability prediction of aluminum sheet alloys under isothermal forming conditions. *J. Manuf. Process.* **2015**, *20*, 406–413. [CrossRef]

159. Neto, D.M.; Martins, J.M.P.; Cunha, P.M.; Alves, J.L.; Oliveira, M.C.; Laurent, H.; Menezes, L.F. Thermo-mechanical finite element analysis of the AA5086 alloy under warm forming conditions. *Int. J. Solids Struct.* **2017**, *151*, 99–117. [CrossRef]

160. Panicker, S.S.; Singh, H.G.; Panda, S.K.; Dashwood, R. Characterization of tensile properties, limiting strains, and deep drawing behavior of AA5754-H22 sheet at elevated temperature. *J. Mater. Eng. Perform.* **2015**, *24*, 4267–4282. [CrossRef]

161. Pourboghrat, F.; Venkatesan, S.; Carsley, J.E. LDR and hydroforming limit for deep drawing of AA5754 aluminum sheet. *J. Manuf. Process.* **2013**, *15*, 600–615. [CrossRef]

162. Wang, W.; Huang, L.; Tao, K.; Chen, S.; Wei, X. Formability and numerical simulation of AZ31B magnesium alloy sheet in warm stamping process. *Mater. Des.* **2015**, *87*, 835–844. [CrossRef]

163. Al-Samman, T.; Gottstein, G. Room temperature formability of a magnesium AZ31 alloy: Examining the role of texture on the deformation mechanisms. *Mater. Sci. Eng. A* **2008**, *488*, 406–414. [CrossRef]

164. Uemori, T.; Katahira, T.; Naka, T.; Tada, N.; Yoshida, F. Cyclic stress and strain responses of AZ31 magnesium alloy sheet metal at elevated temperatures. *Procedia Manuf.* **2018**, *15*, 1792–1799. [CrossRef]

165. Balasubramanian, S.; Anand, L. Plasticity of initially textured hexagonal polycrystals at high homologous temperatures: Application to titanium. *Acta Mater.* **2002**, *50*, 133–148. [CrossRef]

166. Maksoud, I.A.; Ahmed, H.; Rödel, J. Investigation of the effect of strain rate and temperature on the deformability and microstructure evolution of AZ31 magnesium alloy. *Mater. Sci. Eng. A* **2009**, *504*, 40–48. [CrossRef]

167. Liang, S.J.; Liu, Z.Y.; Wang, E.D. Simulation of extrusion process of AZ31 magnesium alloy. *Mater. Sci. Eng. A* **2009**, *499*, 221–224. [CrossRef]

168. Ramezani, M.; Neitzert, T.; Pasang, T.; Sellès, M.A. Characterization of friction behaviour of AZ80 and ZE10 magnesium alloys under lubricated contact condition by strip draw and bend test. *Int. J. Mach. Tools Manuf.* **2014**, *85*, 70–78. [CrossRef]

169. Abu-Farha, F.; Verma, R.; Hector, L.G. High temperature composite forming limit diagrams of four magnesium AZ31B sheets obtained by pneumatic stretching. *J. Mater. Process. Technol.* **2014**, *212*, 1414–1429. [CrossRef]

170. Atrian, A.; Fereshteh-Saniee, F. Deep drawing process of steel/brass laminated sheets. *Compos. Part. B* **2013**, *47*, 75–81. [CrossRef]

171. Rajabi, A.; Kadkhodayan, M.; Manoochehri, M.; Farjadfar, R. Deep-drawing of thermoplastic metal-composite structures: Experimental investigations, statistical analyses and finite element modeling. *J. Mater. Process. Technol.* **2015**, *215*, 159–170. [CrossRef]

172. Gali, O.A.; Riahi, A.R.; Alpas, A.T. The tribological behaviour of AA5083 alloy plastically deformed at warm forming temperatures. *Wear* **2016**, *302*, 1257–1267. [CrossRef]

173. Jaworski, J.; Kluz, R.; Trzepieciński, T. Operational tests of wear dynamics of drills made of low-alloy high-speed hs2-5-1 steel. *Eksploat. i Niezawodn.-Maint. Reliab.* **2016**, *18*, 271–277. [CrossRef]

174. Pelcastre, L. Hot Forming Tribology: Galling of Tools and Associated Problems. Licentiate Thesis, Luleå University of Technology, Lulea, Sweden, 2011.

175. Kondratiuk, J.; Kuhn, P. Tribological investigations on friction and wear behaviour of coatings for hot sheet metal forming. *Wear* **2011**, *270*, 839–849. [CrossRef]

176. Wu, C.; Qu, P.; Zhang, L.; Li, S.; Jiang, Z. A Numerical and experimental study on the interface friction of ball-on-disc test under high temperature. *Wear* **2017**, *376–377*, 433–442. [CrossRef]

177. Venema, J.; Matthews, D.T.A.; Hazrati, J.; Wörmann, J.; van den Boogaard, A.H. Friction and wear mechanisms during hot stamping of AlSi coated press hardening steel. *Wear* **2017**, *380–381*, 137–145. [CrossRef]

178. Uda, K.; Azushima, A.; Yanagida, A. Development of new lubricants for hot stamping of Al-coated 22MnB5 steel. *J. Mater. Process. Technol.* **2016**, *228*, 112–116. [CrossRef]

179. Bay, N.; Azushima, A.; Groche, P.; Ishibashi, I.; Merklein, M.; Morishita, M.; Nakamura, T.; Schmid, S.; Yoshida, M. Environmentally benign tribo-systems for metal forming. *CIRP Ann.* **2010**, *59*, 760–780. [CrossRef]

180. Dohda, K.; Boher, C.; Reza-Aria, F.; Mahayotsanun, N. Tribology in metal forming at elevated temperatures. *Friction* **2015**, *3*, 1–27. [CrossRef]

181. Leszak, E. Apparatus and Process for Incremental Dieless Forming. U.S. Patent 3342051A1, 19 September 1967.

182. Pathak, J. A brief review on incremental sheet metal forming. *Int. J. Latest Eng. Manag. Res.* **2017**, *2*, 35–43.

183. Xu, D.K.; Lu, B.; Cao, T.T.; Zhang, H.; Chen, J.; Hong, H.; Cao, J. Enhancement of process capabilities in electrically-assisted double sided incremental forming. *Mater. Des.* **2016**, *92*, 268–280. [CrossRef]

184. Vahdati, M.; Mahdavinejad, R.; Amini, S. Investigation of the ultrasonic vibration effect in incremental sheet metal forming process. *Proc. Inst. Mech. Eng. Part B J. Eng. Manuf.* **2017**, *231*, 971–998. [CrossRef]

185. Amini, S.; Gollo, A.H.; Paktinat, H. An investigation of conventional and ultrasonic-assisted incremental forming of annealed AA1050 sheet. *Int. J. Adv. Manuf. Technol.* **2017**, *90*, 1569–1578. [CrossRef]

186. Al-Obaidai, A.; Kunke, A.; Kräusel, V. Hot single-point incremental forming of glass-fiber-reinforced polymer(PA6GF47) supported by hot air. *J. Manuf. Process.* **2019**, *43*, 17–25. [CrossRef]

187. Ai, S.; Long, H. A review on material fracture mechanism in incremental sheet forming. *Int. J. Adv. Manuf. Technol.* **2019**, *104*, 33–61. [CrossRef]

188. Xiao, X.; Kim, C.; Lv, X.; Hwang, T.S.; Kim, Y.S. Formability and forming force in incremental sheet forming of AA7075-T6 at different temperatures. *J. Mech. Sci. Technol.* **2019**, *33*, 3795–3802. [CrossRef]

189. Kumar, A.; Gulati, V.; Kumar, P.; Singh, H. Forming force in incremental sheet forming: A comparative analysis of the state of the art. *J. Braz. Soc. Mech. Sci. Eng.* **2019**, *41*, 251. [CrossRef]

190. Fan, G.; Gao, L.; Hussain, G.; Wu, Z. Electric hot incremental forming: A novel technique. *Int. J. Mach. Tool. Manuf.* **2008**, *48*, 1688–1692. [CrossRef]

191. Kurra, S.; Rahman, N.; Regalla, S.; Gupta, A. Modeling and optimization of surface roughness in single point incremental forming process. *J. Mater. Res. Technol.* **2015**, *4*, 304–313. [CrossRef]

192. Rauch, M.; Hascoet, J.; Hamann, J.; Plenel, Y. Tool path programming optimization for incremental sheet forming applications. *Comput. Aided Des.* **2009**, *41*, 877–885. [CrossRef]

193. Oleksik, V. Influence of Geometrical Parameters, Wall Angle and Part Shape on Thickness Reduction of Single Point Incremental Forming. *Procedia Eng.* **2014**, *81*, 2280–2285. [CrossRef]

194. Al-Ghamdi, K.A.; Hussain, G. On the Free-Surface Roughness in Incremental Forming of a Sheet Metal: A Study from the Perspective of ISF Strain, Surface Morphology, Post-Forming Properties, and Process Conditions. *Metals* **2019**, *9*, 553. [CrossRef]

195. Slota, J.; Krasowski, B.; Kubit, A.; Trzepiecinski, T.; Bochnowski, W.; Dudek, K.; Neslušan, M. Residual stresses and surface roughness analysis of truncated cones of steel sheet made by single point incremental forming. *Metals* **2020**, *10*, 237. [CrossRef]

196. Jadhav, S. *Basic Investigations of the Incremental Sheet Metal Forming Process on a CNC Milling Machine*; Shaker Verlag GmbH: Aachen, Germany, 2004.

197. Hirt, G.; Ames, J.; Bambach, M. Basic investigation into the characteristics of dies and support tools used in CNC-incremental sheet forming. In Proceedings of the 25th IDDRG Conference, Porto, Portugal, 19–21 June 2006; pp. 341–348.

198. Martins, P.A.F.; Bay, N.; Skjoedt, M.; Silva, M.B. Theory of single point incremental forming. *CIRP Ann. Manuf. Technol.* **2008**, *57*, 247–252. [CrossRef]

199. Behera, A.K.; de Sousa, R.A.; Ingarao, G.; Oleksik, V. Single point incremental forming: An assessment of the progress and technology trends from 2005 to 2015. *J. Manuf. Process.* **2017**, *27*, 37–62. [CrossRef]

200. Durante, M.; Formisano, A.; Langella, A.; Minutolo, F. The influence of tool rotation on an incremental forming process. *J. Mater. Process. Technol.* **2009**, *209*, 4621–4626. [CrossRef]

201. Obikawa, T.; Satou, S.; Hakutani, T. Dieless incremental micro-forming of miniature shell objects of aluminium foils. *Int. J. Mach. Tool. Manuf.* **2009**, *49*, 906–915. [CrossRef]

202. Duflou, J.R.; Callebaut, B.; Verbert, J.; De Baerdemaeker, D. Laser assisted incremental forming: Formability and accuracy improvement. *CIRP Ann.* **2007**, *56*, 273–276. [CrossRef]

203. Göttmann, A.; Diettrich, J.; Bergweiler, G.; Bambach, M.; Hirt, G.; Loosen, P.; Poprawe, R. Laser-assisted asymmetric incremental sheet forming of titanium sheet metal parts. *Prod. Eng.* **2011**, *5*, 263–271. [CrossRef]

204. Göttmann, A.; Bailly, D.; Bergweiler, G.; Bambach, M.; Stollenwerk, J.; Hirt, G.; Loosen, P. A novel approach for temperature control in ISF supported by laser and resistance heating. *Int. J. Adv. Manuf. Technol.* **2013**, *67*, 2195–2205. [CrossRef]

205. Mohammadi, A.; Vanhove, H.; Van Bael, A.; Duflou, J.R. Towards accuracy improvement in single point incremental forming of shallow parts formed under laser assisted conditions. *Int. J. Mater. Form.* **2016**, *9*, 339–351. [CrossRef]

206. Mohammadi, A.; Qin, L.; Vanhove, H.; Seefeldt, M.; Van Bael, A.; Duflou, J.R. Single point incremental forming of an aged AL-Cu-Mg alloy: Influence of pre-heat treatment and warm forming. *J. Mater. Eng. Perform.* **2016**, *25*, 2478–2488. [CrossRef]

207. Kim, S.W.; Lee, Y.S.; Kang, S.H.; Lee, J.H. Incremental forming of Mg alloy sheet at elevated temperatures. *J. Mech. Sci. Technol.* **2007**, *21*, 1518–1522. [CrossRef]

208. Galdos, L.; Argandona, E.S.D.; Ulacia, I.; Arruebarrena, G. Warm incremental forming of magnesium alloys using hot fluid as heating media. *Key Eng. Mater.* **2012**, *504–506*, 815–820. [CrossRef]

209. Al-Obaidi, A.; Kräusel, V.; Landgrebe, D. Hot single-point incremental forming assisted by induction heating. *Int. J. Adv. Manuf. Technol.* **2015**, *82*, 1163–1171. [CrossRef]

210. Al-Obaidi, A.; Kräusel, V.; Landgrebe, D. Induction Heating Validation of Dieless Single-Point Incremental Forming of AHSS. *J. Manuf. Mater. Process.* **2017**, *1*, 5. [CrossRef]

211. Baharudin, B.; Azpen, Q.; Sulaima, S.; Mustapha, F. Experimental Investigation of Forming Forces in Frictional Stir Incremental Forming of Aluminum Alloy AA6061-T6. *Metals* **2017**, *7*, 484. [CrossRef]

212. Kim, S.W.; Lee, Y.S.; Kwon, Y.N.; Lee, J.H. A Study on Warm Incremental Forming of AZ31 Alloy Sheet. *Trans. Mater. Process.* **2008**, *17*, 373–379.

213. Ji, Y.H.; Park, J.J. Incremental forming of free surface with magnesium alloy AZ31 sheet at warm temperatures. *Trans. Nonferrous Met. Soc. China* **2008**, *18*, s165–s169. [CrossRef]

214. Ji, Y.H.; Park, J.J. Formability of magnesium AZ31 sheet in the incremental forming at warm temperature. *J. Mater. Process. Technol.* **2008**, *201*, 354–358. [CrossRef]

215. Xu, D.; Lu, B.; Cao, T.; Chen, J.; Long, H.; Cao, J. A comparative study on process potentials for frictional stir- and electric hot-assisted incremental sheet forming. *Procedia Eng.* **2014**, *81*, 2324–2329. [CrossRef]

216. Ambrogio, G.; Gagliardi, F. Temperature variation during high speed incremental forming on different lightweight alloys. *Int. J. Adv. Manuf. Technol.* **2015**, *76*, 1819–1825. [CrossRef]

217. Ambrogio, G.; Gagliardi, F.; Chamanfar, A.; Misiolek, W.Z.; Filice, L. Induction heating and cryogenic cooling in single point incremental forming of Ti-6Al-4V: Process setup and evolution of microstructure and mechanical properties. *Int. J. Adv. Manuf. Technol.* **2017**, *91*, 803–812. [CrossRef]

218. Ambrogio, G.; Gagliardi, F.; Bruschi, S.; Filice, L. On the high-speed single point incremental forming of titanium alloys. *CIRP Ann.* **2013**, *62*, 243–246. [CrossRef]

219. Buffa, G.; Campanella, D.; Fratini, L. On the improvement of material formability in SPIF operation through tool stirring action. *Int. J. Adv. Manuf. Technol.* **2013**, *66*, 1343–1351. [CrossRef]

220. Davarpanah, M.A.; Mirkouei, A.; Yu, X.; Malhotra, R.; Pilla, S. Effects of incremental depth and tool rotation on failure modes and microstructural properties in single point incremental forming of polymers. *J. Mater. Process. Technol.* **2015**, *222*, 287–300. [CrossRef]

221. Wang, J.; Li, L.; Jiang, H. Effects of forming parameters on temperature in frictional stir incremental sheet forming. *J. Mech. Sci. Technol.* **2016**, *30*, 2163–2169. [CrossRef]

222. Liu, Z. Friction stir incremental forming of AA7075-O sheets: Investigation on process feasibility. *Procedia Eng.* **2017**, *207*, 783–788. [CrossRef]

223. Uheida, E.H.; Oosthuizen, G.A.; Dimitrov, D.M.; Bezuidenhout, M.B.; Hugo, P.A. Effects of the relative tool rotation direction on formability during the incremental forming of titanium sheets. *Int. J. Adv. Manuf. Technol.* **2018**, *96*, 3311–3319. [CrossRef]

224. Adams, D.; Jeswiet, J. Single point incremental forming of 6061-T6 using electrically assisted forming methods. *Proc. Inst. Mech. Eng. B J. Eng. Manuf.* **2014**, *228*, 757–764. [CrossRef]

225. Bao, W.; Chu, X.; Lin, S.; Gao, J. Experimental investigation on formability and microstructure of AZ31B alloy in electropulse-assisted incremental forming. *Mater. Des.* **2015**, *87*, 632–639. [CrossRef]

226. Honarpisheh, M.; Abdolhoseini, M.J.; Amini, S. Experimental and numerical investigation of the hot incremental forming of Ti-6Al-4V sheet using electrical current. *Int. J. Adv. Manuf. Technol.* **2016**, *83*, 2027–2037. [CrossRef]

227. Khazaali, H.; Fereshteh-Saniee, F. A comprehensive experimental investigation on the influences of the process variables on warm incremental forming of Ti-6Al-4V titanium alloy using a simple technique. *Int. J. Adv. Manuf. Technol.* **2016**, *87*, 2911–2923. [CrossRef]

228. Liu, R.; Lu, B.; Xu, D.; Chen, J.; Chen, F.; Ou, H.; Long, H. Development of novel tools for electricity-assisted incremental sheet forming of titanium alloy. *Int. J. Adv. Manuf. Technol.* **2016**, *85*, 1137–1144. [CrossRef]

229. Najafabady, S.A.; Ghaei, A. An experimental study on dimensional accuracy, surface quality, and hardness of Ti-6Al-4 V titanium alloy sheet in hot incremental forming. *Int. J. Adv. Manuf. Technol.* **2016**, *87*, 3579–3588. [CrossRef]

230. Li, Z.; Lu, S.; Zhang, T.; Zhang, C.; Mao, Z. 1060 Al electric hot incremental sheet forming process: Analysis of dimensional accuracy and temperature. *Trans. Indian Inst. Met.* **2018**, *71*, 961–970. [CrossRef]

231. Li, Z.; Lu, S.; Zhang, T.; Zhang, C.; Mao, Z. Electric assistance hot incremental sheet forming: An integral heating design. *Int. J. Adv. Manuf. Technol.* **2018**, *96*, 3209–3215. [CrossRef]

232. Magnus, C.S. Joule heating of the forming zone in incremental sheet metal forming: Part 1. *Int. J. Adv. Manuf. Technol.* **2017**, *91*, 1309–1319. [CrossRef]

233. Magnus, C.S. Joule heating of the forming zone in incremental sheet metal forming: Part 2. *Int. J. Adv. Manuf. Technol.* **2017**, *89*, 295–309. [CrossRef]

234. Min, J.; Seim, P.; Störkle, D.; Thyssen, L.; Kuhlenkötter, B. Thermal modeling in electricity assisted incremental sheet forming. *Int. J. Mater. Form.* **2017**, *10*, 729–739. [CrossRef]

235. Pacheco, P.A.P.; Silveira, M.E. Numerical simulation of electric hot incremental sheet forming of 1050 aluminum with and without preheating. *Int. J. Adv. Manuf. Technol.* **2017**, *94*, 3097–3108. [CrossRef]

236. Palumbo, G.; Brandizzi, M. Experimental investigations on the single point incremental forming of a titanium alloy component combining static heating with high tool rotation speed. *Mater. Des.* **2012**, *40*, 43–51. [CrossRef]

237. Liu, Z. Heat-assisted incremental sheet forming: A state-of-the-art review. *Int. J. Adv. Manuf. Technol.* **2018**, *98*, 2987–3003. [CrossRef]

238. Albakri, M.; Abu-Farha, F.; Khraisheh, M. A new combined experimental–numerical approach to evaluate formability of rate dependent materials. *Int. J. Mech. Sci.* **2013**, *66*, 55–66. [CrossRef]

239. Wang, L.; Strangwood, M.; Balint, D.; Lin, J.; Dean, T. Formability and failure mechanisms of AA2024 under hot forming conditions. *Mater. Sci. Eng. A* **2011**, *528*, 2648–2656. [CrossRef]

240. Hmida, R.; Thibaud, S.; Gilbin, A.; Richard, F. Influence of the initial grain size in single point incremental forming process for thin sheets metal and microparts: Experimental investigations. *Mater. Des.* **2013**, *45*, 155–165. [CrossRef]

241. Ambrogio, G.; Ciancio, C.; Filice, L.; Gagliardi, F. Theoretical model for temperature prediction in Incremental Sheet Forming-Experimental validation. *Int. J. Mech. Sci.* **2016**, *108*, 39–48.

242. Bhattacharya, A.; Maneesh, K.; Reddy, N.; Cao, J. Formability and surface finish studies in single point incremental forming. *J. Manuf. Sci. Eng.* **2011**, *133*, 061020. [CrossRef]

243. Bagudanch, I.; Garcia-Romeu, M.L.; Centeno, G.; Elías-Zúñiga, A.; Ciurana, J. Forming force and temperature effects on single point incremental forming of polyvinylchloride. *J. Mater. Process. Technol.* **2015**, *219*, 221–229. [CrossRef]

244. Bagudanch, I.; Vives-Mestres, M.; Sabater, M.; Garcia-Romeu, M. Polymer incremental sheet forming process: Temperature analysis using response surface methodology. *Mater. Manuf. Process.* **2017**, *32*, 44–53. [CrossRef]

245. Medina-Sanchez, G.; Garcia-Collado, A.; Carou, D.; Dorado-Vicente, R. Force Prediction for Incremental Forming of Polymer Sheets. *Materials* **2018**, *11*, 1597. [CrossRef] [PubMed]

246. Clavijo-Chaparro, S.L.; Iturbe-Ek, J.; Lozano-Sánchez, L.M.; Sustaita, A.O.; Elías-Zúñiga, A. Plasticized and reinforced poly(methyl methacrylate) obtained by a dissolution-dispersion process for single point incremental forming: Enhanced formability towards the fabrication of cranial implants. *Polym. Test.* **2018**, *68*, 39–45. [CrossRef]

247. Medina-Sánchez, G.; Torres-Jimenez, E.; Lopez-Garcia, R.; Dorado-Vicente, R.; Cazalla-Moral, R. Temperature influence on Single Point Incremental Forming of PVC parts. *Procedia Manuf.* **2017**, *13*, 335–342. [CrossRef]

248. Shim, M.; Park, J. The formability of aluminum sheet in incremental forming. *J. Mater. Process. Technol.* **2001**, *113*, 654–658. [CrossRef]

249. Kim, Y.; Park, J. Effect of process parameters on formability in incremental forming of sheet metal. *J. Mater. Process. Technol.* **2002**, *130*, 42–46. [CrossRef]

250. Hussain, G.; Gao, L.; Hayat, N.; Cui, Z.; Pang, Y.; Dar, N. Tool and lubrication for negative incremental forming of a commercially pure titanium sheet. *J. Mater. Process. Technol.* **2008**, *203*, 193–201. [CrossRef]

251. Lu, B.; Fang, Y.; Xu, D.; Chen, J.; Ou, H.; Moser, N.; Cao, J. Mechanism investigation of friction-related effects in single point incremental forming using a developed oblique roller-ball tool. *Int. J. Mach. Tools Manuf.* **2014**, *85*, 14–29. [CrossRef]

252. Jeswiet, J.; Micari, F.; Hirt, G.; Bramley, A.; Duflou, J.; Allwood, J. Asymmetric single point incremental forming of sheet metal. *CIRP Ann.* **2005**, *54*, 88–114. [CrossRef]

253. Malhotra, R.; Bhattacharya, A.; Kumar, A.; Reddy, N.; Cao, J. A new methodology for multi-pass single point incremental forming with mixed toolpaths. *CIRP Ann.* **2011**, *60*, 323–326. [CrossRef]

254. López, C.; Elías-Zúñiga, A.; Jiménez, I.; Martínez-Romero, O.; Siller, H.R.; Diabb, J.M. Experimental Determination of Residual Stresses Generated by Single Point Incremental Forming of AlSi10Mg Sheets Produced Using SLM Additive Manufacturing Process. *Materials* **2018**, *11*, 2542. [CrossRef] [PubMed]

255. Hajavifard, R.; Maqbool, F.; Schmiedt-Kalenborn, A.; Buhl, J.; Bambach, M.; Walther, F. Integrated Forming and Surface Engineering of Disc Springs by Inducing Residual Stresses by Incremental Sheet Forming. *Materials* **2019**, *12*, 1646. [CrossRef] [PubMed]

256. Zhang, X.; He, T.; Miwa, H.; Nanbu, T.; Murakami, R.; Liu, S.; Cao, J.; Wang, Q.J. A new approach for analyzing the temperature rise and heat partition at the interface of coated tool tip-sheet incremental forming systems. *Int. J. Heat Mass Transf.* **2019**, *129*, 1172–1183. [CrossRef]

257. Araghi, B.T.; Manco, G.L.; Bambach, M.; Hirt, G. Investigation into a new hybrid forming process: Incremental sheet forming combined with stretch forming. *CIRP Ann.* **2009**, *58*, 225–228. [CrossRef]

258. Tandon, P.; Sharma, O.N. Experimental investigation into a new hybrid-forming process: Incremental stretch drawing. *Proc. Inst. Mech. Eng. Part B J. Eng. Manuf.* **2016**, *232*, 475–486. [CrossRef]

259. Jagtab, R.; Kumar, S. An experimental investigation on thinning and formability in hybrid incremental sheet forming process. *Procedia Manuf.* **2019**, *30*, 71–76. [CrossRef]

260. Araghi, B.T.; Göttmann, A.; Bambach, M.; Hirt, G.; Bergweiler, G.; Diettrich, J.; Steiners, M.; Saeed-Akbari, A. Review on the development of a hybrid incremental sheet forming system for small batch sizes and individualized production. *Prod. Eng.* **2011**, *5*, 393–404. [CrossRef]

261. Lu, B.; Zhang, H.; Xu, D.K.; Chen, J. A Hybrid Flexible Sheet Forming Approach towards Uniform Thickness Distribution. *Procedia CIRP* **2014**, *18*, 244–249. [CrossRef]

262. Allwood, J.; Braun, D.; Music, O. The effect of partially cut-out blanks on geometric accuracy in incremental sheet forming. *J. Mater. Proc. Technol.* **2010**, *210*, 1501–1510. [CrossRef]

263. Dong, G.J.; Zhao, C.C.; Cao, M.Y. Flexible-die forming process with solid granule medium on sheet metal. *Trans. Nonferrous Met. Soc. China* **2013**, *23*, 2666–2677. [CrossRef]

264. Koç, M.; Billur, E.; Necati, Ö.N. An experimental study on the comparative assessment of hydraulic bulge test analysis methods. *Mater. Des.* **2011**, *32*, 272–281. [CrossRef]

265. Yang, X.Y.; Lang, L.H.; Liu, K.N.; Liu, B.S. Mechanics analysis of axisymmetric thin-walled part in warm sheet hydroforming. *Chin. J. Aeronaut.* **2015**, *28*, 1546–1554. [CrossRef]

266. Zhou, G.; Wang, Y.N.; Lang, L.H. Accuracy analysis of complex curvature parts based on the rigid-flexible hydroforming. *Int. J. Adv. Manuf. Technol.* **2018**, *99*, 247–254. [CrossRef]

267. Woźniak, D.; Głowacki, M.; Hojny, M.; Pieja, T. Application of CAE systems in forming of drawpieces with use rubber-pad forming processes. *Arch. Metall. Mater.* **2012**, *57*, 1179–1187. [CrossRef]

268. Belhassen, L.; Koubaa, S.; Wali, M.; Dammak, F. Numerical prediction of springback and ductile damage in rubber-padforming process of aluminum sheet metal. *Int. J. Mech. Sci.* **2016**, *117*, 218–226. [CrossRef]

269. Irthiea, I.; Green, G.; Hashim, S.; Kriama, A. Experimental and numerical investigation on micro deep drawing process of stainless steel 304 foil using flexible tools. *Int. J. Mach. Tools Manuf.* **2014**, *76*, 21–33. [CrossRef]

270. Trzepiecinski, T.; Malinowski, T.; Pieja, T. Experimental and numerical analysis of industrial warm forming of stainless steel sheet. *J. Manuf. Process.* **2017**, *30*, 532–540. [CrossRef]

271. Chen, L.; Chen, H.; Guo, W.; Chen, G.; Wang, Q. Experimental and simulation studies of springback in rubber forming using aluminium sheet straight flanging process. *Mater. Des.* **2014**, *54*, 354–360. [CrossRef]

272. Jin, C.K.; Jeong, M.G.; Kang, C.G. Effect of rubber forming process parameters on micro-patterning of thin metallic plates. *Procedia Eng.* **2014**, *81*, 1439–1444. [CrossRef]

273. Liu, Y.; Hua, L.; Lan, J.; Wei, X. Studies of the deformation styles of the rubber-pad forming process used for manufacturing metallic bipolar plates. *J. Power Sources* **2010**, *195*, 8177–8184. [CrossRef]

274. Zhang, Q.; Wang, Z.R.; Dean, T.A. The mechanics of multi-point sandwich forming. *Int. J. Mach. Tools Manuf.* **2008**, *48*, 1495–1503. [CrossRef]

275. Walczyk, D.F.; Hosford, J.F.; Papazian, J.M. Using reconfigurable tooling and surface heating for incremental forming of composite aircraft parts. *J. Manuf. Sci. Eng.* **2003**, *125*, 333–343. [CrossRef]

276. Li, M.Z.; Cai, Z.Y.; Sui, Z.; Yan, Q.G. Multi-point forming technology for sheet metal. *J. Mater. Process. Technol.* **2002**, *129*, 333–338. [CrossRef]

277. Chen, J.J.; Li, M.Z.; Liu, W.; Wang, C.T. Sectional multipoint forming technology fir large-size sheet metal. *Int. J. Adv. Manuf. Technol.* **2005**, *25*, 935–939. [CrossRef]

278. Abosaf, M.; Essa, K.; Alghawail, A.; Tolipov, A.; Su, S.; Pham, D. Optimisation of multi-point forming process parameters. *Int. J. Adv. Manuf. Technol.* **2017**, *92*, 1849–1859. [CrossRef]

279. Nakajima, N. A newly developed technique to fabricate complicated dies and electrodes with wires. *J. Jpn. Soc. Mech. Eng.* **1969**, *72*, 498–506. [CrossRef]

280. Paunoiu, V.; Teodor, V.; Maier, C.; Baroiu, N.; Bercu, G. Study ofthe tool geometry in reconfigurable multipoint forming. *Ann. Dunărea de Jos Univ. Galaţi* **2011**, *11*, 139–144.

281. Sairajan, K.K.; Aglietti, G.S.; Mani, K.M. A review of multifunctional structure technology for aerospace applications. *Acta Astronaut.* **2016**, *120*, 30–42. [CrossRef]

282. Sun, G.Y.; Huo, X.T.; Chen, D.D.; Li, Q. Experimental and numerical study on honey-comb sandwich panels under bending and in-panel compression. *Mater. Des.* **2017**, *133*, 154–168. [CrossRef]

283. Carradò, A.; Faerber, J.; Niemeyer, S.; Ziegmann, G.; Palkowski, H. Metal/polymer/metalhybrid systems: Towards potential formability applications. *Compos. Struct.* **2011**, *93*, 715–721. [CrossRef]

284. Besse, C.C.; Mohr, D. Plasticity of formable all-metal sandwich sheets: Virtual experiments and constitutive modeling. *Int. J. Solids Struct.* **2012**, *49*, 2863–2880. [CrossRef]

285. Cai, Z.Y.; Zhang, X.; Liang, X.B. Multi-point forming of sandwich panels with egg-box-like cores andfailure behaviors in forming process: Analytical models, numerical and experimental investigations. *Mater. Des.* **2018**, *160*, 1029–1041. [CrossRef]

286. Liang, X.B.; Cai, Z.Y.; Zhang, X. Forming characteristics analysis and springback predictionof bi-directional trapezoidal sandwich panels in the multi-point bend-forming. *Int. J. Adv. Manuf. Technol.* **2018**, *98*, 1709–1720. [CrossRef]

287. Zhang, S.H.; Chen, S.F.; Ma, Y.; Song, H.W.; Cheng, M. Developments of new sheet metal forming technology and theory in China. *Acta Metall. Sin.* **2015**, *28*, 1452–1470. [CrossRef]

288. Chen, H.; Güner, A.; Khalifa, N.B.; Tekkaya, A.E. Granular media-based tube press hardening. *J. Mater. Process. Tech.* **2016**, *228*, 145–159. [CrossRef]

289. Dong, G.J.; Zhao, C.C.; Cao, M.Y. Process of back pressure deep drawing with solid granule medium on sheet metal. *J. Cent. South. Univ.* **2014**, *21*, 2617–2626. [CrossRef]

290. Psyk, V.; Risch, D.; Kinsey, B.L.; Tekkaya, A.E.; Kleiner, M. Electromagnetic forming—A review. *J. Mater. Process. Technol.* **2011**, *211*, 787–829. [CrossRef]

291. Kim, D.; Park, H.I.; Lee, J.; Kim, J.H.; Lee, M.-G.; Lee, Y. Experimental study on forming behavior of high-strength steel sheets under electromagnetic pressure. *Proc. Inst. Mech. Eng. Part B J. Eng. Manuf.* **2015**, *229*, 670–681. [CrossRef]

292. Long, A.; Wan, M.; Wang, W.; Wu, X.; Cui, X.; Ma, B. Forming methodology and mechanism of a novel sheet metal forming technology-electromagnetic superposed forming(EMSF). *Int. J. Solids Struct.* **2018**, *151*, 165–180. [CrossRef]

293. Lai, Z.; Cao, Q.; Han, X.; Huang, Y.; Deng, F.; Chen, Q.; Li, L. Investigation on plastic deformation behavior of sheet workpiece during radial Lorentz force augmented deep drawing process. *J. Mater. Process. Technol.* **2017**, *245*, 193–206. [CrossRef]

294. Paese, E.; Geier, M.; Homrich, R.P.; Rosa, P.; Rossi, R. Sheet metal electromagnetic forming using a flat spiral coil: Experiments, modeling, and validation. *J. Mater. Process. Technol.* **2019**, *263*, 408–422. [CrossRef]

295. Thibaudeau, E.; Kinsey, B.L. Analytical design and experimental validation of uniform pressure actuator for electromagnetic forming and welding. *J. Mater. Process. Technol.* **2015**, *215*, 251–263. [CrossRef]

296. Psyk, V.; Kurka, P.; Kimme, S.; Werner, M.; Landgrebe, D.; Ebert, A.; Schwarzendahl, M. Structuring by electromagnetic forming and by forming with an elastomer punch as a tool for component optimisation regarding mechanical stiffness and acoustic performance. *Manuf. Rev.* **2015**, *2*, 23. [CrossRef]

297. Mamalis, A.G.; Manolakos, D.E.; Kladas, A.G.; Koumoutsos, A.K.; Ovchinnikov, S.G. Electromagnetic forming of aluminum alloy sheet using a grooved die: Numerical modelling. *Phys. Met. Metall.* **2006**, *102*, 590–593. [CrossRef]

298. Golovashchenko, S.F. Material formability and coil design in electromagnetic forming. *J. Mater. Eng. Perform.* **2007**, *16*, 314–320. [CrossRef]

299. Balanethiram, V.S.; Hu, X.; Altynova, M.; Daehn, G.S. Hyperplasticity: Enhanced formability at high rates. *J. Mater. Process. Technol.* **1994**, *45*, 595–600. [CrossRef]

300. Cui, X.; Zhang, Z.; Yu, H.; Xiao, X.; Cheng, Y. Springback Calibration of a U-Shaped Electromagnetic Impulse Forming Process. *Metals* **2019**, *9*, 603. [CrossRef]

301. Xiong, W.R.; Wang, W.P.; Wan, M.; Li, X.J. Geometric issues in V-bending electromagnetic forming process of 2024-T3 aluminum alloy. *J. Manuf. Process.* **2015**, *19*, 171–182. [CrossRef]

302. Yu, H.; Chen, J.; Liu, W.; Yin, H.; Li, C. Electromagnetic forming of aluminum circular tubes into square tubes: Experiment and numerical simulation. *J. Manuf. Process.* **2018**, *31*, 613–623. [CrossRef]

303. Su, H.; Huang, L.; Li, J.; Ma, F.; Huang, P.; Feng, F. Two-step electromagnetic forming: A new forming approach to local features of large-size sheet metal parts. *Int. J. Mach. Tools Manuf.* **2018**, *124*, 99–116. [CrossRef]

304. Kamal, M.; Shang, J.; Cheng, V.; Hatkevich, S.; Daehn, G.S. Agile manufacturing of a micro-embossed case by a two-step electromagnetic forming process. *J. Mater. Process. Technol.* **2007**, *190*, 41–50. [CrossRef]

305. Imbert, J.; Worswick, M. Electromagnetic reduction of a pre-formed radius on AA 5754 sheet. *J. Mater. Process. Technol.* **2011**, *211*, 896–908. [CrossRef]

306. Li, J.; Qiu, W.; Huang, L.; Su, H.; Tao, H.; Li, P. Gradient electromagnetic forming (GEMF): A new forming approach for variable-diameter tubes by use of sectional coil. *Int. J. Mach. Tools Manuf.* **2018**, *135*, 65–77. [CrossRef]

307. Centeno, G.; Martínez-Donaire, A.J.; Bagudanch, I.; Morales-Palma, D.; Garcia-Romeu, M.L.; Vallellano, C. Revisiting Formability and Failure of AISI304 Sheets in SPIF: Experimental Approach and Numerical Validation. *Metals* **2017**, *7*, 531. [CrossRef]

308. Centeno, G.; Martínez-Donaire, A.J.; Morales-Palma, D.; Vallellano, C.; Martins, P.A.F. Novel experimental techniques for the determination of the forming limits at necking and fracture. In *Materials Forming and Machining*; Davim, P., Ed.; Woodhead Publishing: Cambridge, UK, 2016; pp. 1–24.

309. Cui, X.H.; Mo, J.H.; Li, J.J.; Zhao, J.; Zhu, Y.; Huang, L.; Li, W.Z.; Zhong, K. Electromagnetic incremental forming (EMIF): A novel aluminum alloy sheet and tube forming technology. *J. Mater. Process. Technol.* **2014**, *214*, 409–427. [CrossRef]

310. Lai, Z.; Cao, Q.; Han, X.; Liu, N.; Li, X.; Huang, Y.; Chen, M.; Cai, H.; Wang, G.; Liu, L.; et al. A comprehensive electromagnetic forming approach for large sheet metal forming. *Procedia Eng.* **2017**, *207*, 54–59. [CrossRef]

311. Tan, J.; Zhan, M.; Gao, P.; Li, H. Electromagnetic forming rules of a stiffened panel with grid ribs. *Metals* **2017**, *7*, 559. [CrossRef]

312. Cao, T.; Lu, B.; Ou, H.; Long, H.; Chen, J. Investigation on a new hole-flanging approach by incremental sheet forming through a featured tool. *Int. J. Mach. Tools Manuf.* **2016**, *110*, 1–17. [CrossRef]

313. Li, C.; Liu, D.; Yu, H.; Ji, Z. Research on formability of 5052 aluminum alloy sheet in a quasi-static-dynamic tensile process. *Int. J. Mach. Tools Manuf.* **2009**, *49*, 117–124. [CrossRef]

314. Kacem, A.; Krichen, A.; Manach, P.Y.; Thuillier, S.; Yoon, J.W. Failure prediction in the hole-flanging process of aluminium alloys. *Eng. Fract. Mech.* **2013**, *99*, 251–265. [CrossRef]

315. Krichen, A.; Kacem, A.; Hbaieb, M. Blank-holding effect on the hole-flanging process of sheet aluminum alloy. *J. Mater. Process. Technol.* **2011**, *211*, 619–626. [CrossRef]

316. Su, H.; Huang, L.; Li, J.; Ma, F.; Ma, H.; Huang, P.; Zhu, H.; Feng, F. Inhomogeneous deformation behaviors of oblique hole-flanging parts during electromagnetic forming. *J. Manuf. Process.* **2020**, *52*, 1–11. [CrossRef]

317. Feng, F.; Li, J.; Yuan, P.; Zhang, Q.; Huang, P.; Su, H.; Chen, R. Application of a GTN Damage Model Predicting the Fracture of 5052-O Aluminum Alloy High-Speed Electromagnetic Impaction. *Metals* **2018**, *8*, 761. [CrossRef]

318. Gillard, A.J.; Golovashchenko, S.F.; Mamutov, A.V. Effect of quasi-static prestrain on the formability of dual phase steels in electrohydraulic forming. *J. Manuf. Process.* **2013**, *15*, 201–218. [CrossRef]

319. Mamutov, A.V.; Golovashchenko, S.F.; Mamutov, V.S.; Bonnen, J.J.F. Modeling of electrohydraulic forming of sheet metal parts. *J. Mater. Process. Technol.* **2015**, *219*, 84–100. [CrossRef]

320. Golovashchenko, S.F.; Gillard, A.J.; Mamutov, A.V. Formability of dual phase steels in electrohydraulic forming. *J. Mater. Process. Technol.* **2013**, *213*, 1191–1212. [CrossRef]

321. Golovashchenko, S.F.; Mamutov, A.V.; Bonnen, J.J.F.; Gillard, A.J. Electrohydraulic forming of sheet metal parts. In Proceedings of the International Conference on the Technology of Plasticity, Aachen, Germany, 25–30 September 2011; pp. 1170–1175.

322. Golovashchenko, S.F.; Bessonov, N.M.; Ilinich, A.M. Two-step method of forming complex shapes from sheet metal. *J. Mater. Process. Technol.* **2011**, *211*, 875–885. [CrossRef]

323. Samei, J.; Green, D.E.; Golovashchenko, S.; Hassannejadasl, A. Quantitative microstructural analysis of formability enhancement in dual phase steels subject to electrohydraulic forming. *J. Mater. Eng. Perform.* **2013**, *22*, 2080–2088. [CrossRef]

324. Maris, C.; Hassannejadasl, A.; Green, D.E.; Cheng, J.; Golovashchenko, S.F.; Gillard, A.J.; Liang, Y. Comparison of quasi-static and electrohydraulic free forming limits for DP600 and AA5182 sheets. *J. Mater. Process. Technol.* **2016**, *235*, 206–219. [CrossRef]

325. Zia, M.; Fazli, A.; Soltananpour, M. Warm electrohydraulic forming: A novel high speed forming process. *Procedia Eng.* **2017**, *207*, 323–328. [CrossRef]

326. Ahmed, K.I.; Gadala, M.S.; El-Sebaie, M.G. Deep spinning of sheet metals. *Int. J. Mach. Tools Manuf.* **2015**, *97*, 72–85. [CrossRef]

327. Xia, Q.; Shima, S.; Kotera, H.; Yasuhuku, D. A study of the one-path deep drawing spinning of cups. *J. Mater. Process. Technol.* **2005**, *159*, 397–400. [CrossRef]

328. Abd-Alrazzaq, M.; Ahmed, M.; Younes, M. Experimental Investigation on the Geometrical Accuracy of the CNC Multi-Pass Sheet Metal Spinning Process. *J. Manuf. Mater. Process.* **2018**, *2*, 59. [CrossRef]

329. Liu, R.; Yu, Z.; Zhao, Y.; Evsyukov, S.A. Formability of Flange Constraint Spinning for Aluminum Cup Part. *J. Shanghai Jiaotong Univ. Sci.* **2019**, *53*, 105–110.

330. Xia, Q.; Xiao, G.; Long, H.; Cheng, X.; Sheng, X. A review of process advancement of novel metal spinning. *Int. J. Mach. Tools Manuf.* **2014**, *85*, 100–121. [CrossRef]

331. Mori, K.I.; Ishiguro, M.; Isomura, Y. Hot shear spinning of cast aluminium alloy parts. *J. Mater. Process. Technol.* **2009**, *209*, 3621–3627. [CrossRef]

332. Wong, C.C.; Dean, T.A.; Lin, J. A review of spinning, shear forming and flow forming processes. *Int. J. Mach. Tools Manuf.* **2003**, *43*, 1419–1435. [CrossRef]

333. Music, O.; Allwood, J.M.; Kawai, K. A review of the mechanics of metal spinning. *J. Mater. Process. Technol.* **2010**, *210*, 3–23. [CrossRef]

334. Żaba, K.; Nowosielski, M.; Puchlerska, S.; Kwiatkowski, M.; Kita, P.; Głodzik, M.; Korfanty, K.; Pociecha, D.; Pieja, T. Investigation of the mechanical properties and microstructure of nickel superalloys processed in shear forming. *Arch. Metall. Mater.* **2015**, *60*, 2637–2644. [CrossRef]

335. Plewiński, A.; Drenger, T. Spinning and flow forming hard-to-deform metal alloys. *Arch. Civ. Mech. Eng.* **2009**, *9*, 101–109. [CrossRef]

336. Chang, S.C.; Huang, C.A.; Yu, S.Y.; Chang, Y.; Han, W.C.; Shieh, T.S.; Chung, H.C.; Yao, H.T.; Shyu, G.D.; Hou, H.Y.; et al. Tube spinnability of AA2024 and 7075 aluminum. *J. Mater. Process. Technol.* **1998**, *80–81*, 676–682. [CrossRef]

337. Żaba, K.; Puchlerska, S.; Kwiatkowski, M.; Nowosielski, M.; Głodzik, M.; Tokarski, T.; Seibt, P. Comparative analysis of properties and microstructure of the plastically deformed alloy Inconel®718 manufactured by plastic working and direct metal laser sintering. *Arch. Metall. Mater.* **2016**, *61*, 143–148. [CrossRef]

338. Polyblank, J.A.; Allwood, J.M. Parametric toolpath design in metal spinning. *CIRP Ann.* **2015**, *64*, 301–304. [CrossRef]

339. Shimizu, I. Asymmetric forming of aluminium sheets by synchronous spinning. *J. Mater. Process. Technol.* **2010**, *210*, 585–592. [CrossRef]

340. Cheng, X.Q.; Xia, Q.; Lai, Z.Y. Investigation on stress and strain distributions of hollow-part with triangular cross-section by spinning. *Int. J. Mater. Prod. Technol.* **2013**, *47*, 162–174. [CrossRef]

341. Xia, Q.; Lai, Z.Y.; Long, H.; Cheng, X.Q. A study of the spinning force of hollow parts with triangular cross sections. *Int. J. Adv. Manuf. Technol.* **2013**, *68*, 2461–2470. [CrossRef]

342. Sugita, Y.; Arai, H. Formability in synchronous multipass spinning using simple pass set. *J. Mater. Process. Technol.* **2015**, *217*, 336–344. [CrossRef]

343. Jia, Z.; Xu, Q.; Han, Z.; Peng, W.F. Precision forming of the straight edge of square section by die-less spinning. *J. Manuf. Sci. Eng.* **2015**, *138*, 011006. [CrossRef]

344. Russo, I.M.; Loukaides, E.G. Toolpath generation for asymmetric mandrel-free spinning. *Procedia Eng.* **2017**, *207*, 1707–1712. [CrossRef]

345. Polyblank, J.A.; Allwood, J.M.; Duncan, S.R. Closed-loop control of product properties in metal forming: A review and prospectus. *J. Mater. Process. Technol.* **2014**, *214*, 2333–2348. [CrossRef]

346. Wang, L.; Long, H. Roller path design by tool compensation in multi-pass conventional spinning. *Mater. Des.* **2013**, *46*, 645–653. [CrossRef]

347. Watson, M.; Long, H. Wrinkling failure mechanics in metal spinning. *Procedia Eng.* **2014**, *81*, 2391–2396. [CrossRef]

348. Wang, L.; Long, H.; Ashley, D.; Roberts, M.; White, P. Effects of the roller feed ratio on wrinkling failure in conventional spinning of a cylindrical cup. *Proc. Inst. Mech. Eng. Part B J. Eng. Manuf.* **2011**, *225*, 1991–2006. [CrossRef]

349. Watson, M.; Long, H.; Lu, B. Investigation of wrinkling failure mechanics in metal spinning by Box–Behnken design of experiments using finite element method. *Int. J. Adv. Manuf. Technol.* **2015**, *78*, 981–995. [CrossRef]

350. El-Khabeery, M.M.; Fattouh, M.; El-Sheikh, M.N.; Hamed, O.A. On the conventional simple spinning of cylindrical aluminium cups. *Int. J. Mach. Tools Manuf.* **1991**, *31*, 203–219. [CrossRef]

351. Russo, I.M.; Cleaver, C.J.; Allwood, J.M. Haptic metal spinning. *Procedia Manuf.* **2019**, *29*, 129–136. [CrossRef]

352. Xia, P. Haptics for Product Design and Manufacturing Simulation. *IEEE Trans. Haptics* **2016**, *9*, 358–375. [CrossRef] [PubMed]

353. Russo, I.M.; Cleaver, C.J.; Allwood, J.M.; Loukaides, E.G. The influence of part asymmetry on the achievable forming height in multi-pass spinning. *J. Mater. Process. Technol.* **2020**, *275*, 116350. [CrossRef]

354. Music, O.; Allwood, J.M. Flexible asymmetric spinning. *CIRP Ann.* **2011**, *60*, 319–322. [CrossRef]

355. Arai, H. Robotic metal spinning—Forming non-axisymmetric products using force control. *J. Robot. Soc. Jpn.* **2006**, *24*, 140–145. (In Japanese) [CrossRef]

356. Lossen, B.; Homberg, W. Friction-spinning – Interesting Approach to Manufacture of Complex Sheet Metal Parts and Tubes. *Procedia Eng.* **2014**, *81*, 2379–2384. [CrossRef]

357. Brummer, C.; Eck, S.; Marsoner, S.; Arntz, K.; Klocke, F. Laser-assisted metal spinning for an efficient and flexible processing of challenging materials. *IOP Conf. Sci. Mater. Sci. Eng.* **2016**, *119*, 012022. [CrossRef]

358. Sim, M.S.; Lee, C.M. A study on the laser preheating effect of Inconel 718 specimen with rotated angle with respect to 2-axis. *Int. J. Precis. Eng. Manuf.* **2014**, *15*, 189–192. [CrossRef]

359. Li, Y.H.; Fan, T.; Zhang, N. Research on Ball Spinning Forming of Superalloy Inconel 718 Thin-Walled Tube. *Adv. Mater. Res.* **2011**, *189–193*, 2742–2745. [CrossRef]

360. Yoshihara, S.; Donald, B.M.; Hasegawa, T.; Kawahara, M.; Yamamoto, H. Design improvement of spin forming of magnesium alloy tubes using finite element. *J. Mater. Proc. Technol.* **2004**, *153–154*, 816–820. [CrossRef]

361. Yang, H.; Huang, L.; Zhan, M. Coupled thermo-mechanical FE smulation of the hot splitting spinning process of magnesium alloy AZ31. *Comput. Mater. Sci.* **2010**, *47*, 857–866. [CrossRef]

362. Jin, K.; Wang, J.; Guo, X.; Dombelsky, J.; Wang, H.; Jin, X.; Ding, R. Experimental analysis of electro-assisted warm spin forming of commercial pure titanium components. *Int. J. Adv. Manuf. Technol.* **2019**, *102*, 293–304. [CrossRef]

363. Jiang, S.S.; Tang, Z.J.; Du, H.; Chen, J.; Zhang, J.T. Research progress of current assisted forming process for titanium alloys. *Precis. Form. Eng.* **2017**, *9*, 7–13.

364. Magargee, J.; Morestin, F.; Cao, J. Characterization of flow stress for commercially pure titanium subjected to electrically assisted deformation. *J. Eng. Mater. Technol.* **2013**, *135*, 041003. [CrossRef]

365. Nguyen-Tran, H.D.; Oh, H.S.; Hong, S.T.; Han, H.N.; Cao, J.; Ahn, S.H.; Chun, D.M. A review of electrically-assisted manufacturing. *Int. J. Precis. Eng. Manuf.-Green Technol.* **2015**, *2*, 365–376. [CrossRef]

366. Homberg, W.; Hornjak, D.; Beerwald, C. Manufacturing of complex functional graded workpieceswith the frictio-spinning process. *Int. J. Mater. Form.* **2010**, *3*, 843–946. [CrossRef]

367. Zhan, M.; Yang, H.; Guo, J.; Wang, X.X. Review on hot spinning for difficult-to-deform lightweight metals. *Trans. Nonferrous Met. Soc. China* **2015**, *25*, 1732–1743. [CrossRef]

368. Childerhouse, T.; Long, H. Processing maps for wrinkle free and quality enhanced parts by shear spinning. *Procedia Manuf.* **2019**, *29*, 137–144. [CrossRef]

369. Hatori, S.; Sekiguchi, A.; Özerc, A. Conceptual design of multipurpose forming machine and experiments on force-controlled shear spinning of truncated cone. *Procedia Manuf.* **2018**, *15*, 1255–1262. [CrossRef]

370. Han, Z.R.; Fan, Z.J.; Xiao, Y.; Jia, Z. The constant temperature control system of multi-pass and die-less shear spinning by flame heating. *Int. J. Adv. Manuf. Technol.* **2018**, *97*, 2439–2446. [CrossRef]

371. Lee, H.S.; Song, Y.B.; Hong, S.S. Shear Spinning of Ti-6Al-4V Alloy at Hot Working Temperature. *Trans. Mater. Process.* **2011**, *20*, 432–438. [CrossRef]

372. Niklasson, F. Shear Spinning of Nickelbased Super-Alloy 718. In *The Minerals, Metals & Materials Series*; Ott, E., Liu, X., Andersson, J., Bi, Z., Bockenstedt, K., Dempster, I., Groh, J., Heck, K., Jablonski, P., Kaplan, M., et al., Eds.; Springer: Cham, Switzerland, 2018; pp. 769–778.

373. Prakash, R.; Singhal, R.P. Shear spinning technology for manufacture of long thin wall tubes of small bore. *J. Mater. Process. Technol.* **1995**, *54*, 186–192. [CrossRef]

374. Sekiguchi, A.; Arai, H. Control of wall thickness distribution by oblique shear spinning methods. *J. Mater. Process. Technol.* **2012**, *212*, 786–793. [CrossRef]

375. Suzuki, N.; Tokuhiro, S.; Takeuchi, O. Double cylinder forming by die-less shear spinning for air intake lip skin for aero jest engine nacelle. *Procedia Manuf.* **2018**, *15*, 1270–1277. [CrossRef]

376. Jia, Z.; Li, D.C.; Han, Z.R.; Xiao, Y. Mechanics of double-sheet die-less shear spinning: A novel potential method for wall thickness control in spinning process. *J. Braz. Soc. Mech. Sci. Eng.* **2019**, *41*, 497. [CrossRef]

377. Bylya, O.I.; Khismatullin, T.; Blackwell, P.; Vasin, R.A. The effect of elasto-plastic properties of materials on their formability by flow forming. *J. Mater. Process. Technol.* **2018**, *252*, 34–44. [CrossRef]

378. Sivanandini, M.; Dhami, S.S.; Pabla, B.S. Flow forming of tubes-A review. *Int. J. Sci. Eng. Res.* **2012**, *3*, 1–11.

379. Zhao, M.J.; Wu, Z.L.; Chen, Z.R.; Huang, X.B. Analysis of flow control forming of magnesium alloy wheel. *IOP Conf. Ser. Mater. Sci. Eng.* **2017**, *170*, 012006. [CrossRef]

380. Wang, X.; Gao, P.; Zhan, M.; Yang, K.; Dong, Y.; Li, Y. Development of microstructural inhomogeneity in multi-pass flow forming of TA15 alloy cylindrical parts. *Chin. J. Aeronaut.* **2019**. [CrossRef]

381. Singh, A.K.; Nrasimhan, K.; Singh, R. Finite element modeling of backward flow forming of Ti6Al4V alloy. *Mater. Today Proc.* **2018**, *5*, 24963–24970. [CrossRef]

Article

A Comparative Assessment of Six Machine Learning Models for Prediction of Bending Force in Hot Strip Rolling Process

Xu Li [1,*], Feng Luan [2,*] and Yan Wu [3]

[1]   The State Key Laboratory of Rolling and Automation, Northeastern University, Shenyang 110819, China
[2]   School of Computer Science and Engineering, Northeastern University, Shenyang 110169, China
[3]   School of Metallurgy, Northeastern University, Shenyang 110819, China; wuy@smm.neu.edu.cn
*   Correspondence: lixu@ral.neu.edu.cn (X.L.); luanfeng@mail.neu.edu.cn (F.L.); Tel.:+86-24-8368-1808 (X.L.)

Received: 12 April 2020; Accepted: 20 May 2020; Published: 22 May 2020

Abstract: In the hot strip rolling (HSR) process, accurate prediction of bending force can improve the control accuracy of the strip crown and flatness, and further improve the strip shape quality. In this paper, six machine learning models, including Artificial Neural Network (ANN), Support Vector Machine (SVR), Classification and Regression Tree (CART), Bagging Regression Tree (BRT), Least Absolute Shrinkage and Selection operator (LASSO), and Gaussian Process Regression (GPR), were applied to predict the bending force in the HSR process. A comparative experiment was carried out based on a real-life dataset, and the prediction performance of the six models was analyzed from prediction accuracy, stability, and computational cost. The prediction performance of the six models was assessed using three evaluation metrics of root mean square error (RMSE), mean absolute error (MAE), and coefficient of determination (R^2). The results show that the GPR model is considered as the optimal model for bending force prediction with the best prediction accuracy, better stability, and acceptable computational cost. The prediction accuracy and stability of CART and ANN are slightly lower than that of GPR. Although BRT also shows a good combination of prediction accuracy and computational cost, the stability of BRT is the worst in the six models. SVM not only has poor prediction accuracy, but also has the highest computational cost while LASSO showed the worst prediction accuracy.

Keywords: bending force prediction; hot strip rolling (HSR); comparative assessment; machine learning; regression

---

1. Introduction

In recent years, with the development of hot strip rolling (HSR) technology, product users continuously call for increased requirements. These increased requirements include strip variety, specifications, and strip shape quality. A good strip shape quality produced by the HSR process has a desired crown and flatness, and it is also an important factor to determine the competitiveness of strip in the market. Therefore, strip shape quality has become a hot topic of many scholars [1,2].

There are many factors that affect the strip shape quality, which are mainly related to the roller, strip, and rolling conditions in the HSR process. However, the field environment of the rolling process is very complex, and there are many factors that affect the strip shape quality. There is still no perfect solution to the strip shape quality problem in the world. In order to improve the strip shape quality, most scholars mainly studied the following two aspects. The research on production equipment is the first thought of researchers. In order to improve strip shape quality, it is necessary to control roll crown effectively. Therefore, it can be achieved by replacing the work rolls with ultra-high strength and ultra-high hardness to reduce the flexural deformation of the rolls. Secondly, the research on rolling technology has been carried out. In order to improve the precision of the preset model and compensate the influence of external factors on strip shape measurement precision, various factors affecting shape control model were studied. For example, in the process of strip production, if the detection accuracy of the roller is too low, it will directly affect the adjustment ability of the strip shape control mechanism, so the strip shape quality could not be improved [1,3].

Hydraulic roll bending control is one of the main methods to control the shape of hot rolled strip. The hydraulic roll bending system is more and more widely used in shape control of rolling mill because of its fast response and convenient real-time control. As shown in Figure 1, the principle of the hydraulic roller bending control system is that the bending force generated by the hydraulic cylinder is applied to the roller neck between the working roller and the supporting roller to change the deflection of the working roller instantaneously. Therefore, the shape of the gap of the load rollers is changed and the strip shape is controlled [4].

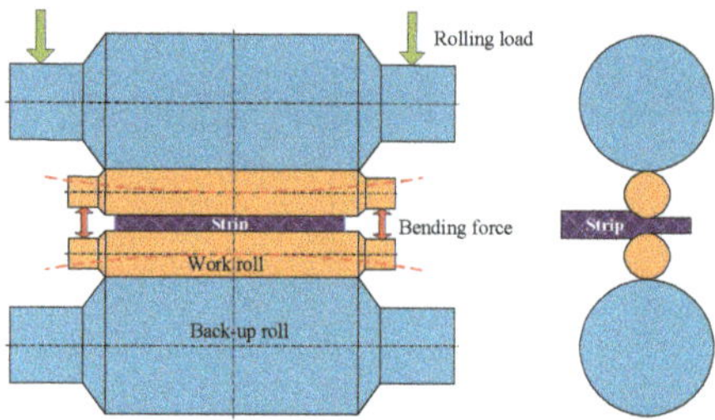

Figure 1. The schematic diagram of hydraulic roll bending technique.

When the rolling process and equipment parameters are changed, the preset value of bending force needs to be adjusted in time. As a result of the adjustment, the roll gap shape is consistent with the cross-sectional shape of the strip, so that the shape of the strip rolled by the rolling mill can meet the requirements. In the production process, the setting value of bending force needs to be adjusted constantly with the requirement of the rolling process. Therefore, the bending force is usually calculated according to the rolling factors such as temperature, thickness, width, rolling force, material, thermal expansion of rolls, wears of rolls, and so on, aiming at the convexity and flatness of the strip. Due to the multivariable, strong coupling, nonlinear, and time-varying characteristics of rolling factors, the calculation model of hot rolling bending force is extremely complicated [5,6].

The traditional mathematical model considers that all the rolling factors related to bending force have linear or approximate linear effects on bending force, and the coupling relationship between the rolling factors is weak in the model. Therefore, the mathematical model established according to the traditional theory is difficult to achieve the ideal prediction effect of bending force because of the limitations of its own structure [7].

Since the 1990s, artificial intelligence methods have been widely applied to rolling processes. Furthermore, Artificial Neural Networks (ANN) have been extensively studied and applied in the fields of mechanical property prediction [8–10], rolling force prediction [11–14], roughing mill temperature prediction [15], strip shape and crown prediction [16–20]. For the first time, Wang et al. [21] used the ANN model optimized by genetic algorithm to predict the bending force in the rolling process. The accuracy of the model is verified by actual factory data, which shows that the model can be flexibly used for on-line control and rolling schedule optimization.

These studies reveal that ANNs have been shown to perform nonlinear data well and have better predictive quality as compared to traditional mathematical models due to its good learning ability. However, they also have some shortcomings, such as the unexplained nature of relationships between the input and output parameters of the process, the need for higher calculation times, and the tendency of over-fitting, which leads to poor performance [22]. In recent years, besides ANN modeling, some new machine learning methods have emerged, such as Support Vector Machine (SVM), Classification and

Regression Tree (CART), Bagging Regression Tree (BRT), Least Absolute Shrinkage and Selection Operator (LASSO), and Gaussian Process Regression (GPR). For the prediction research in the rolling field, some scholars have realized that better prediction results and prospects can be obtained by adopting new machine learning methods [21,23,24], and so far, there are few literature reports on bending force prediction.

Therefore, this research is motivated to investigate the application of SVM, CART, BRT, LASSO, GPR and ANN on bending force prediction in hot rolling process, and to comprehensively analyze and evaluate the prediction performance of these models from prediction accuracy, stability and computational cost. Through the comprehensive evaluation results of these models, a prediction model of bending force with high prediction accuracy and good stability can be proposed, and finally the profile quality of strip can be improved.

Inspired by this motivation, this paper first provides the basic principles of the six models; verifies the predictability of the bending force using these models based on real-life dataset of a 1580-mm hot rolling process in a steel factory. All the gauges used in this research were calibrated, and the measurement results are reliable and valid. The remaining part of the paper is organized as follows. Section 2 briefly describes the HSR process, the influencing factors of bending force, the acquisition, and processing of experimental data. Section 3 gives a brief description of literature review and basic theories of the six machine learning models. In addition, the three evaluation metrics are also given in this section. Sections 4 and 5 report the experimental results and discussion, respectively. In Section 6, we draw the conclusions.

2. Case Study and Data

2.1. Hot Rolling Technology and Bending Force

Figure 2 shows the complete rolling process in a typical HSR process. The HSR process consists of 6 key parts: the reheating furnace, the roughing mill, the hot coil box and flying shear, the finishing mill, the laminar cooling, and the coiler. The key equipment of the production line is a finishing mill group composed of 8 groups of stands, which determines the final shape of the strip. Each group of stand consists of a pair of work rolls and a pair of backup rolls. The spacing between the stand is 5.5 m. The whole line is equipped with work roll shifting and hydraulic roll bending systems to control flatness and plate crown.

A single batch consists of a coil of rough steel, which enters the reheating furnace to be reheated to the appropriate temperature. Next, the strip passes through the roughing mill, where its thickness and width are reduced to close to the desired value. Then, the strip enters the finishing mill section, where the strip is carefully milled to the required width and thickness. The profile of the strip can be controlled by changing the bending forces between the two work rolls [25]. The strip thickness and flatness are measured in real time by an X-ray gauge at the end of the finishing stands as shown in Figure 2. Measuring the final dimensions of the strip is vital for the mill controllers. The controllers adjust mill parameters in real time with feedback from the gage to minimize strip flatness. Next, the strip is cooled by water to an appropriate final temperature. Finally, the strip is coiled and is ready for shipment.

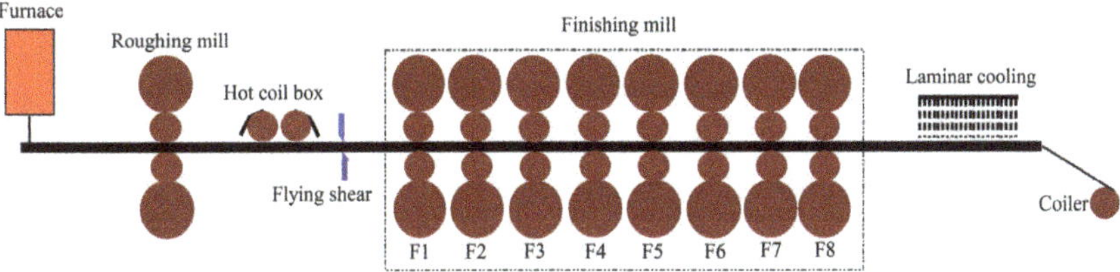

Figure 2. Schematic layout of hot strip rolling (HSR).

2.2. Data Collection and Analysis

In this paper, the final stand rolling data of a 1580-mm HSR process in a steel factory are collected for experiments. The purpose of models in hot rolling is to predict strip characteristics prior to rolling the strip based on information about the mill. For the proposed prediction model of bending force, the input variables are entrance temperature (°C), entrance thickness (mm), exit thickness (mm), strip width (mm), rolling force (kN), rolling speed (m/s), roll shifting (mm), yield strength (MPa), and target profile (μm). The output variable of the model is the bending force (kN). The information of the detection equipment for these parameters is shown in Table 1. In order to ensure the validity of the parameters, all the gauges used in this research were calibrated. The fractal dimension visualization diagram of the collected dataset is shown in Figure 3. Obviously, the input data vary considerably in different dimensions. Table 2 shows the data distributions for each input variables. In order to eliminate the difference between the numbers of different dimensional data, avoid prediction error increase because of the big difference between input and output data, and update the weights and biases conveniently in the modeling process. It is necessary to scale data to a small interval in a certain proportion. Normalization is required prior to data entry into the model [26]. The following formula is used to normalize the data:

$$y_i' = \frac{y_i - y_{min}}{y_{max} - y_{min}} \tag{1}$$

where y_i', y_i, y_{min}, and y_{max} are the normalized data, original data, maximal data, and minimal data, respectively.

Table 1. Parameter information.

Parameter	Detection Equipment	Specifications	Brand
Entrance temperature	Infrared thermometer	SYSTEM4	LAND
Exit thickness	X-ray thickness gauge	RM215	TMO
Strip width	Width gauge	ACCUBAND	KELK
Rolling force	Load Cell	Rollmax	KELK
Rolling speed	Incremental Encoder	FGH6	HUBNER
Roll shifting	Position Sensor	Tempsonics	MTS
Target profile	Profile Gauge	RM312	TMO
Bending force	Pressure Transducer	HDA3839	HYDAC

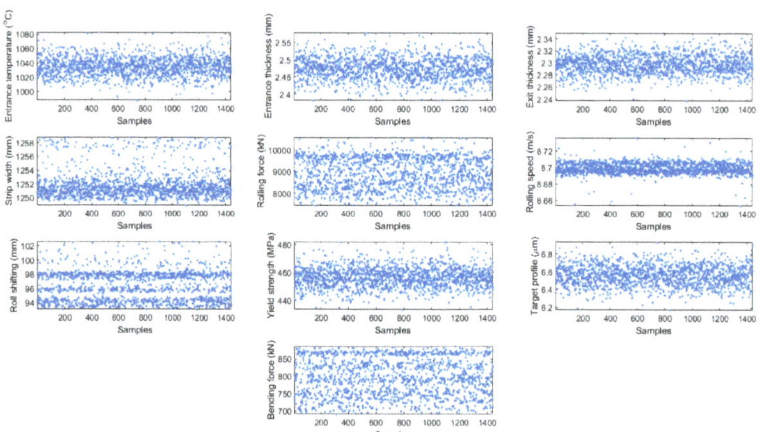

Figure 3. The fractal dimension visualization diagram of the data.

Table 2. Input parameters and the values.

Parameter	Unit	Mean	Range of Value
Entrance temperature	°C	1035.2	988.05~1082.8
Entrance thickness	mm	2.4750	2.3827~2.5782
Exit thickness	mm	2.2981	2.2374~2.3494
Strip width	mm	1252.0	1248.9~1258.8
Rolling force	kN	8940.4	7438.4~10,590
Rolling speed	m/s	8.6995	8.6535~8.7362
Roll shifting	mm	96.436	93.125~102.625
Yield strength	MPa	456.71	433.74~482.02
Target profile	μm	65.613	61.740~69.315

In the present study, the K-fold cross validation method was used and 1440 pairs of measured bending force data were divided into five subsets. Four subsets were employed to train the machine learning models and the remaining one for testing the models. Furthermore, the measurement data should be processed with z-score normalization to the same scale to reduce the impact of different magnitudes and dimensions.

3. Methodology

3.1. Artificial Neural Network (ANN)

ANNs are complex computational models inspired by the human nervous system, which are capable of machine learning and pattern recognition. ANN includes a wide range of learning algorithms that have been developed in statistics and artificial intelligence. It uses analogy with biological neurons to generate general solutions to the problem. Since ANNs are nonlinear classification techniques and also composed of an interconnected group of artificial neurons, they have the ability to learn complex relationships between input and output variables [17]. ANN is the earliest prediction model applied in the rolling field, including mechanical property prediction, rolling force prediction, roughing mill temperature prediction, flatness, and crown prediction [7–21]. Therefore, ANN is the most widely applied model and the basic model for comparison.

3.2. Support Vector Machine (SVM)

Support Vector Machine (SVM) is a supervised learning method developed from statistical learning theory to analyze data and pattern recognition, which can be used to classify and regression data [27]. SVM as a regression technique (SVR) is a nonlinear algorithm and the basic principle is to map the data to a high-dimensional feature space using a nonlinear mapping, and then construct the regression estimation function in the high-dimensional feature space and then map back to the original space, and this nonlinear transformation is achieved by defining the appropriate kernel function. Many machine learning algorithms follow the principle of empirical error minimization, while SVR follows the principle of structural risk minimization, so it can obtain better generalization performance [28]. SVRs are prominent in research and practice, due to their use of linear optimization techniques to find optimal solutions to nonlinear predictive problems in higher-dimensional feature spaces. Therefore, it has been widely employed for regression and forecasting in the fields of agriculture, hydrology, the environment, and metallurgy [29–31]. This encourages us to apply an SVR to the prediction of the HSR process.

3.3. Classification and Regression Tree (CART)

Decision trees (DT) is an important algorithm for machine learning. The classification and regression tree methodology, also known as the CART was introduced in 1984 by Breiman et al. [32]. CART has low computational complexity because of its recursive computation. To predict a response, follow the decisions in the tree from the root (beginning) node down to a leaf node. The leaf node contains the response. So CART is non-parametric and can find complex relationships between input and output variables. Therefore, CART also has the advantage of discovering nonlinear structures and variables interactions in the training samples [33]. Regression tree is a data mining algorithm widely used in regression problems of biology [34], environment [35] and material processing [36]. We selected CART to prediction bending force because they are an explanatory technique, able to reveal data structure, identify important characteristics, and develop rules.

3.4. Bagging Regression Tree (BRT)

Bagging (short for bootstrap aggregating) is a simple and very powerful ensemble method. Bagging is one of the simplest techniques, which can reduce variance when combined with the base learner generation, with surprisingly good performance [37]. Bagging Regression Tree (BRT) is the application of the bootstrap procedure to RT. The basic idea underlying BRT is the recognition that part of the output error in a single regression tree is due to the specific choice of the training dataset. Therefore, if several similar datasets are created by resampling with replacement (that is, bootstrapping) and regression trees are grown without pruning and averaged, the variance component of the output error is reduced [38,39]. The BRT has been widely used in the fields of biostatistics [40], remote sensing [41], and material processing [42] due to its flexibility and interpretability to high-order nonlinear modeling. Therefore, it is reasonable to compare and evaluate BRT as one of the optional models.

3.5. Least Absolute Shrinkage and Selection Operator (LASSO)

LASSO stands for Least Absolute Shrinkage and Selection Operator [43]. The idea behind the LASSO algorithm is to achieve a minimization of the residual sum of squares while regularizing the sum of the absolute value of the coefficients being less than a given constant. LASSO technique has been successfully developed in recent years, combining shrinkage and highly correlated variables. LASSO regression is characterized by variable selection and regularization while fitting the generalized linear model. Regularization is to control the complexity of the model through a series of parameters so as to avoid over fitting. LASSO has been widely used in temperature prediction [44], the wavelength analysis [45], and streamflow prediction [46]. In view of its wide application in industry, LASSO is also taken as one of the research models in this paper.

3.6. Gaussian Process Regression (GPR)

Gaussian Process Regression (GPR) is a new machine learning regression method developed in recent years, and it is also a non-parametric model algorithm based on Bayesian network. The GPR algorithm can adaptively determine the number of model parameters according to the information of the training samples, and add the prior knowledge of the existing objects in the modeling process, and then combine the actual experimental data to obtain the posterior Gauss process model [47]. When GPR is applied to practical problems, GPR can give a confidence interval while outputting the mean value, making the validity of the prediction result continuously enhanced. In addition, because the GPR can quantitatively model Gaussian noise, it has excellent prediction accuracy [48,49]. Because of its good predictive ability, GPR has been widely used in data-driven modeling of various problems in industry [50–53], so GPR has also become an optional scheme in this paper.

3.7. Model Information

All models were implemented in Matlab (Version 2015b, MathWorks, Natick, MA, USA) under a computer with a hardware configuration of Intel Core i7-7500U CPU 2.7 GHz, and 8 GB of RAM. CART, BRT, LASSO, and GPR were carried out with treefit, fitensemble, lasso, and fitrgp functions, respectively. These four functions are included in Matlab's Statistics and Machine Learning Toolbox. The parameters of these models were automatically optimized by Matlab function according to the training dataset. For SVR, the parameter C (trade-off parameter between the minimization of errors and smoothness of the solution), and the parameter σ (the width of the RBF kernel function) are needed to determine the process of model establishment. In order to reveal the effect of C and σ on prediction results in training dataset. Ten logarithmically, equally spaced points were generated between 1 and 1000 for C. Twenty logarithmically, equally spaced points were generated between 10 and 1000 for σ. The optimized C and σ were determined to be 100 and 69.5, respectively. For ANN, the transfer function of hidden layer and output layer are "tansig" and "purelin", respectively. The "tainlm" (Levenberg-Marquardt) was chosen as the optimal learning algorithm. In this paper, the performance of neural networks with 2–30 neurons was investigated considering the single hidden layer. The number of neurons in the network hidden layer is determined to be 6, and the ANN has the best performance.

3.8. Comparison of Model and Statistical Error Analysis

The accuracy and performance of the studied models for bending force prediction were evaluated and compared using three commonly used statistical metrics, which were root mean square error (RMSE, Equation (2)), mean absolute error (MAE, Equation (3)), and coefficient of determination (R^2, Equation (4)). The mathematical equations of the statistical indicators are described below.

$$\text{RMSE} = \sqrt{\frac{\sum_{i=1}^{N}\left(y_i - y_i^*\right)^2}{N}} a = 1 \tag{2}$$

$$\text{MAE} = \frac{1}{N}\sum_{i=1}^{N}\left|y_i - y_i^*\right| \tag{3}$$

$$R^2 = 1 - \frac{\sum_{i=1}^{N}\left(y_i - y^*\right)^2}{\sum_{i=1}^{N}\left(y_i - \overline{y}\right)^2} \tag{4}$$

where y_i and y_i^* are the measured values and predictive values respectively, N is the total number of predicted data. Higher values of R^2 are preferred, i.e., closer to 1 means better model performance and regression line fits the data well. On the contrary, the lower the RMSE and MAE values are, the better the model performs.

4. Results

4.1. Prediction Accuracy of Various Models

Table 3 shows the results of the ANN, SVM, CART, BRT, LASSO, and GPR models performing 30 trials in training and testing dataset. It can be seen that the predicted bending force varies considerably depending on the model selection. In training dataset, no matter which evaluation metrics are used, BRT shows the best prediction performance with the highest R^2 and lowest RMSE and MAE. In testing dataset, GPR shows the best prediction performance with the highest R^2 and lowest RMSE and MAE. The prediction performance of BRT follows that of GPR. On the contrary, LASSO has the worst prediction accuracy not only in the training dataset but also in the testing dataset.

Table 3. Accuracy statistical results of the six machine learning models performing 30 trials in training and testing dataset.

Model		Training Dataset			Testing Dataset		
		RMSE (kN)	MAE (kN)	R^2	RMSE (kN)	MAE (kN)	R^2
ANN	Max	10.3392	8.0444	0.9678	11.1894	8.6467	0.9621
	Min	9.3885	7.2853	0.9611	10.1867	7.8557	0.9543
	Mean	9.7603	7.6325	0.9653	10.5676	8.2064	0.9590
SVM	Max	13.0896	10.3777	0.9400	13.4531	10.7890	0.9371
	Min	12.8499	9.9659	0.9378	13.1528	10.3986	0.9328
	Mean	12.9733	10.1730	0.9389	13.2755	10.5367	0.9356
CART	Max	9.1048	6.8205	0.9749	10.5625	7.8368	0.9640
	Min	8.3138	6.3089	0.9699	9.9445	7.2870	0.9591
	Mean	8.7054	6.5526	0.9724	10.2547	7.5449	0.9615

Table 3. *Cont.*

Model		Training Dataset			Testing Dataset		
		RMSE (kN)	MAE (kN)	R^2	RMSE (kN)	MAE (kN)	R^2
BRT	Max	6.1908	4.7867	0.9865	9.7559	7.6485	0.9674
	Min	6.1018	4.7222	0.9861	9.4643	7.4094	0.9653
	Mean	6.1552	4.7561	0.9862	9.6441	7.5587	0.9660
LASSO	Max	13.5908	11.2526	0.9345	14.5408	11.9976	0.9330
	Min	13.4307	11.0978	0.9329	13.5506	11.1993	0.9222
	Mean	13.5261	11.1914	0.9336	14.0047	11.5624	0.9283
GPR	Max	7.4269	5.7891	0.9805	8.6418	6.6802	0.9745
	Min	7.3335	5.7243	0.9800	8.3605	6.5054	0.9726
	Mean	7.3887	5.7580	0.9802	8.5121	6.6137	0.9735

Note: Abbreviations: root mean square error, RMSE; mean absolute error, MAE; coefficient of determination, R^2; Artificial Neural Network, ANN; Support Vector Machine, SVR; Classification and Regression Tree, CART; Bagging Regression Tree, BRT; Least Absolute Shrinkage and Selection operator, LASSO; Gaussian Process Regression, GPR.

Figure 4 shows the accuracy ranking results of the six models in training and testing dataset. The prediction accuracy value of the model can be read out from the ordinate, and the ranking results of these models are showed with numbers above the color bars. As can be seen from Figure 4, first of all, the accuracy ranking results show slight difference in training and testing dataset. In training dataset, the accuracy ranking results is consistent with the three evaluation metrics, and the rank order is: BRT, GPR, CART, ANN, SVM, LASSO. However, the accuracy ranking result changes slightly in testing dataset. In addition, with different accuracy evaluation metrics, the accuracy ranking results are also different. The accuracy rank with the metrics of RMSE and R^2 in descending order is: GPR, BRT, CART, ANN, SVM, LASSO; and that of MAE in descending order is: GPR, CART, BRT, ANN, SVM, LASSO. Based on the comprehensive performance of the two datasets, it can be considered that GPR model shows the best prediction accuracy and LASSO model has the worst prediction accuracy.

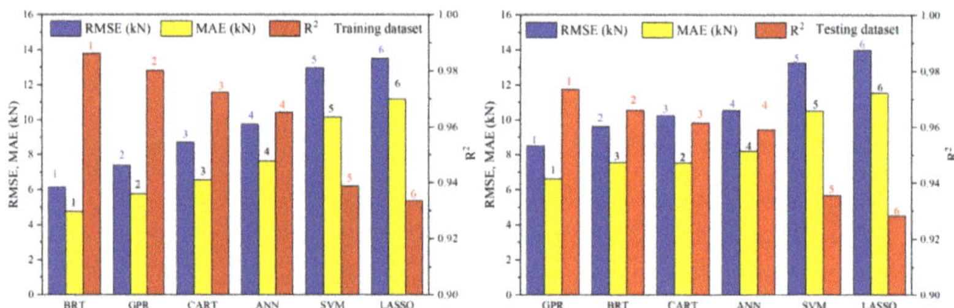

Figure 4. Prediction accuracy and ranking results of the six machine learning models performing 30 trials.

The scatter plot of the bending force values measured by the factory and the values predicted by the six machine learning models in training and testing dataset are presented in Figure 5. Scattered points of different colors in the figure represent the predicted values by different models. In training and testing dataset, the predicted values of BRT and GPR models are closely distributed on both sides of the straight line y = x. The results show that the predicted bending force of the two models have a better correlation with the measured bending force value, and the two models are superior to the other four models for bending force prediction. On the contrary, LASSO has the worst prediction accuracy with the most scattered predicted value around the straight line y = x, indicating that the predicted values are much different from the measured values. The maximum error of all other data points predicted by the six models is within 10%. Therefore, these six models have achieved good prediction performance.

Figure 6 shows the measured bending force and predicted bending force by the six models in training and testing dataset, and also shows prediction errors below. It clearly shows that the BRT has the best prediction performance in training dataset, the maximum positive error is 23.60 kN and the maximum negative error is −25.28 kN. In testing dataset, GPR has the best prediction performance with the maximum positive error as 31.84 kN and the maximum negative error as −26.50 kN. The errors of BRT and GPR are more concentrated in the range of 0 kN, which means that the number of samples with large error values is smaller. On the contrary, the LASSO performs worst in the six models of two datasets. In training dataset, the maximum positive error is 41.07 kN and the maximum negative error is −44.56 kN. In testing datasets, the maximum positive error is 42.99 kN and the maximum negative error is −27.04 kN.

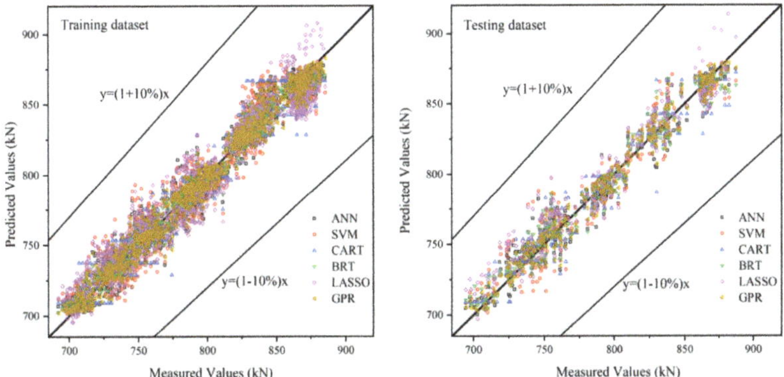

Figure 5. Scatter plots of the measured bending force values and the predicted crown values by the six machine learning models.

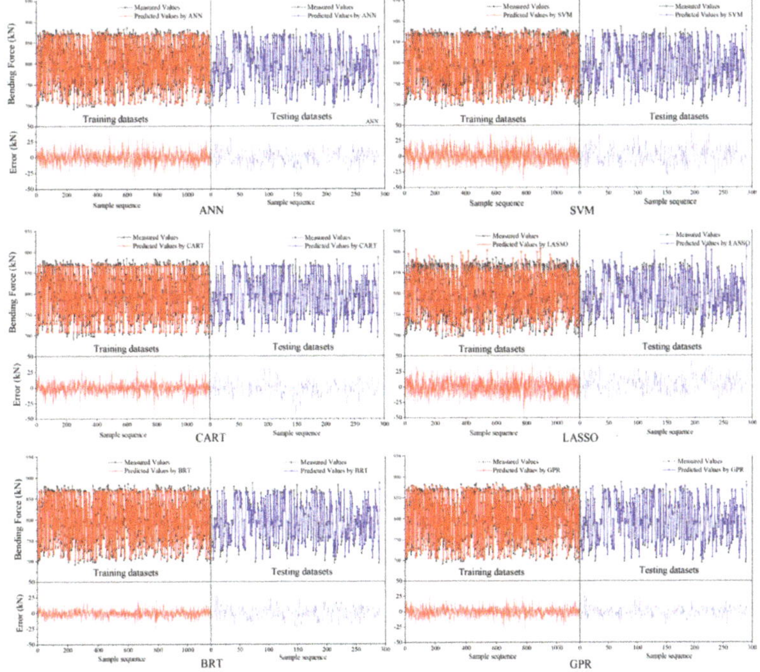

Figure 6. Comparison results and absolute errors between measured values and predicted values by six machine learning models.

Figure 7 shows the histograms and distribution curves of the errors. All the error distribution curves have a bell shape of normal distribution, which indicates that the prediction errors of all models are normal distribution. Whether in training or in testing datasets, GPR, BRT, and CART models perform relatively well, and their normal distribution curves are higher and narrower, which indicate that more prediction values with smaller errors are obtained. In addition, it is also found that the dataset centers (the highest point of normal distribution curve) of most models are close to the zero point of errors. The dataset center represents the average value of errors, indicating that the probabilities of positive error and negative error are almost equal. However, the normal distribution curve of LASSO shifts to right obviously in testing dataset, which indicates that the positive error of LASSO model is much more than the negative error and the predicted values are higher.

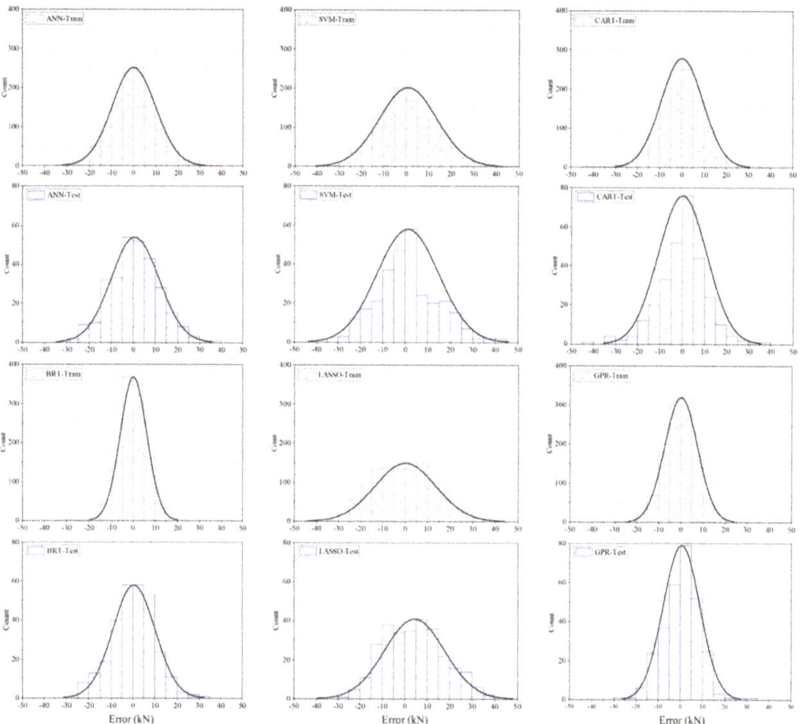

Figure 7. Histograms and normal distribution curves of the error of six machine learning models performing 30 trials.

4.2. Stability of Various Models

The prediction accuracy results show that the prediction accuracy in testing dataset of all models is lower than that in the training dataset. In addition, in training dataset, the prediction accuracy of BRT model is better than the GPR. However, in testing dataset, the prediction accuracy of GPR model is better than the BRT (showed in Figure 4). The difference of prediction accuracy between in training and testing dataset can be regarded as the stability of the model. The stability of machine learning model is also an important factor affecting the prediction performance, which should be taken into account when evaluating the reliability of predicted result. The stability of the machine learning model is the relative change percentage of the evaluation metrics (including RMSE, MAE, and R^2) of the model in training and testing datasets [31]. The smaller the relative change percentage, the higher the stability of the model. The relative change percentage of evaluation metrics under the two datasets can be described by $\delta_{i,N}$ and calculated by the following formula:

$$\delta_{i,N} = \left| \frac{\delta_{i,test} - \delta_{i,train}}{\delta_{i,train}} \right| \times 100\% \tag{5}$$

where i represent the evaluation metrics (RMSE, MAE, or R^2) and N represent the one of the six models.

Figure 8 shows the $\delta_{i,N}$ from three evaluation metrics of the six models performing 30 trials. It shows that the stability rankings of ANN, SVR, CART, BRT, LASSO, and GPR models are slightly different with different evaluation metrics. With the evaluation metrics of RMSE and R^2, SVR shows the most stable performance with the lowest $\delta_{i,N}$ values of 2.33% and 0.35%, respectively. LASSO shows the most stable performance with the lowest $\delta_{i,N}$ values of 2.33% in the evaluation metrics of MAE. However, no matter which evaluation metrics is used, the BRT shows the most unstable performance with the highest $\delta_{i,N}$. The $\delta_{i,N}$ are 56.68%, 58.93%, and 2.05% calculated by RMSE, MAE, and R^2, respectively. This unstable performance reveals that when new input data is used, it will lead to a significant reduction in prediction accuracy. This is because the BRT model has a large number of hyper-parameters, which need to be carefully optimized for model application [31].

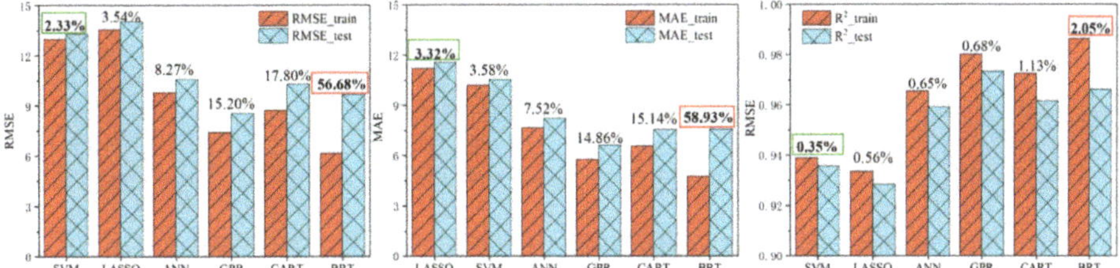

Figure 8. Difference of prediction accuracy metrics (RMSE, MAE, and R^2) between training datasets and testing datasets for 30 trials (the $\delta(i,N)$ of the models are shown above the bars).

The distribution of the three evaluation metrics obtained from six machine learning models performing 30 trials in training and testing dataset are illustrated in Figure 9 using a boxplot. It represents the degree of spread for the prediction accuracy with its respective quartile. In training dataset, the suspected outliers of the prediction accuracy only appear in ANN and CART models. In testing dataset, suspected outliers appear in ANN, SVM, CART, and BRT models. At the same time, it is found that the quartile distance of testing dataset is increasing compared with that in training dataset, which indicates that the degree of dispersion of prediction accuracy became larger. Although Figure 8 shows that both SVM and LASSO models have the most stable performance. However, the variation of the quartile distance of the LASSO model is most obvious in the two datasets. Considering comprehensively, it can be considered that SVM is the most stable model.

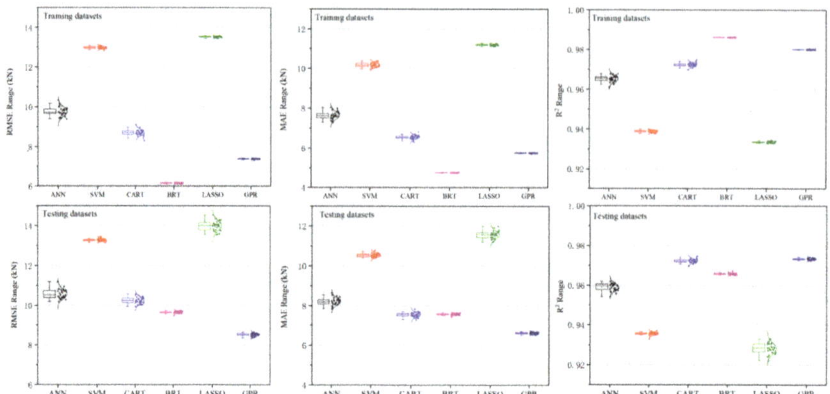

Figure 9. Boxplots of the prediction accuracy (RMSE, MAE, and R^2) obtained from six machine learning models performing 30 trials.

4.3. Computational Costs of Various Models

Table 4 and Figure 10 shows the computational cost (time used for computation) of the six machine learning models. The CART and LASSO show the lowest computational costs of 0.59 s and 0.35 s, respectively. BRT and ANN also show smaller computational costs of 2.11 s and 9.35 s, respectively. Compared with the above four models, the computational cost of GPR increases to 63.25 s. Furthermore, the computational cost of SVM reaches the maximum value, which is 305.09 s.

Table 4. Comparison of computational costs (time used for computation) of the six machine learning models.

Model	Computational Cost (s)		
	Max	**Min**	**Mean**
ANN	11.66	8.12	9.35
SVM	321.59	295.24	305.09
CART	0.71	0.56	0.59
BRT	2.64	1.87	2.11
LASSO	0.51	0.31	**0.35**
GPR	64.55	62.06	63.25

Note: The lowest computational cost among all models are marked in bold.

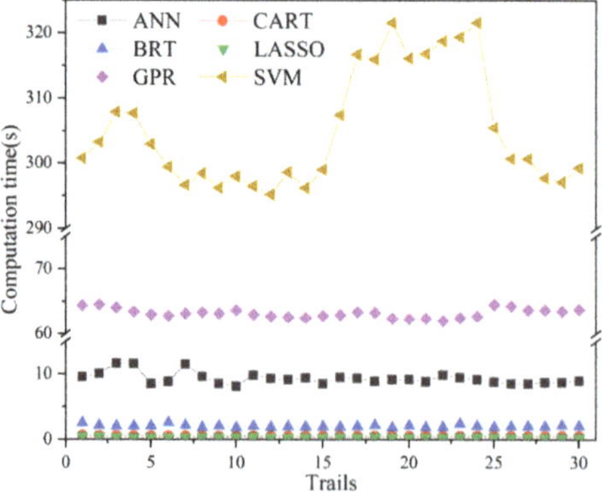

Figure 10. Comparison of computational cost (time used for computation) of the six machine learning models.

4.4. Comprehensive Evaluation of Various Models

Based on the above results, six machine learning models comprehensively evaluated from prediction accuracy, stability, and computational cost, and the results are shown in Figure 11. It must be pointed out that the prediction accuracy here is in testing dataset. Figure 11 shows that GPR provides the best combination of prediction accuracy, stability, and computational cost. The prediction accuracy of BRT, CART, and ANN models is slightly worse than that of GPR. For the three models, the prediction accuracy and computational cost of them are not much different, but BRT has the worst stability. In addition, SVM does not perform well in terms of prediction accuracy and computational cost. LASSO has good stability and computational cost, but the prediction accuracy is the worst.

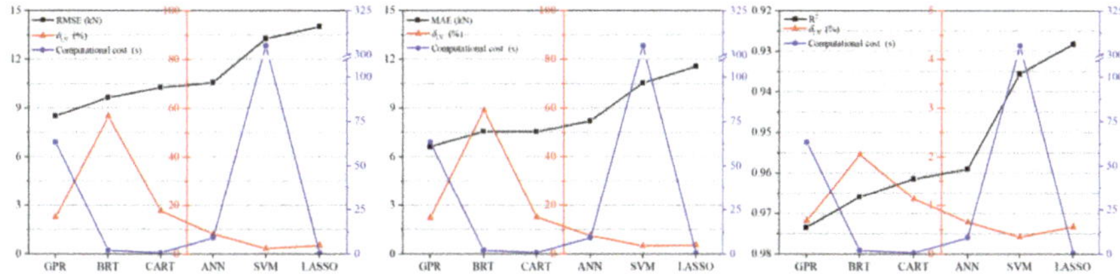

Figure 11. Comprehensive evaluation results of the six machine learning models.

5. Discussion

With the development of technology, today, in the production process of a series of steel, the equipment will maintain a stable operation state, so the rolling process is also carried out stably. Therefore, for specific strip specification, most rolling processes will obtain relatively stable datasets without large variability. Then, the key technology of rolling parameter prediction is how to improve the prediction accuracy. The stable data can better reflect the normal rolling process. The purpose of this paper is to discuss the comparison results of prediction accuracy when different models are applied to bending force prediction. Therefore, in order to reflect the most essential characteristics of the model and obtain a fair comparison result, the machine learning models used in this paper were not over optimized by combining with other intelligent optimization algorithms (such as Genetic Algorithm (GA), Particle Swarm Optimization (PSO), etc.). We believe that this paper can be used as a reference and basis for other similar prediction applications and research to select machine learning models, and it is more practical to select basic original models and parameters.

6. Conclusions

In this paper, we applied six machine learning models, including ANN, SVR, CART, BRT, LASSO, and GPR, to predict the bending force in the HSR process. A comparative experiment was carried out based on real-life dataset, and the prediction performance of the six models was analyzed from prediction accuracy, stability, and computational cost. All the gauges used in this research were calibrated to ensure the validity of the data and reliability of the results. The prediction performance of the six models was assessed using three evaluation metrics of RMSE, MAE, and R^2.

(1)   The comparison results of prediction accuracy show that the accuracy ranking results in testing dataset are slightly different under the three evaluation metrics. However, considering that GPR performs best, followed by BRT, CART, ANN, SVM, and LASSO respectively. The bending force measured by experiment is 690~890 kN, while the prediction error of GPR is only 8.51 kN (RMSE) and 6.61 kN (MAE).

(2)   The ranking results of stability show inconsistency in the three evaluation metrics. However, considering comprehensively, SVM shows the most stable performance with the γ of 2.33% (RMSE), 0.32% (MAE) and 0.35% (R^2). The stability decreases in the order of LASSO, ANN,

GPR, CART, and BRT. BRT shows the most unstable performance with the γ of 56.68% (RMSE), 58.93% (MAE), and 2.05% (R^2).

(3) The computational cost of the six models presents three levels. The computational costs of LASSO, CART, BRT, and ANN are increasing gradually, but they are all within ten seconds. The computational cost of GPR model is slightly higher, at about 63 s. However, the computational cost of SVM has reached more than 300 s.

(4) Comprehensively considering the prediction accuracy, stability, and computational cost of the six models, GPR can be considered the most promising machine learning model for predicting bending force. The prediction accuracy and stability of CART and ANN is slightly lower than GPR, but the computational cost is relatively small, so it can also be used as an alternative. In addition, BRT also shows the better combination of prediction accuracy and computational cost, but the stability of BRT is the worst among the six models. SVM not only performs poorly in prediction accuracy, but also has the greatest computational cost. While LASSO has good stability and small computational cost, but it also has the worst prediction accuracy.

Author Contributions: Conceptualization, funding acquisition as well as supervision of this research project were done by X.L.; the experimental investigations were carried out by F.L. with the support of X.L.; Y.W. took the lead in writing the manuscript; X.L. and F.L. revised the paper. All authors have read and agreed to the published version of the manuscript.

Funding: This research was funded by the National Key R&D Program of China (No. 2017YFB0304100), National Natural Science Foundation of China (No. 51634002), and the Fundamental Research Funds for the Central Universities (Nos. N180708009 and N171604010).

Conflicts of Interest: The authors declare no conflict of interest.

References

1. Takahashi, R.R. State of the art in hot rolling process control. *Control Eng. Pract.* **2001**, *9*, 987–993. [CrossRef]
2. Zhu, H.T.; Jiang, Z.Y.; Tieu, A.K.; Wang, G.D. A fuzzy algorithm for flatness control in hot strip mill. *J. Mater. Process. Technol.* **2003**, *140*, 123–128. [CrossRef]
3. Wang, Q.L.; Sun, J.; Liu, Y.M.; Wang, P.F.; Zhang, D.H. Analysis of symmetrical flatness actuator efficiencies for UCM cold rolling mill by 3D elastic-plastic FEM. *Int. J. Adv. Manuf. Technol.* **2017**, *92*, 1371–1389. [CrossRef]
4. Jia, C.Y.; Shan, X.Y.; Cui, Y.C.; Bai, T.; Cui, F.J. Modeling and Simulation of Hydraulic Roll Bending System Based on CMAC Neural Network and PID Coupling Control Strategy. *J. Iron Steel Res. Int.* **2013**, *20*, 17–22. [CrossRef]
5. Zhang, S.H.; Deng, L.; Zhang, Q.Y.; Li, Q.H.; Hou, J.X. Modeling of rolling force of ultra-heavy plate considering the influence of deformation penetration coefficient. *Int. J. Mech. Sci.* **2019**, *159*, 373–381. [CrossRef]
6. Zhang, S.H.; Song, B.N.; Gao, S.W.; Guan, M.; Zhao, D.W. Upper bound analysis of a shape-dependent criterion for closing central rectangular defects during hot rolling. *Appl. Math. Model.* **2018**, *55*, 674–684. [CrossRef]
7. Zhang, W.; Wang, Y.; Sum, M. Modeling and Simulation of Electric-Hydraulic Control System for Bending Roll System. In Proceedings of the 2008 IEEE Conference on Robotics, Automation and Mechatronics (RAM), Chengdu, China, 21–24 September 2008; pp. 1–4. [CrossRef]
8. Sterjovski, Z.; Nolan, D.; Carpenter, K.R.; Dunne, D.P.; Norrish, J. Artificial neural networks for modeling the mechanical properties of steels in various applications. *J. Mater. Process. Technol.* **2015**, *170*, 536–544. [CrossRef]
9. Wang, P.; Huang, Z.Y.; Zhang, M.Y.; Zhao, X.W. Mechanical Property Prediction of Strip Model Based on PSO-BP Neural Network. *J. Iron Steel Res. Int.* **2008**, *15*, 87–91. [CrossRef]
10. Ozerdem, M.S.; Kolukisa, S. Artificial Neural Network approach to predict mechanical properties of hot rolled, nonresulfurized, AISI 10xx series carbon steel bars. *J. Mater. Process. Technol.* **2008**, *199*, 437–439. [CrossRef]

11. Bagheripoor, M.; Bisadi, H. Application of artificial neural networks for the prediction of roll force and roll torque in hot strip rolling process. *Appl. Math. Model.* **2013**, *37*, 4593–4607. [CrossRef]

12. Lee, D.; Lee, Y. Application of neural-network for improving accuracy of roll-force model in hot-rolling mill. *Control Eng. Pract.* **2002**, *10*, 473–478. [CrossRef]

13. Mahmoodkhani, Y.; Wells, M.A.; Song, G. Prediction of roll force in skin pass rolling using numerical and artificial neural network methods. *Ironmak. Steelmak.* **2017**, *44*, 281–286. [CrossRef]

14. Yang, Y.Y.; Linkens, D.A.; Talamantes-Silva, J.; Howard, I.C. Roll force and torque prediction using neural network and finite element modelling. *ISIJ Int.* **2003**, *43*, 1957–1966. [CrossRef]

15. Laurinen, P.; Röning, J. An adaptive neural network model for predicting the post roughing mill temperature of steel slabs in the reheating furnace. *J. Mater. Process. Technol.* **2005**, *168*, 423–430. [CrossRef]

16. John, S.; Sikdar, S.; Swamy, P.K.; Das, S.; Maity, B. Hybrid neural-GA model to predict and minimise flatness value of hot rolled strips. *J. Mater. Process. Technol.* **2008**, *195*, 314–320. [CrossRef]

17. Deng, J.F.; Sun, J.; Peng, W.; Hu, Y.H.; Zhang, D.H. Application of neural networks for predicting hot-rolled strip crown. *Appl. Soft Comput.* **2019**, *78*, 119–131. [CrossRef]

18. Kim, D.H.; Lee, Y.; Kim, B.M. Application of ANN for the dimensional accuracy of workpiece in hot rod rolling process. *J. Mater. Process. Technol.* **2002**, *130–131*, 214–218. [CrossRef]

19. Alaei, H.; Salimi, M.; Nourani, A. Online prediction of work roll thermal expansion in a hot rolling process by a neural network. *Int. J. Adv. Manuf. Technol.* **2016**, *85*, 1769–1777. [CrossRef]

20. Sikdar, S.; Kumari, S. Neural network model of the profile of hot-rolled strip. *Int. J. Adv. Manuf. Technol.* **2009**, *42*, 450–462. [CrossRef]

21. Wang, Z.H.; Gong, D.Y.; Li, X.; Li, G.; Zhang, D. Prediction of bending force in the hot strip rolling process using artificial neural network and genetic algorithm (ANN-GA). *Int. J. Adv. Manuf. Technol.* **2017**, *93*, 3325–3338. [CrossRef]

22. Laha, D.; Ren, Y.; Suganthan, P.N. Modeling of steelmaking process with effective machine learning techniques. *Expert Syst. Appl.* **2015**, *42*, 4687–4696. [CrossRef]

23. Hu, Z.Y.; Wei, Z.H.; Sun, H.; Yang, J.M.; Wei, L.X. Optimization of Metal Rolling Control Using Soft Computing Approaches: A Review. *Arch. Comput. Methods Eng.* **2019**, *11*, 1–17. [CrossRef]

24. Shardt, Y.A.W.; Mehrkanoon, S.; Zhang, K.; Yang, X.; Suykens, J.; Ding, S.X.; Peng, K. Modelling the strip thickness in hot steel rolling mills using least-squares support vector machines. *Can. J. Chem. Eng.* **2018**, *96*, 171–178. [CrossRef]

25. Peng, K.X.; Zhong, H.; Zhao, L.; Xue, K.; Ji, Y.D. Strip shape modeling and its setup strategy in hot strip mill process. *Int. J. Adv. Manuf. Technol.* **2014**, *72*, 589–605. [CrossRef]

26. Leaman, R.; Dogan, R.I.; Lu, Z.Y. Dnorm: Disease name normalization with pairwise learning to rank. *Bioinformatics* **2013**, *29*, 2909–2917. [CrossRef]

27. Cortes, C.; Vapnik, V. Support-vector networks. *Mach. Learn.* **1995**, *20*, 273–297. [CrossRef]

28. Smola, A.J.; Schölkopf, B. A Tutorial on Support Vector Regression. *Stat. Comput.* **2004**, *14*, 199–222. [CrossRef]

29. Shrestha, N.K.; Shukla, S. Support vector machine based modeling of evapotranspiration using hydro-climatic variables in a sub-tropical environment. *Agric. Forest Meteorol.* **2015**, *200*, 172–184. [CrossRef]

30. Ghorbani, M.A.; Shamshirband, S.; Haghi, D.Z.; Azani, A.; Bonakdari, H.; Ebtehaj, I. Application of firefly algorithm-based support vector machines for prediction of field capacity and permanent wilting point. *Soil Tillage Res.* **2017**, *172*, 32–38. [CrossRef]

31. Fan, J.; Yue, W.; Wu, L.; Zhang, F.; Cai, H.; Wang, X.; Lu, X.; Xiang, Y. Evaluation of SVM, ELM and four tree-based ensemble models for predicting daily reference evapotranspiration using limited meteorological data in different climates of China. *Agric. Forest Meteorol.* **2018**, *263*, 225–241. [CrossRef]

32. Gordon, A.D. Classification and Regression Trees. In *Biometrics*; Breiman, L., Friedman, J.H., ROlshen, A., Stone, C.J., Eds.; International Biometric Society: Washington, DC, USA, 1984; Volume 40, p. 874. [CrossRef]

33. Brezigar-Masten, A.; Masten, I. CART-based selection of bankruptcy predictors for the logic model. *Expert Syst. Appl.* **2012**, *39*, 10153–10159. [CrossRef]

34. Vondra, M.; Touš, M.; Teng, S.Y. Digestate evaporation treatment in biogas plants: A techno-economic assessment by Monte Carlo, neural networks and decision trees. *J. Clean. Prod.* **2019**, *238*, 117870. [CrossRef]

35. Günay, M.E.; Türker, L.; Tapan, N.A. Decision tree analysis for efficient CO_2 utilization in electrochemical systems. *J. CO_2 Util.* **2018**, *28*, 83–95. [CrossRef]

36. Madhusudana, C.K.; Kumar, H.; Narendranath, S. Fault Diagnosis of Face Milling Tool using Decision Tree and Sound Signal. *Mater. Today Proc.* **2018**, *5*, 12035–12044. [CrossRef]

37. Wang, G.; Hao, J.; Mab, J.; Jiang, H. A comparative assessment of ensemble learning for credit scoring. *Exp. Syst. Appl.* **2011**, *38*, 223–230. [CrossRef]

38. Breiman, L. Bagging predictors. *Mach. Learn.* **1996**, *24*, 123–140. [CrossRef]

39. Bühlmann, P.; Yu, B. Analyzing Bagging. *Ann. Stat.* **2002**, *30*, 927–961. [CrossRef]

40. Hothorn, T.; Lausen, B.; Benner, A.; Radespiel-Troger, M. Bagging survival trees. *Stat. Med.* **2004**, *23*, 77–91. [CrossRef]

41. Chan, C.W.; Huang, C.; Defries, R. Enhanced algorithm performance for land cover classification from remotely sensed data using bagging and boosting. *IEEE Trans. Geosci. Remote* **2001**, *39*, 693–695. [CrossRef]

42. Wang, X.J.; Yuan, P.; Mao, Z.Z.; You, M.S. Molten steel temperature prediction model based on bootstrap Feature Subsets Ensemble Regression Trees. *Knowl.-Based Syst.* **2016**, *1011*, 48–59. [CrossRef]

43. Tibshirani, R. Regression shrinkage and selection via the lasso. *J. R. Stat. Soc. B Met.* **1996**, *58*, 267–288. [CrossRef]

44. Spencer, B.; Alfandi, O.; Al-Obeidat, F. A Refinement of Lasso Regression Applied to Temperature Forecasting. *Procedia Comput. Sci.* **2018**, *130*, 728–735. [CrossRef]

45. Zhang, R.Q.; Zhang, F.Y.; Chen, W.C.; Yao, H.M.; Guo, J.; Wu, S.C.; Wu, T.; Du, Y.P. A new strategy of least absolute shrinkage and selection operator coupled with sampling error profile analysis for wavelength selection. *Chemometr. Intell. Lab.* **2018**, *17515*, 47–54. [CrossRef]

46. Chu, H.B.; Wei, J.H.; Wu, W.Y. Streamflow prediction using LASSO-FCM-DBN approach based on hydro-meteorological condition classification. *J. Hydrol.* **2020**, *580*, 124253. [CrossRef]

47. Sniekers, S.; Vaart, A.V.D. Adaptive Bayesian credible sets in regression with a Gaussian process prior. *Statistics* **2015**, *9*, 2475–2527. [CrossRef]

48. Liu, H.T.; Cai, J.F.; Ong, Y.S.; Wang, Y. Understanding and comparing scalable Gaussian process regression for big data. *Knowl.-Based Syst.* **2019**, *164*, 324–335. [CrossRef]

49. Kong, D.; Chen, Y.; Li, N. Gaussian process regression for tool wear prediction. *Mech. Syst. Signal Pr.* **2018**, *104*, 556–574. [CrossRef]

50. Wang, B.; Chen, T. Gaussian process regression with multiple response variables. *Chemometr. Intell. Lab.* **2015**, *142*, 159–165. [CrossRef]

51. Zhang, C.; Wei, H.; Zhao, X.; Liu, T.; Zhang, K. A Gaussian process regression based hybrid approach for short-term wind speed prediction. *Energy Convers. Manag.* **2016**, *126*, 1084–1092. [CrossRef]

52. Aye, S.A.; Heyns, P.S. An integrated Gaussian process regression for prediction of remaining useful life of slow speed bearings based on acoustic emission. *Mech. Syst. Signal Pr.* **2017**, *84*, 485–498. [CrossRef]

53. Liu, Y.Q.; Pan, Y.P.; Huang, D.P.; Wang, Q.L. Fault prognosis of filamentous sludge bulking using an enhanced multi-output gaussian processes regression. *Control Eng. Pract.* **2017**, *62*, 46–54. [CrossRef]

 *metals*

Article

Fatigue Life Assessment of Refill Friction Stir Spot Welded Alclad 7075-T6 Aluminium Alloy Joints

Andrzej Kubit [1,*]**, Mateusz Drabczyk** [2]**, Tomasz Trzepiecinski** [3]**, Wojciech Bochnowski** [4]**, Ľuboš Kaščák** [5] **and Jan Slota** [5]

[1]   Department of Manufacturing and Production Engineering, Rzeszow University of Technology, al. Powst. Warszawy 8, 35-959 Rzeszów, Poland
[2]   Institute of Technology, Faculty of Mathematics and Natural Sciences, University of Rzeszów, al. Rejtana 16c, 35-959 Rzeszow, Poland; mat.drabczyk@gmail.com
[3]   Department of Materials Forming and Processing, Rzeszow University of Technology, al. Powst. Warszawy 8, 35-959 Rzeszów, Poland; tomtrz@prz.edu.pl
[4]   Centre for Innovative Technologies, University of Rzeszów, ul. Pigonia 1, 35-959 Rzeszów, Poland; wobochno@ur.edu.pl
[5]   Institute of Technology and Material Engineering, Faculty of Mechanical Engineering, Technical University of Košice, Mäsiarska 74, 040 01 Košice, Slovakia; lubos.kascak@tuke.sk (Ľ.K.); jan.slota@tuke.sk (J.S.)
*   Correspondence: akubit@prz.edu.pl; Tel.: +48-17-743-2019

Received: 3 April 2020; Accepted: 12 May 2020; Published: 13 May 2020

Abstract: Refill Friction Stir Spot Welding (RFSSW) shows great potential to be a replacement for single-lap joining techniques such as riveting or resistance spot welding used in the aircraft industry. In this paper, the fatigue behaviour of RFSSW single-lap joints is analysed experimentally in lap-shear specimens of Alclad 7075-T6 aluminium alloy with different thicknesses, i.e., 0.8 mm and 1.6 mm. The joints were tested under low-cycle and high-cycle fatigue tests. Detailed observations of the fatigue fracture characteristics were conducted using a scanning electron microscope (SEM) with energy dispersive X-ray spectroscopy (EDS). The locations of fatigue failure across the weld, fatigue crack initiation, and propagation behaviour are discussed on the basis of the SEM analysis. The possibility of predicting the propagation of fatigue cracks in RFSSW joints is verified based on Paris's law. Two fatigue failure modes are observed at different load levels, including shear fracture mode transverse crack growth at high stress-loading conditions and at low load levels, and destruction of the lower sheet due to stretching as a result of low stress-loading conditions. The analysis of SEM micrographs revealed that the presence of aluminium oxides aggravates the inhomogeneity of the material in the weld nugget around its periphery and is a source of crack nucleation. The results of the fatigue crack growth rate predicted by Paris's law were in good agreement with the experimental results.

Keywords: aircraft industry; aluminium alloy; friction stir spot welding; mechanical engineering; single-lap joints

---

1. Introduction

In recent years, Refill Friction Stir Spot Welding (RFSSW) technology [1], which is a derivative of Friction Stir Spot Welding (FSSW), is one the main solid-state joining techniques, which are being extensively studied in the literature. RFSSW is seen as an attractive technique for joining difficult-to-weld aluminium alloys in aircraft structures [2–5]. One of the crucial assessment qualities in the aircraft industry is the fatigue performance of structural components and, therefore, many efforts have been made to investigate the fatigue properties of various grades of friction stir welded joints of aluminium alloys. Fatigue loading is the main cause of failure of weld joints and is evaluated either by fatigue crack propagation behaviour [6,7] or stress vs. the number of cycles to failure (S–N) behaviour [8,9]. The results of the investigations of many authors revealed that the fatigue strength of friction stir welded joints is superior to fusion welded joints [10,11] and is equal to or less than base metal [12,13].

Various results on the influence of welding conditions on the fatigue strength of friction spot welded joints are reported in the literature. James et al. [14] reported that lower welding speed exhibits higher fatigue strength. Cavaliere et al. [9] reported an increase in fatigue strength with increase in welding speed. In contrast, Ericsson and Sandstrom [15] reported that fatigue properties are relatively independent of welding speed.

Most of the previous studies investigated the fatigue properties of joints between similar aluminium alloys fabricated by friction stir welding. Sivaraj et al. [16] evaluated the fatigue crack growth of 7075-T651 aluminium alloy joints. They found that the fatigue life of joints is appreciably lower than that of the unwelded base metal, which is attributed to the dissolution of precipitates in the weld zone. The effect of base metal temper conditions on the fatigue behaviour of friction stir welded joints of AA7039 aluminium alloy was studied by Sharma et al. [17]. It has been found that better tensile properties and superior fatigue strength are observed in AA7039 aluminium alloy joints in W-temper conditions. Effertz et al. [18] studied the fatigue life of FSSW 7075-T76 aluminium alloy using the Weibull distribution. They found that, for a relatively low load, corresponding to 10% of the maximum supported by the joint, the number of cycles surpasses 1×10^6. Hence, an infinite life can be attributed to the service component.

Many of the previous studies have focused on determining the influence of welding parameters on the failure modes of joints under quasi-static loading conditions and the weld zone microstructure. Hassanifard et al. [19] introduced a new method for enhancing the life and strength of FSSW joints using a localised plasticity process. The S-N data obtained revealed that the cold expansion method could improve the fatigue life of FSSW joints by up to six times. The microstructures and failure modes of friction spot welds in lap-shear specimens of aluminium 6111-T4 sheets were studied by Lin et al. [20]. The micrographs show that, under quasi-static loading conditions, the failure mainly starts from the necking of the upper sheet outside the weld. Under cyclic loading conditions, the micrographs revealed two types of fatigue cracks: growths into the lower sheet outside the stir zone and initiation of the crack from the bend surface of the upper sheet outside the weld. Finding a gap in the literature, Singh et al. [21] investigated the effect of the environment and welding speed on the fatigue properties of the FSSWed Al-Mg-Cr alloy. The study showed that the presence of oxides in the weld nugget zone in the case of liquid nitrogen cooling causes a significant increase in the fatigue crack growth rate, while the fatigue crack growth rate is lower in the case of welding. The observations of fatigue cracks of 6061-T6 aluminium lap-shear specimens under cyclic loading conditions suggest that the fatigue crack is initiated near the possible original notch tip in the stir zone and propagates along the circumference of the nugget. Then it propagates through the sheet thickness and finally grows in the width direction to cause a final fracture [22]. Moreover, the results indicate that the fatigue life predictions based on Paris's law agree well with the experimental results. Wang et al. [23] proposed improving the mechanical properties of 2024 aluminium alloy joints by using graphene nanosheets to strengthen the tip of the hook defect. According to the authors, the improvement of fatigue life originates from the combined effect of crack deflection, which results from its tight incorporation between the aluminium matrix and the graphene nanosheets. Kubit et al. [24,25] analysed the effect of structural defects on the fatigue strength of RFSSW joints using C-scan scanning acoustic microscopy and SEM. The observations of the fatigue fractures reveal that the clad layer at the bottom of the weld is a source of initiation of fatigue cracking. Insufficient plasticisation of sheet material is a crucial defect that influences the number of destructive cycles. It was found that the fracture mechanism depends on the value of load amplitude. Shahani and Farrahi [26] investigated the mechanical behaviour of the 6061-T6 aluminium alloy FSSW joints. It was concluded that the stress intensity factors and kink angle change during the growth of the fatigue crack. Therefore, the contact kink angle assumed in the literature is not an appropriate assumption. Investigations presented by Chowdhury et al. [27]

were aimed at evaluating the fatigue properties of FSSWed 5754-O aluminium alloys. At higher cyclic loads, nugget pull-out failure occurred since the fatigue crack propagated circumferentially around the nugget, while, at lower cyclic loads, fatigue failure occurred perpendicularly to the loading direction. Failure modes of spot friction welds in cross-tension specimens of 6061-T6 aluminium alloy sheets have been investigated by Lin et al. [28]. Paris's law for fatigue crack propagation has been successfully used for accurate development of the fatigue crack growth model. An overview of the fatigue mechanism, fatigue life assessment, crack growth rate, and factors influencing the fatigue of friction stir welded aluminium alloy joints has been provided by Li et al. [29].

The RFSSW solid-state technique eliminates the disadvantages usually observed in other spot-like joining technologies, such as the difficulty of automation, weight penalty, requirement to drill holes (riveted and bolted joints), and the presence of a keyhole originated by the FSSW [18,30]. The advantages of the RFSSW technique such as no additional material being required to fabricate the joint mean that it is ideal for joining structural components made of lightweight alloys in the aircraft industry. Lightweight aluminium alloys are characterised by high fatigue strength [31,32]. However, the welding process causes non-uniformity in the weld structure, which produces a stir zone (SZ), a thermo-mechanically affected zone (TMAZ), and a heat affected zone (HAZ). In recent work [24,25], the authors have shown that the structure of the joint in the Alclad sheet is characterised by the presence of a bonded ligament causing discontinuity in the weld structure. The presence of weld defects affects the fatigue failure mechanism. In the case of linear friction stir welding (LFSW), many scientific papers have been written on the fatigue strength of LFSW joints [33–35]. However, the issue of the fatigue cracking mechanism for RFSSW joints is still not adequately described, and only a few scientific studies deal with this issue [35,36].

In this paper, the RFSSW technique is used to join Alclad 7075-T6 aluminium alloy plates of various thicknesses, i.e., 1.6 mm and 0.8 mm. The single-lap joints were subjected to a fatigue test. Typical failure mechanisms for low stress-loading conditions and high-stress loading conditions are presented and discussed. The location of fatigue crack initiation and the propagation behaviour are also discussed based on the scanning electron microscopy (SEM) analysis. Furthermore, one of the main aims of the work was to verify the possibility of predicting the propagation of fatigue cracks in RFSSW joints using Paris's law.

2. Experimental Methodology

2.1. Material

Alclad 7075-T6 aluminium alloy sheets were selected as the test material. Sheets of two thicknesses corresponding to the thickness of the stringers (1.6 mm) and skin plate (0.8 mm) of the fuselage of an aircraft were joined using RFSSW. The main alloying elements of the 7075-T6 aluminium alloy are zinc, magnesium, and copper. While the alloy is characterised by high static and fatigue strength, it is difficult to cold form because the value of its yield stress is close to its tensile strength [7]. In addition, 7075-T6 aluminium alloy is difficult to weld and is characterised by a relatively low corrosion resistance [5,7]. However, the basic feature of this alloy is its combination of high mechanical strength and low weight, which makes it a material commonly used in the manufacturing of aircraft structures. Due to its relatively high susceptibility to intergranular corrosion, sheets used in the construction of thin-walled aircraft structures are usually clad, i.e., covered with a thin layer of pure aluminium (approximately 5% of the thickness of the base metal) in the hot rolling process.

The table lists the basic mechanical properties of the sheets tested, which were determined in the uniaxial tensile test using a Z100 universal testing machine (Zwick/Roell, Ulm, Germany). The following parameters were determined through the tensile test: yield stress $R_{p0.2}$ = 413.7 MPa, ultimate tensile stress R_m = 482.6 MPa, and elongation at failure A = 7%. Three specimens were tested and the average values of each specific parameter were determined. The chemical composition of 7075-T6 aluminium alloy is listed in Table 1.

Table 1. The chemical composition (in weight percentage) of 7075-T6 material.

Al	Zn	Cr	Cu	Mg	Mn	Fe	Si	Ti	Other	
									Each	Total
87.1–91.4	5.1–6.1	0.18–0.28	1.2–2.0	2.1–2.9	max. 0.3	max. 0.5	max. 0.4	max. 0.2	0.05	0.15

2.2. Welding Process

The fatigue tests were conducted on the single-lap RFSSW joints presented in Figure 1. The welding was conducted using an RPS100 spot welder by Harms & Wende GmbH & Co KG (Hamburg, Germany). Before welding, the workpieces were cleaned with acetone on the faying surfaces. The RFSSW process consists of the following stages.

- The clamping ring is fixed on the top surface of the upper sheet, and the tool stays there for a certain amount of time to produce the initial frictional pre-heating (Figure 2a),
- the sleeve plunges the sheet to the desired depth, and, at the same time, the pin moves in the opposite direction (Figure 2b),
- after reaching the desired plunge depth, the directions of movement of both the sleeve and pin begin to reverse (Figure 2c),
- the weld cycle is completed by removing the tool from the surfaces of the sheets (Figure 2d).

The values of the welding parameters are crucial in assuring the fabrication of welds without defects such as voids in the weld corner, bonding ligament, and structural discontinuities. The RFSSW parameters were selected based on the authors' experience and the results of previous investigations of the authors presented in the paper [37]. The first stage of the poly-optimisation of the values of welding process parameters was focused on the analysis of the tool plunge depth g on the load capacity of the joint. For this purpose, the welds were fabricated at a tool plunge depth equal to 1.3, 1.5, 1.7, and 1.9 mm, with three different tool rotational speeds n = 2000, 2400, and 2800 rpm. The load capacity of the joints obtained ranged from 5510 N to 8000 N. In the next part of the research, the focus was put on determining the influence of tool rotational speed and the duration of welding on the load capacity of joints. For the durations of welding t = 1.5 s and 2.5 s, an increase of tool rotational speed causes an increase in the load capacity of the joint. For the duration t = 1.5 s, an increase in the tool rotational speed from 2000 to 2800 rpm results in an increase in load capacity by 17.9% (from 6870 to 8100 N), while, for the welding time t = 2.5 s, it does this by 11.5%. During the investigations, the hypothesis that the factor analysed did not have an influence on the load capacity of the connection that was tested. For the analysis, a confidence interval of 95% and α = 5% (the probability of rejecting the hypothesis tested, assuming it is true) was used. The dependent variable was the maximum shear load, while the process parameters were set as the independent variables. The values of the parameters that were selected assured the most favourable joint quality in terms of the highest static strength and quality of the weld structure (minimum number of structural discontinuities, structural notches, as small as the heat affected zone, etc.) [37].

The values of the welding parameters are as follows: tool rotational speed n = 2400 rpm, duration of welding t = 2.5 s, and tool plunge depth g = 1.55 mm.

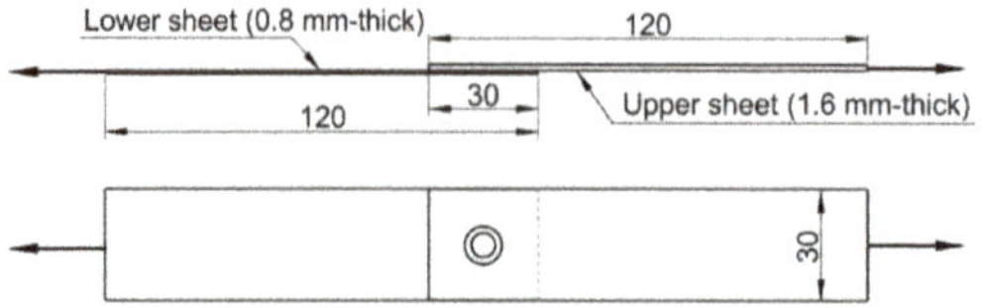

Figure 1. Geometry and dimensions of the specimen for the tensile test.

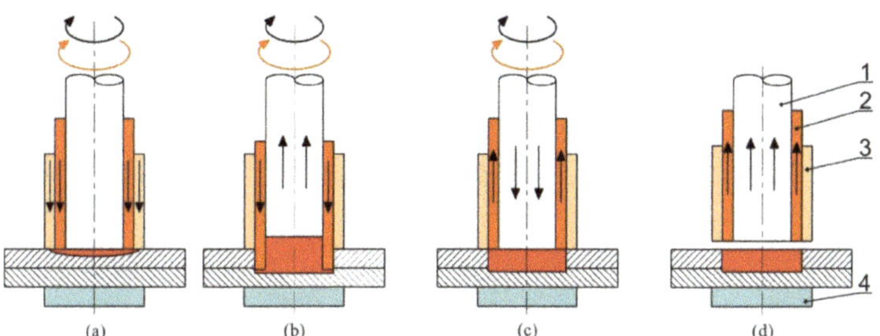

Figure 2. Stages of the Refill Friction Stir Pot Welding (RFSSW) process: touchdown and preheating (**a**), plunging (**b**), refilling (**c**), and retreating (**d**): 1-pin, 2-sleeve, 3-clamping ring, and 4-bracking anvil.

2.3. Fatigue Test

Fatigue strength tests were carried out on an Instron ElectroPulsTM E10000 testing machine (Instron, Norwood, MA, USA) designed for dynamic and static testing on a wide range of components and materials. Fatigue tests were carried out at room temperature with the following conditions: a frequency of 50 Hz and a limited number of cycles equal to 2×10^6. The coefficient of the stress cycle of $R = 0.1$ used is often used in aircraft component testing [38] and corresponds to a tension-tension cycle in which $\sigma_{min} = 0.1\sigma_{max}$. Four specimens were tested for each level of amplitude.

2.4. Characterisation of Fracture Surfaces

Microstructure and fracture morphologies of selected specimens were analysed using a S-3400N (Hitachi, Chiyoda, Japan) variable pressure scanning electron microscope (SEM). The samples were etched using Keller's solution (2 mL HF, 3 mL HCL, 5 mL HNO_3, and 190 mL H_2O). When analysing fatigue fractures, an analysis of the chemical composition was performed on selected areas of the fracture surface to verify fracture areas and identify impurities in the weld. The chemical composition analysis was conducted using energy dispersive spectrometry (EDS). Energy-dispersive X-ray (EDX) spectra were collected on the Quanta 3D 200i (FEI Company, Hillsboro, OR, USA) scanning electron microscope.

3. Results and Discussion

3.1. Microstructure Analysis

The cross-section of the RFSSW joint (Figure 3) can be divided into four regions in terms of the microstructural characteristics of the joint in sequence from the stir zone (SZ) toward the base material (BM), the thermo-mechanically affected zone (TMAZ), and the heat-affected zone (HAZ). In the SZ, a homogenous fine-grained microstructure was found, which was characterised by fully dynamically recrystallised equiaxial grains. In the HAZ, the grains are similar to those of the BM, but, in the direction of the TMAZ, the grains become slightly coarse. Within the centre of the SZ in the RFSSW process, a fine-grained microstructure is observed. In the case of the TMAZ, a high gradient of a grain size change is revealed. On the boundary of these grains, ultrafine precipitates were observed. Literature data suggest that these are likely particles of Al_7Cu_2Fe, $Mg\,(Zn, Cu, Al)_2$, and also $MgZn_2$ strengthening phases that occur in the 7xxx series aluminium alloys [25].

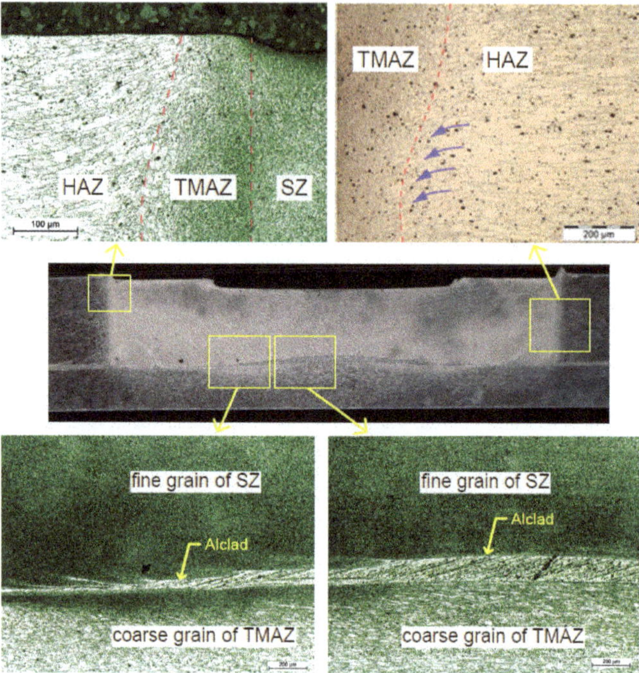

Figure 3. Cross-sectional view of the weld made with the parameters n = 2400 rpm, t = 2.5 s, and g = 1.55 mm.

The weld made using RFSSW consists of two main parts: the nugget of the joint and the HAZ. The nugget of the weld has a fine grain structure. This indicates the occurrence of the recrystallization caused by both high heat and pressure. In the bottom part of the weld, a plated coating (Alclad) is observed. Alclad within the weld nugget, which covered the sheet metals to protect them against corrosion, causes heterogeneity in the structure of the weld. These heterogeneities reduce the weld strength and clearly indicate the direction of flow of the material. The Alclad has a higher thermal conductivity (229 W/mK) where BM (134 W/mK) transfers the heat out of the joint very quickly. The Alclad layer also exists in the lower zone of the joint which, despite reaching a much higher temperature, is not degraded.

3.2. Fatigue Diagram

A fatigue diagram with areas corresponding to the specific failure mechanism is shown in Figure 4. Based on the results of shear fatigue testing of the joints, two basic mechanisms of fatigue failure were observed. A limited number of cycles to failure of 10^5 is adopted to be the boundary between low-stress and high-stress fatigue loading. In the area of high-stress fatigue strength, the joints were mainly damaged by shearing in the plane in which the sheets were joined. High-stress fatigue strength was characterised by relatively high plastic deformation of the joint. The weld structure is weakened by a clad layer called a bonding ligament, which is composed of pure aluminium compressed at the bottom of the weld (Figure 5a). The bonding ligament originating from the Alclad at the original lap interface has been deformed and redistributed due to plasticised metal flow during the welding process. The bonding ligament, especially near the axis of the weld, is an inherent problem of the RFSSW joints [24,37].

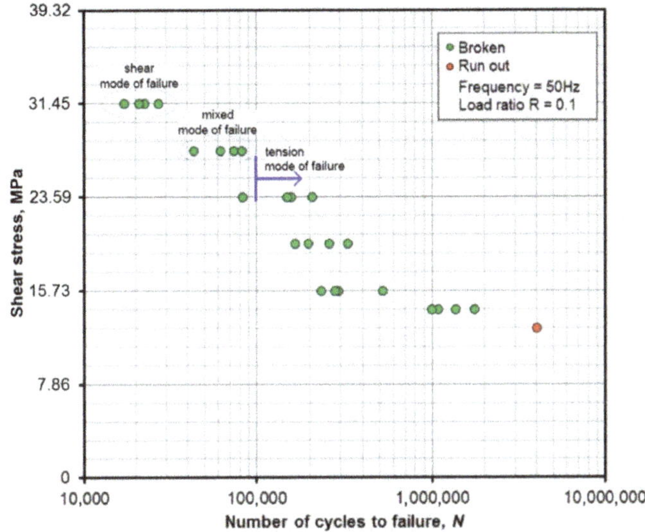

Figure 4. Fatigue diagram for Refill Friction Stir Spot Welding (RFSSW) joints subjected to high-cycle fatigue strength with a limited number of cycles equal to 4×10^6.

3.3. Fatigue with a High Value of Stress

Plastic deformations of the weld resulting from fatigue cycles are the cause of stress concentration at the boundary of the clad material in the weld structure and the fine-grained base metal (BM) in the SZ zone. Moreover, a thin and brittle layer of oxides was formed in the welding process as a result of thermo-mechanical changes in the BM. The presence of aluminium oxides aggravates the inhomogeneity of the weld structure around its periphery and is a source of crack nucleation. The oxides are considered to contaminate the weld. Aluminium and its alloys are very reactive with atmospheric oxygen and form a layer of oxide on its surface [39]. The oxide layer can cause entrapped oxide flaws in the stir zone. The existence of oxides in the nugget zone is confirmed by EDS analysis of the fracture surface of the joint in the area of the upper sheet that underwent fatigue shearing after 22,348 cycles with a maximum variable force value of 2,000 N. Figure 5b shows a fragment of the fracture surface with three characteristic areas of BM, a clad layer, and the boundary surface between the clad and BM containing a significant amount of oxides. The chemical composition in these areas is revealed by EDS spectra: EDS_1 (Figure 5c), EDS_2 (Figure 5d), and EDS_3 (Figure 5e), respectively.

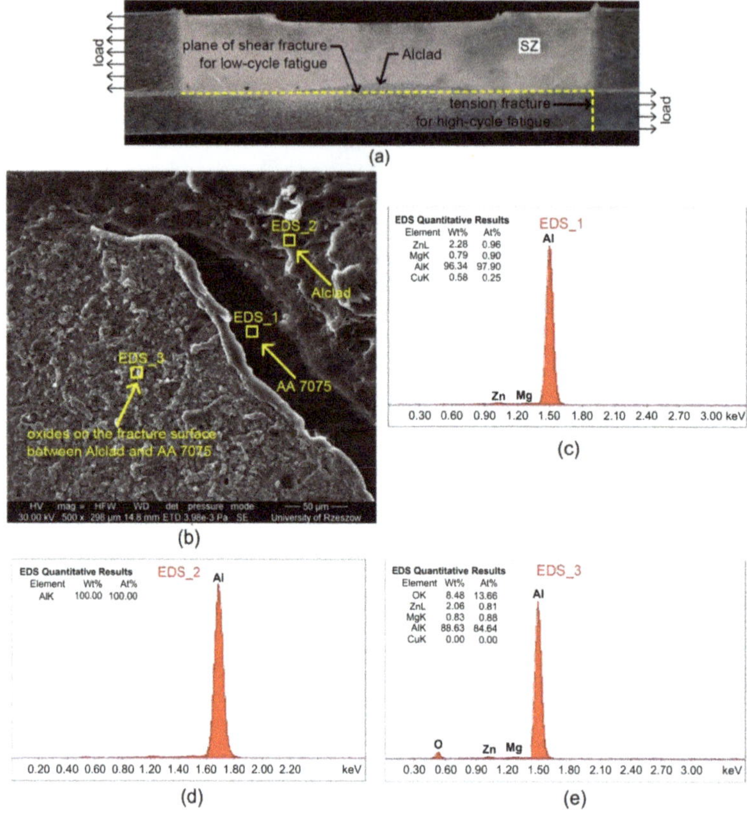

Figure 5. (**a**) Cross-sectional view of RFSSW joint structure, (**b**) scanning electron microscope (SEM) micrograph of fatigue fracture for a joint that was destroyed after 22,348 cycles at a stress level of 31.45 MPa and energy dispersive X-ray spectroscopy (EDS) spectra of weld material in areas EDS_1 (**c**), EDS_2 (**d**), and EDS_3 (**e**) in Figure 4b.

For a variable force with a maximum value of 2000 N, all the test samples exhibited the shear model of failure. Figure 6a shows the fracture surface in the shear plane of the lower sheet for a joint that was destroyed after loading more than 22,348 cycles. The presence of oxides on the surface of the clad material in the nugget zone was demonstrated on the basis of EDS spectra of fatigue fractures (Figure 6b–d). A high content of oxides on the fracture surface was also observed in the case of other samples that were destroyed as a result of shearing. Figure 7 presents a SEM micrograph of a sample that was destroyed after loading more than 26,913 cycles.

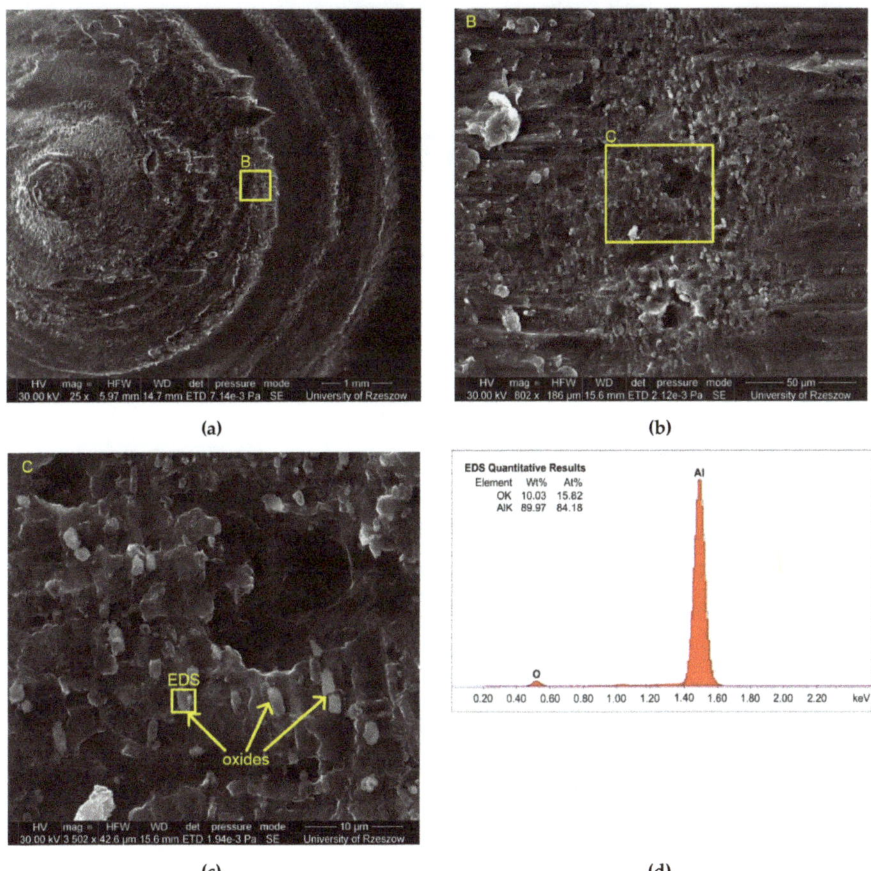

Figure 6. (**a**) Scanning electron microscope (SEM) micrograph of a fracture surface for a joint that was destroyed after 22,348 cycles. At a stress level of 31.45 MPa (**b**) and (**c**) magnified views in areas A and B, (**d**) EDS spectra of the aluminium oxides shown in Figure 5c.

In the case of selected specimens, especially those characterised by higher fatigue life from among those loaded with a variable force with a maximum value of 2000 N, SEM analysis has shown that, despite the fact that the samples were destroyed by shearing, the second mechanism of fatigue cracking progressed in parallel to the shear fracture. In the lower sheet, transverse cracks were disclosed, which were the result of stretching this sheet. This type of crack was usually initiated at the boundary of the SZ and TMAZ in the vicinity of the axis of the joint, and propagated toward the edges of the sheets. In the first phase of fatigue loading, the fatigue crack propagated along the weld perimeter, after which it took a linear direction. Figure 8a shows the fatigue fracture for a joint, which led to damage after 22,348 load cycles.

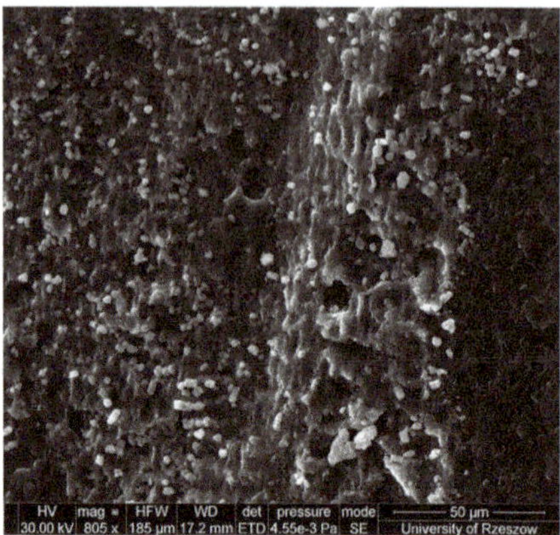

Figure 7. SEM micrograph of a fracture surface for a joint that was destroyed after 26,913 load cycles at a stress level of 31.45 MPa.

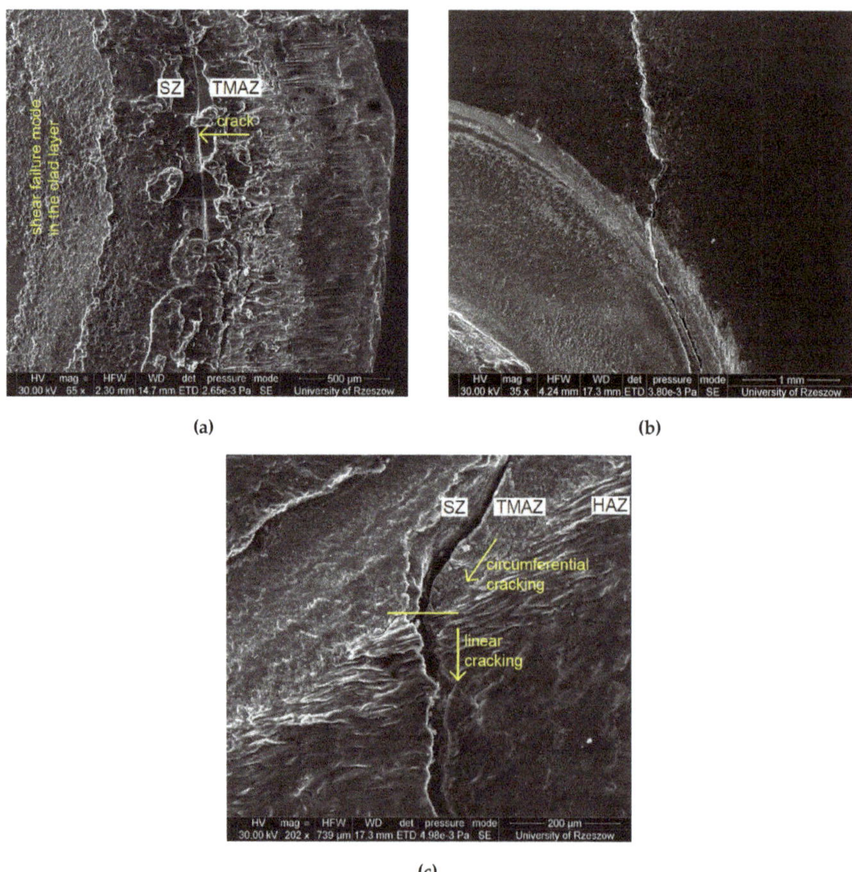

(a)

(b)

(c)

Figure 8. SEM micrographs of a fracture surface for joints that were destroyed after (**a**) 22,348 cycles at a stress level of 31.45 MPa, (**b**) 26,913 cycles at a stress level of 31.45 MPa, and (**c**) 42,639 cycles at a stress level of 27.52 MPa with a clear transverse fatigue crack in the lower sheet.

The joint was destroyed as a result of shearing at the boundary of the clad material and through the clad material. However, the crack of the lower plate was simultaneously initiated along the weld perimeter. Figure 8b shows the propagation of the crack as a result of tensile stress. The change in the direction of crack propagation from peripheral to transverse is clearly visible. A similar phenomenon was observed for a joint that had been destroyed as a result of shearing more than 42,639 cycles with a variable force with a maximum value of 1,750 N (Figure 8c). The results obtained are consistent with the study of Venukumar et al. [40] who tested the fatigue behaviour of friction stir spot welding refilled by the friction forming process (FSSW-FFP) lap-shear specimens, and concluded that, under high-stress loading conditions with lower load ranges, more transverse cracks were observed when compared to those with higher load ranges.

3.4. Fatigue with a Low Value of Stress

In the low stress range of fatigue strength, determined by the number of 10^5 cycles with variable force, all test specimens were destroyed as a result of stretching the lower sheet (Figure 9), which was thinner than the upper sheet. In the low stress range of loading conditions, there are no plastic deformations caused by subsequent load cycles with a variable load. Hence, there is no crack initiation at the boundary between the clad material lying at the bottom of the weld and the fine-grained material.

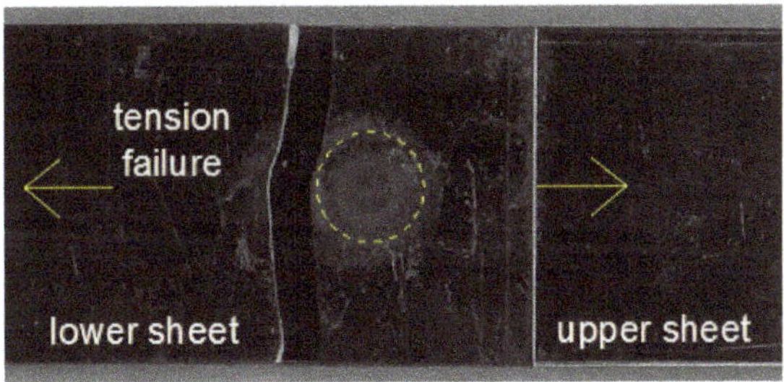

Figure 9. A typical pattern of destruction of the weld under high-cycle loading conditions.

3.5. Fatigue Crack Growth Rate

Due to the long-term operation of the RFSSW structure, knowledge about their high-cycle fatigue life is very important. An accurate description of fracture mechanisms due to fatigue during long-term operation, and, thus, the prediction of crack propagation, is an important issue. Therefore, it was decided to attempt to predict the speed of crack propagation in RFSSW joints based on well-established hypotheses. The authors propose the use of Paris's law in order to describe the propagation of fatigue cracks in the lap-shear specimens considered in this article. Paris's law for crack propagation has been successfully adopted to predict the fatigue crack growth rate of FSSW aluminium alloy joints [22,41]. In this paper, it was assumed that the initiation of the fatigue crack occurred at the boundary of the SZ and TMAZ zones. In addition, the initial crack propagation in the peripheral direction of the weld is not described by Paris's law.

The SZ, TMAZ, and HAZ have different properties in relation to the native material. In the RFSSW process, the material in the previously mentioned zones were subjected to thermo-mechanical change. This caused grain refining and modification of the mechanical properties of the material as a result of high temperature. Therefore, the mechanisms accompanying fatigue cracking and its initiation in the area are omitted in this analysis. The analysis was started when the direction of crack propagation was already linear and was propagating toward the edge of the sheets. Paris's law was valid for the linear cracking in the quasi-transverse direction to the load direction. Figure 10 shows in schematic fashion the area of validity of Paris's law (or Paris-Erdogan law) for the lap-shear specimens analysed. In this graph, the rate of crack growth da/dN is expressed on a logarithmic scale, and the amplitude of the stress intensity factor ΔK is expressed on a linear scale. The ΔK is a material constant at a given stress ratio R. The length of the initial crack was adopted in the analysis as "pre-crack" (Figure 11). The threshold region in the propagation in Figure 10 does not occur for the ΔK-values below the threshold level. The Paris region corresponds to stable macroscopic crack growth. The final failure region is associated with the high crack growth rate prior to failure.

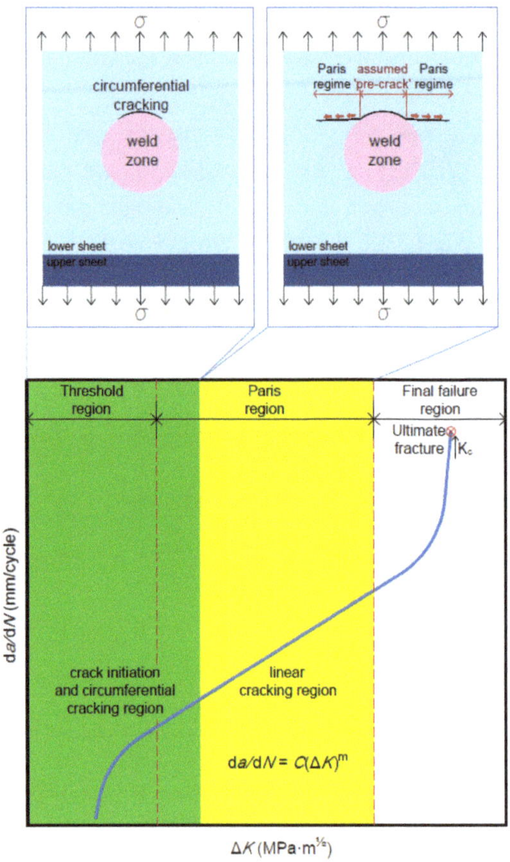

Figure 10. Rate of fatigue-crack growth with the assumed range of validity of Paris Law for lap-shear RFSSW specimens.

Paris and Erdogan [42] proposed that the relationship between the rate of fatigue-crack growth and the stress-intensity range is as follows.

$$\frac{da}{dN} = C \cdot (\Delta K)^m \tag{1}$$

$$\Delta K = \Delta \sigma \cdot \sqrt{\pi \cdot a} \tag{2}$$

$$\Delta \sigma = \sigma_{max} - \sigma_{min} \tag{3}$$

$$N = \int_{a_1}^{a_2} \frac{1}{C \cdot \left(\Delta \sigma \cdot \sqrt{\pi \cdot a} \right)^m} da \tag{4}$$

where da/dN is the fatigue crack growth rate, N is the life or number of cycles, a is crack length, ΔK is the stress intensity factor range, and C and m are constants proposed to incorporate the effects of material, mean load, loading frequency, and environment.

The fatigue crack growth rate is defined as the increase in its length during one loading cycle, and the case depends on the gap length, working stress, and material properties. Changes in the loading stress σ during the load cycle cause corresponding changes in the stress in the area close to the edge of the crack, and, thus, changes in the stress intensity factor range ΔK (Equation (2)). Changes in the stress near the crack edge determine its behaviour.

The detailed procedure for obtaining feasible and comparable constants in Paris's law is defined in standards BS 6835:1988 and ASTM 647-95a. The values of constants C and m for 7075-T6 aluminium alloy used in this research are as follows: $C = 7 \times 10^{-11}$, $m = 4$.

Experimental verification was carried out thanks to which it was possible to confirm the accuracy of Paris's law for the selected fatigue area of the RFSSW joint. For selected specimens subjected to fatigue loading, measurements of the distances between the ends of cracks (Figure 11) were conducted during the tests. These measurements were carried out when the direction of crack propagation in the lower sheet changes from peripheral to quasi-transverse. These distances were assumed as 'pre-cracks' and then based on Paris's law. The theoretical courses of crack propagation were determined according to Equations (1)–(4).

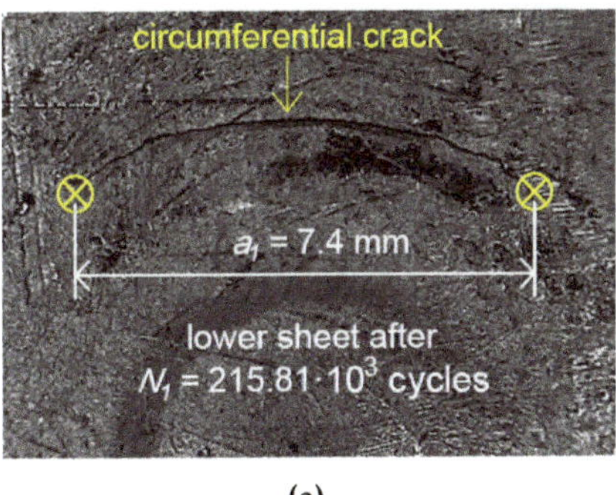

(a)

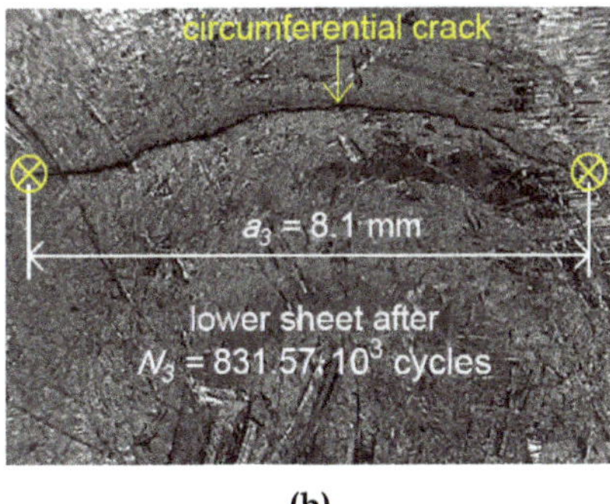

(b)

Figure 11. Views of fatigue cracks observed during the test when the crack changes its direction of propagation from peripheral to quasi-transverse for joints that were destroyed after (**a**) 215,810 cycles and (**b**) 831,570 cycles

The predicted evolution of crack length according to Paris's law is shown in Figure 12. In general, the experimental results are well predicted using the Paris equation for specific loading conditions and geometrical configurations. This conclusion is in accordance with the results of Theos [43] who conducted fatigue analysis of welded aircraft wing panels. The highest consistency was obtained for the joint, which cracked at the number of cycles of $N = 289.82 \times 10^3$. The weld was resistant to destruction when the Paris law underestimates the number of cycles to failure N. The Paris equation has some limitations, i.e., asymptotic behaviour in both the threshold and final failure region is not described with this equation. Moreover, the stress ratio is not accounted for the crack growth. A significant portion of the fatigue life is occupied in the subcritical crack growth region described by Paris's law, particularly structures constructed from sheet or plates.

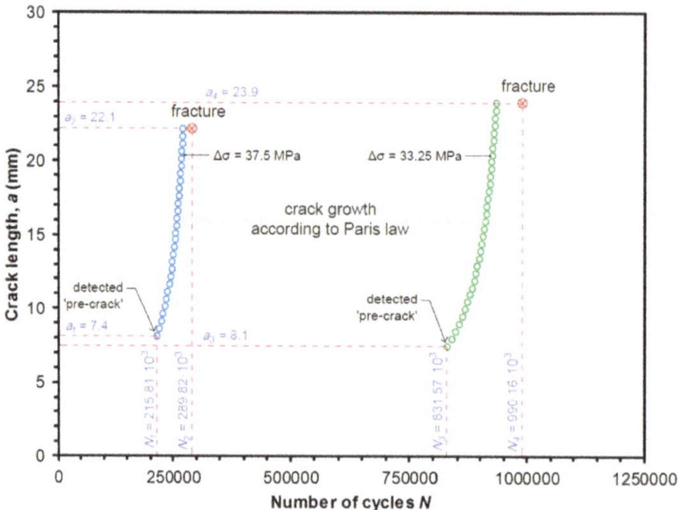

Figure 12. Fatigue crack growth behaviour with indication of the area of validity of Paris law for a lap-shear RFSSW joint loaded at a shear stress level of 15.73 MPa (blue points) and of 14.15 MPa (green points).

Analysis of SEM micrographs of fatigue fractures (Figure 13) shows that, in the area of the lower sheet, the propagation of the fatigue crack was approximately uniform throughout its thickness. This may prove that the assumption of Paris's law in the area of linear cracking of the joint is correct. Good consistency of the experimental results with the predicted number of cycles at which the joint would be destroyed allows one to conclude that the assumptions are correct. The considerations presented may form the basis for broader analyses, which predict crack propagation in RFSSW structures.

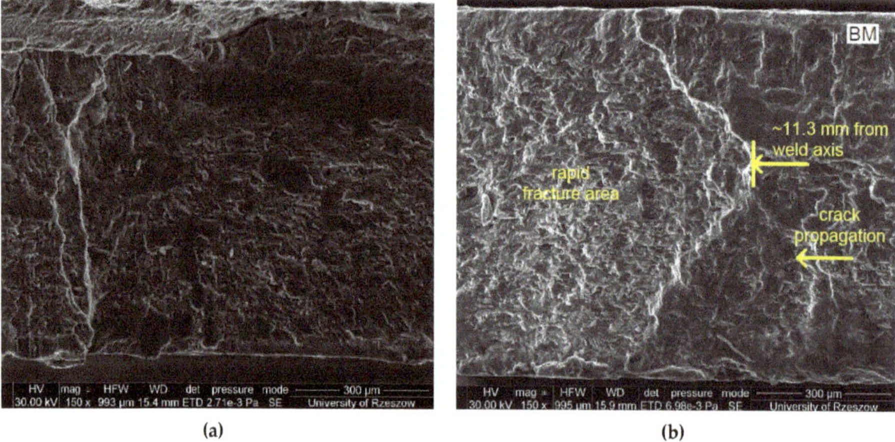

Figure 13. Fatigue fractures of lower sheets of joints that were destroyed as a result of stretching after (**a**) 231,794 cycles at a stress level of 15.73 MPa and (**b**) 1,363,043 cycles at a stress level of 14.15 MPa.

4. Residual Stresses Induced by Friction Stir Welding

Residual stresses are defined as self-equilibrating forces that remain in a material once all external forces have been removed. They arise due to complex thermal-mechanical-metallurgical interactions during welding. Residual stresses (RS) are results of complex thermal-mechanical- metallurgical interactions during welding. It may result in distortion of the weldments, which can lead to fitting issues and/or failures. Friction stir welding (FSW) also leads to the introduction of RS in welded structures. According to the Williams and Steuwer [44], residual stresses can be classified in terms of length scales on which they act over.

- type-I: extend over macroscopic areas and are averaged over several grains,
- type-II: extend between grains or sub-regions of grains and averaged over these areas,
- type-III: act on the inter-atomic level such as around inclusions or dislocation.

Several approaches for evaluating FSW-induced residual stresses have been proposed in literature based on the experimental techniques and numerical approaches. Residual stresses induced by friction stir welding are considered to be a major driving force for crack propagation if friction stir welded the structures [45]. The dominant role of RS on crack growth rates in 2024-T351 aluminium alloy joints was explicitly pinpointed by Bussu and Irving [46]. Their analysis underlined that the secondary role is played by the local microhardness and microstructure. The trend of crack propagation behaviour in aluminium alloys based on residual stress measurement has been discussed by Hong et al. [47]. The role of residual stresses and microstructure on crack propagation in 6063-T5 aluminium alloy joints has been experimentally investigated by Tra et al. [48]. It was found that crack propagation is mainly driven by the inhomogeneous microstructure in and around the welded area, whereas the effect of RS is not significant. A completely opposite conclusion was pointed out by Pouget and Reynolds [49]. They concluded that accurate growth rate predictions can be achieved by including a residual stress effect into the calculation. The key role of process-induced residual stresses on crack growth in FSW joints has also been shown by Citarella et al. [50] and Sepe et al. [51]. Difficulties in experimentally obtaining all the components of residual stress tensor significantly reduce the potential application of experimental methods [46].

Recently, some attempts to numerically assess the fatigue behaviour in FSW structures were discussed in literature. A numerical investigation on the influence of residual stress, induced by the FSW process, on fatigue crack growth in 2024-T3 aluminium friction stir welded butt joints has been proposed by Carlone et al. [45]. They found that, if the initial crack starts from the weld line, the process-induced opening stresses have an accelerating effect on the crack propagation. The implemented finite element method-dual boundary element method (FEM-DBEM) proved to be able to effectively predict multiple crack growth in the presence of residual stresses induced by the fabrication process. The influence of FSW process parameters on the residual stresses distribution has been also considered in recent investigations by Carlone et al. [52]. Results of FEM-DBEM-based computations revealed that the FSW process induced compressive residual stresses that play a delaying effect on the crack propagation if the initial notch is far enough from the weld line. Zadeh et al. [53] analysed three dimensional simulations of fatigue crack growth in friction stir welded joints of 2024-T351 aluminium alloy. Results of computations verified with experiments and the analytical method show 90% accuracy in terms of fracture mechanics and fatigue life prediction.

The effective numerical approaches able to predict the residual stresses and crack growth in friction stir welded joints are highly desirable. Therefore, further research will address the need for experimental and numerical analysis of the residual stresses in RFSSW joints.

5. Conclusions

This paper presents the results of fatigue life assessment of RFSSW single-lap joints subjected to high-cycle fatigue tests. The following conclusions are drawn from the research.

- As far as fatigue strength in low cycle conditions is concerned, the joints were mainly damaged by shearing in the plane in which the sheets were joined.

- The bonding ligament is the main element of the RFSSW joint weakening the fatigue strength of the joint.

- The RFSSW joint of Alclad sheets is contaminated by a high content of aluminium oxides. The presence of aluminium oxides aggravates the heterogeneity of the material in the weld nugget around its periphery and is a source of crack nucleation. Oxides could partially break down and be released from the outside of the weld area during the welding process.

- As far as fatigue strength is concerned in low stress-loading conditions, determined by the number of 10^5 cycles with a variable force, all test specimens were destroyed as a result of stretching the lower sheet.

- Paris's law for crack propagation has been successfully adopted to predict the fatigue crack growth of lap-shear RFSSW specimens. Although some assumptions have been made, the comparison of the analytical and experimental fatigue crack growth rate confirms the potential of Paris's law to analyse the crack growth in RFSSW joints.

Author Contributions: Conceptualization, A.K., M.D., and W.B. Investigation, A.K., M.D., and W.B. Formal analysis, A.K., M.D., and W.B. Validation, A.K., M.D., and W.B. Interpretation of the results, all authors contributed equally. Writing–original draft, A.K., and T.T. Writing–review & editing, T.T. Project administration, L'.K. and J.S. Funding acquisition, L'.K. and J.S. All authors have read and agreed to the published version of the manuscript.

Funding: The Slovak Research and Development Agency under grant number APVV-17-0381 – Increasing the efficiency of forming and joining parts of hybrid car bodies – funded this research. And the Grant Agency of the Ministry of Education, Science, Research, and Sport of the Slovak Republic grant number VEGA 1/0259/19.

Acknowledgments: The authors are grateful for the support of experimental works to the Slovak Research and Development Agency under project APVV-17-0381 – Increasing the efficiency of forming and joining parts of hybrid car bodies, and the Grant Agency of the Ministry of Education, Science, Research, and Sport of the Slovak Republic grant number VEGA 1/0259/19. The present work was carried out with the generous support of the Belgian Welding Institute (Zwijnaarde—Ghent, Belgium). The authors of this paper would like to kindly thank ir. Koen Faes for preparing the welded joints.

Conflicts of Interest: The authors declare no conflict of interest.

References

1. Schilling, C.; dos Santos, J. Method and device for linking at least two adjoining work pieces by friction welding. US Patent 6722556B2, 20 April 2004.

2. Jata, K.V.; Sankaran, K.K.; Ruschau, J.J. Friction-stir welding effects on microstructure and fatigue of aluminum alloy 7050-T7451. *Metall. Mater. Trans. A* **2000**, *31*, 2181–2192. [CrossRef]

3. Zhao, Y.; Dong, C.; Wang, C.; Miao, S.; Tan, J.; Yi, Y. Microstructures evolution in refill friction stir spot welding of Al-Zn-Mg-Cu alloy. *Metals* **2020**, *10*, 145. [CrossRef]

4. Schmal, C.; Meschut, G.; Buhl, N. Joining of high strength aluminum alloys by refill friction stir spot welding (III-1854-18). *Weld. World* **2019**, *63*, 541–550. [CrossRef]

5. Shen, Z.; Yang, X.; Zhang, Z.; Cui, L.; Li, T. Microstructure and failure mechanisms of refill friction stir spot welded 7075-T6 aluminum alloy joints. *Mater. Des.* **2013**, *44*, 476–486. [CrossRef]

6. Fratini, L.; Pasta, S.; Reynolds, A.P. Fatigue crack growth in 2024-T351 friction stir welded joints: Longitudinal residual; stress and microstructural effect. *Int. J. Fatigue* **2009**, *31*, 495–500. [CrossRef]

7. Hatamleh, O.; Lyons, J.; Forman, R. Laser and shot peening effects on fatigue crack growth in friction stir welded 7075-T7351 aluminum alloy joints. *Int. J. Fatigue* **2007**, *29*, 421–434. [CrossRef]

8. Jana, S.; Mishra, R.S.; Baumann, J.B.; Grant, G. Effect of stress ratio on the fatigue behavior of a friction stir processed cast Al–Si–Mg alloy. *Scripta Mater.* **2009**, *61*, 992–995. [CrossRef]

9. Cavaliere, P.; Santis, A.D.; Panella, F.; Squillace, A. Effect of welding parameters on mechanical and microstructural properties of dissimilar AA6082-AA2024 joints produced by friction stir welding. *Mater. Des.* **2009**, *30*, 609–616. [CrossRef]

10. Zhou, C.; Yang, X.; Luan, G. Effect of root flaws on the fatigue property of frictionstir welds in 2024-T3 aluminium alloys. *Mater. Sci. Eng. A* **2006**, *418*, 155–160. [CrossRef]

11. Wang, X.; Wang, K.; Shen, Y.; Hu, K. Comparison of fatigue property between friction stir and TIG welds. *J. Univ. Sci. Technol. B.* **2008**, *15*, 280–284. [CrossRef]

12. Uemastu, Y.; Tokaji, K.; Shibata, H.; Tozaki, Y.; Ohmune, T. Fatigue behavior of friction stir welds without neither welding flash nor flaw in several aluminium alloys. *Int. J. Fatigue* **2009**, *31*, 1443–1453.

13. Di, S.; Yang, X.; Luan, G.; Jian, B. Comparative study on fatigue properties betweenAA2024-T4 friction stir welds and base materials. *Mater. Sci. Eng. A* **2006**, *435–436*, 389–395. [CrossRef]

14. James, M.N.; Hattingh, D.G.; Bradley, G.R. Weld tool travel speed effects on fatigue life of friction stir welds in 5083 aluminium. *Int. J. Fatigue* **2003**, *25*, 1389–1398. [CrossRef]

15. Ericsson, M.; Sandstrom, R. Influence of welding speed on the fatigue of friction stir welds, and comparison with MIG and TIG. *Int. J. Fatigue* **2003**, *25*, 1379–1387. [CrossRef]

16. Sivaraj, P.; Kanagarajan, D.; Balasubramanian, V. Fatigue cract growth behaviour of friction stir welded AA7075-T651 aluminium alloy joints. *Trans. Nonferr. Met. Soc. Chin.* **2014**, *24*, 2459–2467. [CrossRef]

17. Sharma, C.; Dweredi, D.K.; Kumar, P. Fatigue behavior of friction stir weld joints of Al–Zn–Mg alloy AA7039 developed using base metal in different temper condition. *Mater. Des.* **2014**, *64*, 334–344. [CrossRef]

18. Effertz, P.S.; Infante, V.; Quintino, L.; Suhuddin, U.; Hanke, S.; dos Santos, J.F. Fatigue life assessment of friction spot welded 7075-T76 aluminium alloy using Weibull distribution. *Int. J. Fatigue* **2016**, *87*, 381–390. [CrossRef]

19. Hassanifard, S.; Mohammadpour, M.; Rashid, H.A. A novel method for improving fatigue life of friction stir spot welded joints using localized plasticity. *Mater. Des.* **2014**, *53*, 962–971. [CrossRef]

20. Lin, P.C.; Lin, S.H.; Pan, J.; Pan, T. Failure modes and fatigue life estimations of spot friction welds in lap-shear specimens of aluminum 6111–T4 sheets. Part 1: Welds made by a concave tool. *Int. J. Fatigue* **2008**, *30*, 74–89. [CrossRef]

21. Singh, R.K.R.; Prasad, R.; Pandey, S.; Sharma, S.K. Effect of cooling environment and welding speed on fatigue properties of friction stir welded Al-Mg-Cr alloy. *Int. J. Fatigue* **2019**, *127*, 551–563. [CrossRef]

22. Wang, D.A.; Chen, C.H. Fatigue lives of friction stir spot welds in aluminium 6061-T6 sheets. *J. Mater. Proc. Tech.* **2009**, *209*, 367–375. [CrossRef]

23. Wang, S.; Wei, X.; Xu, J.; Hong, J.; Song, X.; Yu, C.; Chen, J.; Chen, X.; Lu, H. Strengthening and toughening mechanisms in refilled friction stir spot welding of AA2014 aluminum alloy reinforced by graphene nanosheets. *Mater. Design* **2020**, *186*, 108212. [CrossRef]

24. Kubit, A.; Trzepiecinski, T.; Faes, K.; Drabczyk, M.; Bochnowski, W.; Korzeniowski, M. Analysis of the effect of structural defects on the fatigue strength of RFSSW joints using C-scan scanning acoustic microscopy and SEM. *Fatigue Fract. Eng. Mater.* **2019**, *42*, 1308–1321. [CrossRef]

25. Kubit, A.; Trzepiecinski, T.; Bochnowski, W.; Drabczyk, M.; Faes, K. Analysis of the mechanism of fatigue failure of the refill friction stir spot welded overlap joints. *Arch. Civ. Mech. Eng.* **2019**, *19*, 1419–1430. [CrossRef]

26. Shahani, A.R.; Farrahi, A. Experimental investigation and numerical modeling of the fatigue crackgrowth in friction stir spot welding of lap-shear specimen. *Int. J. Fatigue* **2019**, *125*, 520–529. [CrossRef]

27. Chowdhury, S.H.; Chen, D.L.; Bhole, S.D.; Cao, X.; Wanjara, P. Lap shear strength and fatigue life of friction stir spot welded AZ31magnesium and 5754 aluminum alloys. *Mater. Sci. Eng. A* **2012**, *556*, 500–509. [CrossRef]

28. Lin, P.C.; Su, Z.M.; He, R.Y.; Lin, Z.L. Failure modes and fatigue life estimations of spot friction welding cross-tension specimens of aluminum 6061-T6 sheets. *Int. J. Fatigue* **2012**, *38*, 25–35. [CrossRef]

29. Li, H.; Gao, J.; Li, G. Fatigue of friction stir welded aluminum alloy joints: A review. *Appl. Sci.* **2018**, *8*, 2626. [CrossRef]

30. Campanelli, L.; Uceu, S.; dos Santos, J.; Alcântara, N. Parameters optimization for friction spot welding of AZ31 magnesium alloy by Taguchi method. *Soldagem Inspeção* **2011**, *17*, 26–31. [CrossRef]

31. Rosendo, T.; Barra, B.; Tier, M.A.D.; da Silva, A.A.M.; dos Santos, J.F.; Strohaecker, T.T.; Alcântara, N.G. Mechanical and microstructure investigation of friction spot welded AA6181-T4 aluminum alloy. *Mater. Des.* **2011**, *32*, 1094–1100. [CrossRef]

32. Uematsu, Y.; Tokaj, K. Comparison of fatigue behavior between resistance spot and friction stir spot welded aluminum alloy sheets. *Sci. Technol. Weld. Joi.* **2009**, *14*, 62–67. [CrossRef]

33. Azevedo, J.; Infante, V.; Quintino, L.; dos Santos, J. Fatigue behaviour of friction stir welded steel joints. *Adv. Mater. Res.* **2014**, *891–892*, 1488–1493. [CrossRef]
34. Toumpis, A.; Galloway, A.; Molter, L.; Polezhayeva, H. Systematic investigation of the fatigue performance of a friction stir welded low alloy steel. *Mater. Design* **2015**, *80*, 116–128. [CrossRef]
35. Uematsu, Y.; Tokaji, K.; Tozaki, Y.; Kurita, T.; Murata, S. Effect of re-filling probe hole on tensile failure and fatigue behavior of friction stir spot welded joints in Al–Mg–Si alloy. *Int. J. Fatigue* **2008**, *30*, 1956–1966. [CrossRef]
36. Wang, X.J.; Wang, X.L.; Zhang, Z.K.; Wang, L.; Zhao, Q.S.; Deng, X.B. Fatigue behavior of friction spot stir welding with no-keyhole of aluminum alloy. *Adv. Mater. Res.* **2014**, *1052*, 509–513. [CrossRef]
37. Kluz, R.; Kubit, A.; Trzepiecinski, T.; Faes, K. Polyoptimisation of the refill friction stir spot welding parameters applied in joining 7075-T6 Alclad aluminium alloy sheets used in aircraft components. *Int. J. Adv. Manuf. Tech.* **2019**, *103*, 3443–3457. [CrossRef]
38. Roylance, D. *Fatigue*; Massachusetts Institute of Technology: Cambridge, MA, USA, 2001; Available online: http://web.mit.edu/course/3/3.11/www/modules/fatigue.pdf (accessed on 15 March 2020).
39. Chen, H.B.; Wang, J.F.; Zhen, G.D.; Chen, S.B.; Lin, T. Effects of initial oxide on microstructural and mechanical properties of friction stir welded AA2219 alloy. *Mater. Des.* **2015**, *86*, 49–54.
40. Venukumar, S.; Muthukumaran, S.; Yalagi, S.G.; Kailas, S.V. Failure modes and fatigue behavior of conventional and refilled friction stir spot welds in AA 6061-T6 sheets. *Int. J. Fatigue* **2014**, *61*, 93–100. [CrossRef]
41. Su, Z.M.; He, R.Y.; Lin, P.C.; Dong, K. Fatigue analyses for swept friction stir spot welds in lap-shear specimens of Alclad 2024-T3 aluminum sheets. *Int. J. Fatigue* **2014**, *61*, 129–140. [CrossRef]
42. Paris, P.C.; Erdogan, F. A critical analysis of crack propagation laws. *J. Basic Eng.* **1963**, *85*, 528–533. [CrossRef]
43. Theos, A. Design and Analysis of Welded Aircraft Wing Panels. Ph.D. Thesis, Cranfield Univesity, Bedford, UK, 5 December 2005.
44. Williams, S.W.; Steuwer, A. Residual stresses in friction stir welding. In *Friction Stir Welding*; Lohwasser, D., Chen, Z., Eds.; CRC Press: Boca Raton, DC, USA, 2010; pp. 215–244.
45. Carlone, P.; Citarella, R.; Sonne, M.R.; Hattel, J.H. Multiple crack growth prediction in AA2024-T3 friction stir welded joints, including manufacturing effects. *Int. J. Fatigue* **2016**, *90*, 69–77. [CrossRef]
46. Bussu, G.; Irving, P.E. The role of residual stress and heat affected zone properties on fatigue crack propagation in friction stir welded 2024-T351 aluminium joints. *Int. J. Fatigue* **2003**, *25*, 77–88. [CrossRef]
47. Hong, S.; Kim, S.; Lee, C.G.; Kim, S.J. Fatigue crack propagation behaviour of friction stir welded Al-Mg-Si alloy. *Scripta Mater.* **2006**, *55*, 1007–1010. [CrossRef]
48. Tra, T.H.; Okazaki, M.; Suzuki, K. Fatigue crack propagation behaviour in friction stir welding of AA6063-T5: roles of residual stress and microstructure. *Int. J. Fatigue* **2012**, *43*, 23–29. [CrossRef]
49. Pouget, G.; Reynolds, A.P. Residual stress and microstructure effects on fatigue crack growth in AA2050 friction stir welds. *Int. J. Fatigue* **2008**, *30*, 463–472. [CrossRef]
50. Citarella, R.; Carlone, P.; Lepore, M.; Palazzo, G.S. Numerical-experimental crack growth analysis in AA2024-T3 FSWed butt joint. *Adv. Eng. Software* **2015**, *80*, 47–57. [CrossRef]
51. Sepe, R.; Armentani, E.; di Lascio, P.; Citarella, R. Crack growth behaviour of welded stiffened panel. *Procedia Eng.* **2015**, *109*, 473–483. [CrossRef]
52. Carlone, P.; Citarella, R.; Lepore, M.; Palazzo, G.S. A FEM-DBEM investigation of the influence of process parameters on crack growth in aluminium friction stir welded butt joint. *Int. J. Mater. Form.* **2015**, *8*, 591–599. [CrossRef]
53. Zadeh, M.; Ali, A.; Golestaneh, A.F.; Sahari, B.B. Three dimensional simulation of fatigue crack growth in friction stir welded joints of 2024-T351 Al alloy. *J. Sci. Ind. Res.* **2009**, *68*, 775–782.

Article

Effect of Lubrication on Friction in Bending under Tension Test-Experimental and Numerical Approach

Tomasz Trzepiecinski [1] **and Hirpa G. Lemu** [2,*]

[1] Department of Materials Forming and Processing, Rzeszow University of Technology, 35-959 Rzeszów, Poland; tomtrz@prz.edu.pl
[2] Faculty of Science and Technology, University of Stavanger, N-4036 Stavanger, Norway
* Correspondence: hirpa.g.lemu@uis.no; Tel.: +47-518-32173

Received: 29 March 2020; Accepted: 21 April 2020; Published: 23 April 2020

Abstract: This paper is aimed to determine the value of coefficient of friction (COF) at the rounded edge of the die in the sheet metal forming operations using the bending under tension (BUT) test. The experimental part of the investigations is devoted to the study of the frictional resistances of low alloy steel sheet under different strains of the specimen, surface roughnesses of the tool and for different lubrication conditions. Three oils are destined for different conditions of duties in the stamping process. Numerical modeling of the material flow in the BUT test has been conducted in the MSC.Marc program. One of the objectives of the numerical computations is to know the type of the contact pressure acting on the cylindrical surface countersample in the BUT test by assuming the anisotropic properties of the metallic sheet. It has been found that the COF in the rounded edge of the die does not vary with increasing sheet elongation. Taking into account that normal pressure increases with increasing specimen elongation and workpiece material is subjected to strain hardening phenomenon, the COF value is very stable during the friction test. The effectiveness of the lubrication depends on the balance between two mechanisms accompanied by friction process: roughening of workpiece asperities and adhesion of the contacting surfaces. In the case of high surface roughness of tool due to a dominant share of ploughing, all of the lubricants used were not able to decrease the COF in a sufficient extent. The used lubricants were able to reduce the value of friction coefficient approximately by 3–52% in relation to the surface roughness of rolls.

Keywords: bending under tension test; BUT; coefficient of friction; friction; material properties; mechanical engineering; sheet metal forming

---

1. Introduction

The presence of friction in sheet metal forming (SMF) processes is generally an unfavorable phenomenon and causes changes in the force and energy parameters: the total work and power needed for the process increases, which also causes the forces and unit pressures on the tool surfaces to increase. In a stamping tool, frictional resistance is an important process parameter that controls the material flow in the tool and the surface finish of produced parts [1,2]. At high pressures, the occurrence of material adhesion to the tool and the formation of accretions affect the quality of the product surface [3,4]. Thermal dissipation of friction work at high values of unit pressure and relative speeds leads to a significant increase in temperature in areas adjacent to the contact surface. Limiting of the sliding speed and thus reducing the efficiency of the process may be necessary. The appearance of tangential stresses on the contact surface causes changes in the stress–strain state in the entire volume of the formed material. Friction is one of the reasons for the formation of heterogeneous deformation, which leads to the formation of heterogeneous material properties and undesirable deformations of free surfaces. So, friction has a number of negative effects in SMF processes [5–7]. Therefore, it is intended to reduce the value of friction forces. There are extensive studies on correlation of the microstructure of the tool material with the coefficient of friction (COF). However, as found by Kirkhorn et al. [8], increased or decreased carbide content could, however, not be directly correlated to an alteration of the friction coefficient when comparing the broad range of different tool materials.

There are many tribological approaches to measure the frictional phenomena existing at the die–workpiece interface. The methods used to represent the friction conditions in conventional SMF have been summarized by Trzepiecinski and Lemu [9], where the main advantages and disadvantages of the modeling methods of the friction phenomena in specific areas of the material are also presented. Experimental prediction of the frictional resistances is a key issue in the suitable setting of the numerical models of sheet metal forming process [10–12].

The basic way to reduce friction forces in plastic forming is proper lubrication. Grease lubrication forms layers when separated on a surface of a deformed object or tool, separating both surfaces partially or completely [13]. The basic properties of the lubricant include viscosity and surface activity, which may be essential for maintaining the lubricant under high unit pressure. Due to the wide variety of conditions for plastic forming processes, there are no universal lubricants. Each process requires a separate approach to the lubrication problem and the use of appropriate lubricants [14,15]. The results of the investigations of Hol et al. [16] show that the friction coefficient varies in space and time, and depend on local process conditions such as the nominal contact pressure and lubrication in the sheet material. There is extensive development of friction models destined for the description of the frictional phenomena for different lubrication conditions [17].

Friction occurring on the rounded surface of the die has an adverse effect on the value of the possible limit deformations of the sheet. One of the basic tests provided for the experimental determination of the friction resistance values on the rounded edge of the die is the bending under tension (BUT) test developed by Littlewood and Wallace [18]. In the BUT test, the tribological conditions in the die entry zone can be simulated by drawing a sheet strip over a die shoulder with superimposed back tension on the strip. Many efforts have been made in the literature to recognize the effect of many process parameters on the contact pressure and friction resistance in the rounded region of the die or the punch. Nanayakkara et al. [19] quantitatively determined the effect of roller radius and the tooling pressure on the COF. It was found that the tool radii have a direct effect in the contact pressure. The second cognitive conclusion is that there is a clear relationship between the contact pressure and the COF. To understand the sheet–die interaction, the bending and unbending response of the sheet as it contacts the cylindrical die, Coubrough et al. [20] conducted investigations on the BUT test. The actual contact angle is found to be less than the geometric angle-of-wrap and increases with increasing strip tension. Fratini et al. [1] used the BUT test to measure friction under lubricated and dry friction conditions. The authors concluded that the effect of lubricant in the reduction of the COF is rather small; this is due to the loose of lubricant, which occurred during the tests because of the contact pressure at the sheet metal–tool interface. However, the lubricant reduced the stick–slip phenomena, which are dangerous during SMF operations, since they can induce the cracks in the surface layer of workpiece. Lemu and Trzepiecinski [21] studied the effect of the amount of plastic deformation of the brass, aluminum alloy and steel specimens on the value of COF. It was found that the use of tools with low surface roughness value to reduce the COF is unfounded because the increased real contact area decreases the effectiveness of the lubricant and increases the interatomic interaction of surfaces. Moreover, the results indicated that use of machine oil reduces the friction coefficient value to a lower degree for lower roughness values. Hoffmann et al. [22] conducted a study where they compared the wear occurred in the die radius for different combinations of die and sheet material. The highest wear of die occurred in the regions where the contact pressure between die and workpiece are the highest. Berglund et al. [23] evaluated the correlation between machining finish, punch material and the COF result. The texture characterization parameters measured after the BUT test exhibit strong correlations with the friction and all are related to the areas and inclinations of the surface. Pereira et al. [24] demonstrated that, in the sheet–die interface in BUT test, there is a transient region in the contact pressure that corresponds to the beginning of sheet deformation, after which the pressure stabilizes. It has been specified that the yield stress and relationship between sheet thickness and radius of the die greatly influence the values of contact pressure. In a later study, Pereira et al. [25] analyzed numerically through finite element simulation of the evolution of contact pressure in the die radius during the stamping process. It has been found that the same pressure peaks occur during bending of the sheet over the radius of the die, causing the central part of these peaks to lose contact and greatly reducing the pressure in this region. Ceron and Bay [26] investigated how the BUT test combined with a classical analytical modeling may lead to very large errors in estimation of the COFs. The numerical simulations of the normal stress distributions demonstrated in comparative analysis of an industrial, multistage deep drawing that finite element modeling provides appropriate estimates.

Most of the investigations on the numerical analysis of the tribological test have been conducted assuming the isotropic properties of the material. Moreover, due to bending of the sheet over a rounded die, the assumption that the specimen is subjected to plane strain state is also unjustified, which has been confirmed by authors in the previous papers devoted to the draw bead friction test [27,28]. Thus, one objective of this work reported in this paper is to better understand the contact pressure acting on the rounded edge of the cylindrical countersample in the BUT test by assuming the anisotropic properties of the metallic sheet. Another aim of the paper is to calculate the COF that acts during the BUT test for different strains of the sheet metal. Frictional properties of low alloy steel sheet have been experimentally evaluated for different lubricants.

2. Experimental

2.1. Material

In the investigations, cold-rolled low carbon DC04 grade steel sheets with a thickness of 1 mm were used. These steel sheets are characterized by super deep-drawing properties manifested by high plasticity and high susceptibility to strain hardening. According to the standard EN 10130:2009, DC04 steel sheet is suitable for cold forming of complicated outer and inner components of automobile bodies and other pressworks, especially for high deformation speed. The chemical composition of the DC04 steel sheet is shown in Table 1.

Table 1. Chemical composition of DC04 steel sheet (in wt. %).

C	Mn	P	S	Fe
≤0.08	≤0.4	≤0.03	≤0.03	remainder

Mechanical properties of the sheets have been determined in a uniaxial tensile test according to ISO 6892-1:2016 [29]. Flat specimens have been cut from the sheet metal in three directions: along rolling direction (RD) and at angles of 45° and 90° to the RD. Z100 (Zwick Roell, Ulm, Germany) uniaxial tensile test machine was used in the investigations of the mechanical properties of the sheets (Table 2). The following parameters have been determined: ultimate tensile strength R_m, yield stress $R_{p0.2}$ and anisotropy factor r (Lankford's coefficient). Moreover, the strain hardening properties have been determined by approximation of the true stress–true strain relation using the Hollomon equation:

$$\sigma = K \times \varepsilon^n \tag{1}$$

where σ is the true stress, ε is the true strain, K is the strength coefficient and n is the strain hardening exponent.

Table 2. Selected mechanical properties of the DC04 steel sheet.

Specimen Orientation	$R_{p0.2}$ (MPa)	R_m (MPa)	K (MPa)	n	r
0°	172	306	513	0.17	1.49
45°	179	319	502	0.19	1.32
90°	184	210	524	0.20	1.58

Surface roughness parameters (Table 3) of the tested sheet were determined using a Talysurf CCI Lite 3D optical profiler (Taylor Hobson, Leicester, UK). The following basic parameters have been determined: root mean square roughness Sq, average roughness Sa, maximum pit depth Sv, 10-point peak-valley surface roughness Sz, total height St and the highest peak of the surface Sp.

Table 3. Surface roughness parameters of the DC04 steel sheet.

Sa (µm)	Sq (µm)	Sv (µm)	Sz (µm)	St (µm)	Sp (µm)
1.46	1.72	5.89	12.42	11.39	8.87

2.2. Friction Simulator

Experimental BUT tests have been carried out using a special friction simulator (Figure 1), which is mounted on a universal tensile test machine. The friction simulator consists of a frame in which the horizontal tension member with load cell is mounted. One end of the specimen was mounted at the end of the horizontal tension member, while the second end was mounted in the vertical tension member with load cell. The specimens for friction test were cut along the rolling direction of the sheet metal. They were stretched with speed of 0.25 mm·s^{-1}. The strip specimen with width of $w = 10$ mm and length of $L = 135$ mm was wrapped around a cylindrical fixed roll with radius of $R = 20$ mm.

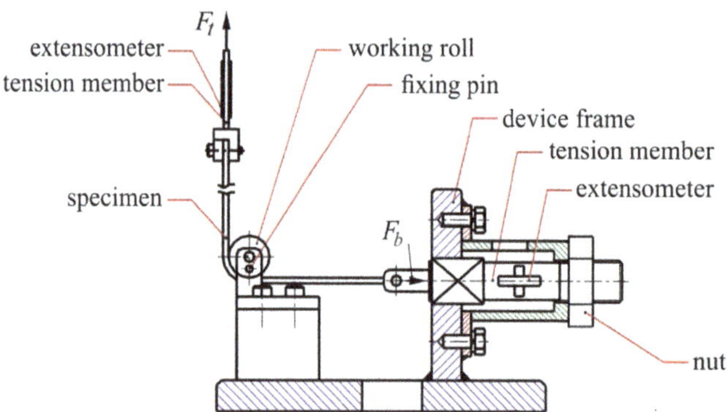

Figure 1. Schematic view of the testing device.

The device was equipped with three rolls differing in their surface roughness. To characterize the surface roughness of cylindrical countersamples, the roughness average Ra measured parallel with the roll axis was assumed. The rolls made of cold-worked tool steel had the roughness qualities $Ra = 0.32$, 0.63 and 1.25 µm. Before testing, all the specimens were degreased with acetone. For lubricated conditions three oils, namely machine oil (MO), deep-drawing oil (DDO) and heavy-drawing oil (HDO) with the following specifications were used:

- Machine oil LAN-46 (Orlen Oil): kinematic viscosity 43.9 mm^2·s^{-1} (at 40 °C), viscosity index 94, flow temperature −10 °C and ignition temperature 232 °C,
- Deep-drawing oil L (Orlen Oil): kinematic viscosity 330 mm^2·s^{-1} (at 40 °C), freezing point −29 °C, flash point 238 °C and weld point 500 daN,

- Heavy-Draw 1150 oil (Lamson Oil): density 975 kg·m^{-3} (at 20 °C); viscosity 1157 mm^2·s^{-1} (at 40 °C) and flash point 277 °C.

In SMF technology, when stamping components with complex shapes, where extremely different friction conditions, contact pressures and slip speeds occur, different types of lubricants should be used. The lubricants were selected in such a way that the first one is used for typical deep-drawing applications, while the second one is used for super deep-drawing applications and Heavy-Draw 1150 oil is heavy duty stamping oil is used in specially difficult forming applications. All lubricants were distributed uniformly on the surface of the samples using a shaft.

During the test, front tension force F_t and back tension force F_b were simultaneously measured. Assuming that the wrap angle α (Figure 2) is constant during the test, the equilibrium equation of the elementary sector of the strip $d\alpha$ can be shown as:

$$F + q\mu wRd\gamma - (F + dF) = 0 \tag{2}$$

$$qwRd\gamma - F\sin\frac{d\gamma}{2} - (F + dF)\sin\frac{d\gamma}{2} = 0 \tag{3}$$

where μ is the COF, w is the width of the strip and q is unit contact pressure determined according to the equation:

$$q = \frac{F_b + F_t}{2wR} \tag{4}$$

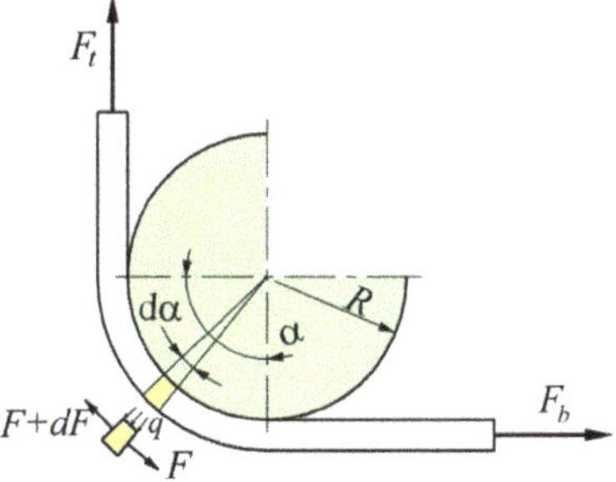

Figure 2. Forces acting on an elemental cut of the strip.

For a very small $d\alpha$ it can be assumed that $\sin\frac{d\gamma}{2} \approx \frac{d\gamma}{2}$, furthermore dF is much smaller that F. Thus, combining Equations (2) and (3) gives

$$\mu d\alpha = \frac{dF}{F} \tag{5}$$

Taking into account in the Equation (4) that $\alpha = \frac{\pi}{2}$ the COF is determined to be

$$\mu = \frac{2}{\pi}\ln\left(\frac{Ft}{F_b}\right) \tag{6}$$

The occurrence of friction between the roll and specimen causes that $F_t > F_b$. The friction test is realized until the elongation of the specimen becomes equal to 20%. It should be noted that the tensile force F_t includes the deformation resistance related with the bending of the specimen around the roll. In this way, Equation (6) does not include explicitly the bending force [21].

3. Numerical Modeling

3.1. Description of the FE-Based Model

Finite element (FE) based numerical modeling of the BUT test was carried out using the MSC.Marc program (MSC.Software, Newport Beach, CA, USA). The geometrical model of the countersample and specimen correspond to the experimental conditions. Due to the fact that strength of the roll material is considerably higher than the strength of specimen material, it was assumed that no deformation exists in the roll. So, the roller surface was modeled as perfectly rigid. The end of the specimen at the back tension side was fixed (Figure 3). The numerical modeling of the BUT test consisted of two steps: (1) bending the sheet around the roll surface (Figure 3a), and (2) applying the tensile force to the upper end of the specimen (Figure 3b). The drawing speed corresponds to the experimental conditions.

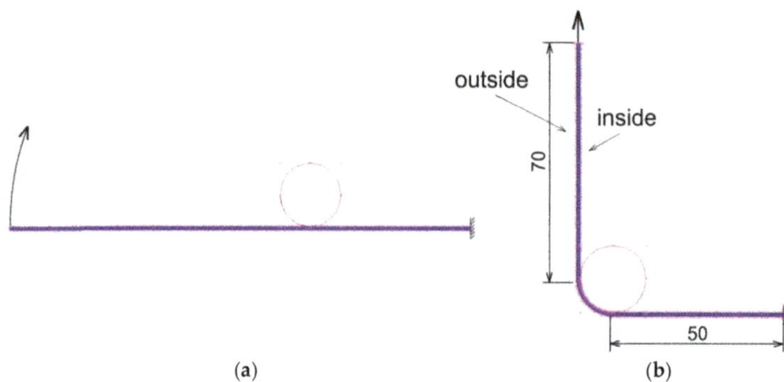

Figure 3. Geometrical model of the bending under tension (BUT) test: (**a**) initial configuration and (**b**) configuration at the beginning of the friction test.

3.2. FE Mesh

The geometric model of the specimen has been discretized by using 8-node isoparametric hex-8 brick elements [30] with the assumed strain formulation, which improve the bending characteristic of the finite elements. According to the program documentation [31], this can significantly improve the solution accuracy though the computational costs of assembling the stiffness matrix may increase. The hex-8 element has been successfully used by Trzepiecinski and Fejkiel [27] to study the material flow of a sheet specimen through a drawbead.

3.3. Material Model

An elastic–plastic material model approach was considered to build the material model. The plastic behavior of the sheet metal was established using Hill (1948) [32] yield criterion, which can be applied for a material description of steel sheet metals [33–35]. Work hardening with power-type law hardening law has also been incorporated in the finite element method (FEM) based on the average values of material parameters K and n determined for three directions, according to the Table 2. The Hill (1948) formulation can be expressed in terms of rectangular Cartesian stress components as:

$$\bar{\sigma} = \sqrt{(F(\sigma_{22} - \sigma_{33})^2 + G(\sigma_{33} - \sigma_{11})^2 + H(\sigma_{11} - \sigma_{22})^2 + 2L\sigma_{23}^2 + 2M\sigma_{31}^2 + 2N\sigma_{12}^2}} \tag{7}$$

where $\bar{\sigma}$ is the equivalent stress, and indices 1, 2 and 3 represent the rolling, transverse and normal directions to the sheet surface, respectively.

Parameters (constants) F, G, H, L, M and N define anisotropy state of material and are equal to:

$$F = \tfrac{1}{2}\left(\tfrac{1}{R_{22}^2} + \tfrac{1}{R_{33}^2} - \tfrac{1}{R_{11}^2}\right), G = \tfrac{1}{2}\left(\tfrac{1}{R_{11}^2} + \tfrac{1}{R_{33}^2} - \tfrac{1}{R_{22}^2}\right), H = \tfrac{1}{2}\left(\tfrac{1}{R_{11}^2} + \tfrac{1}{R_{22}^2} - \tfrac{1}{R_{33}^2}\right),$$
$$L = \tfrac{3}{2R_{23}^2}, M = \tfrac{3}{2R_{13}^2}, N = \tfrac{3}{2R_{12}^2}, \tag{8}$$

Parameters R_{11}, R_{22}, R_{33}, R_{12}, R_{13} and R_{23} are defined in form of user input consisting of ratios of yield stress in different directions with respect to a reference stress according to Equation (8).

$$R_{11} = \frac{\sigma_{11}}{\sigma_0}, R_{22} = \frac{\sigma_{22}}{\sigma_0}, R_{33} = \frac{\sigma_{33}}{\sigma_0}, R_{12} = \frac{\sigma_{12}}{\tau_0}, R_{13} = \frac{\sigma_{13}}{\tau_0}, R_{23} = \frac{\sigma_{23}}{\tau_0}, \tag{9}$$

The automatic algorithm for the evaluation of parameters R_{11}–R_{23} has been built in MSC.Marc program.

The elastic material behavior of sheet is specified using the following properties: Young's modulus $E = 2.1$ GPa and Poisson's ratio $\nu = 0.3$.

3.4. Contact Conditions

To describe contact conditions the classical Coulomb friction law was assumed in Equation (10):

$$f_t = f_n \cdot \mu \cdot T \cdot \frac{2}{\pi} arctan \frac{\|v_r\|}{\delta} \tag{10}$$

where f_t-tangential (friction) force, f_n-normal force, $\|v_r\|$—relative sliding velocity, δ-value of the relative velocity below which sticking occurs and T-tangential vector in the direction of the relative velocity.

The value of δ determines how closely the mathematical friction model represents the step function given as:

$$f_t = f_n \cdot \mu \cdot sign \, \|v_r\| \tag{11}$$

A small value of relative velocity δ results in a reduced value of the effective friction. A very small value may result in poor convergence of contact algorithm. It is recommended [27] that the value of relative velocity should be 1–10% of a typical relative sliding velocity $\|v_r\|$. In this paper, the value of $\delta = 5\%$ was used.

3.5. Mesh Sensitivity Analysis

Mesh sensitivity analysis (MSA) is a crucial part of each FE-based analysis. The finite element size should be balanced between assurance accurate results and the computational cost. MSA was carried out for three different element sizes: 1 mm × 1 mm × 1 mm, 0.5 mm × 0.5 mm × 0.5 mm and 0.25 mm × 0.25 mm × 0.25 mm. A more dense mesh was created by dividing elements into two equal parts in all directions (Figure 4). The value of the front tensile force at a sheet elongation of 5%, 10% and 15% was adopted as the parameter constituting the basis for the selection of mesh size. Only for the MSA, constant value of COF ($\mu = 0.3$) was assumed. The parameters of the numerical models and the results of MSA are shown in Table 4.

Table 4. Parameters of the mesh and mesh sensitivity analysis (MSA) results.

Model No.	Element Size, mm	Number of Elements	Number of Nodes	Tensile Front Force, N at Specimen Elongation			Computation Time, s
				5%	10%	15%	
N1	1 × 1 × 1	1380	3058	2709	2979	3096	220
N2	0.5 × 0.5 × 0.5	11,040	17,451	2697	2976	3089	2008
N3	0.25 × 0.25 × 0.25	88,320	113,365	2695	2976	3086	24,547

The reduction of the element edge size from 1 to 0.5 mm resulted in a change in the front tensile force value of 0.1–0.44%. At the same time, the computation time increased almost ten times. Further reduction of the element size from 0.5 to 0.25 mm caused the value of the front tensile force to change by less than 0.1%. Due to considerably extended computation time and at the same time a slight change in the value of the front tensile force model N3 might be rejected. Therefore, the N2 model with mesh 0.5 mm × 0.5 mm × 0.5 mm is a reasonable solution.

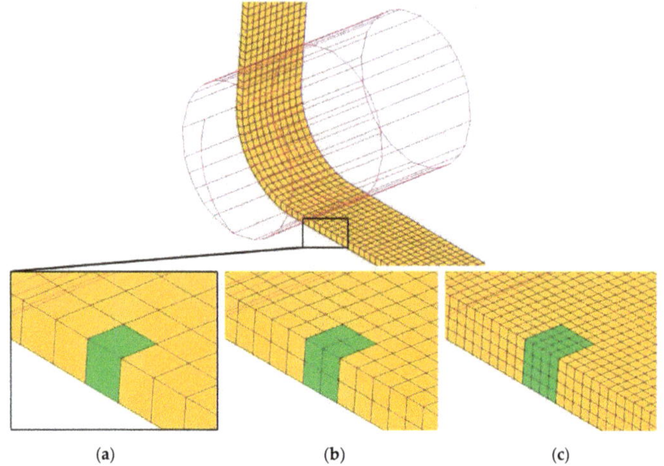

Figure 4. View of the mesh in the MSA: (**a**) 1 mm × 1 mm × 1 mm, (**b**) 0.5 mm × 0.5 mm × 0.5 mm and (**c**) 0.25 mm × 0.25 mm × 0.25 mm.

4. Results and Discussion

4.1. Experimental

4.1.1. Effect of Surface Roughness of Tools

As a result of experimental investigations, the relations between tensile forces and percentage elongation of specimen were obtained. Figure 5 presents an example of such relation registered for the specimen tested using a roll with roughness of $Ra = 0.32$ μm at HDO lubrication. Increasing the tensile forces can be simply attributed to the strain hardening phenomenon. Although the value of both tensile forces increased, the value of the friction coefficient was very stable during the test.

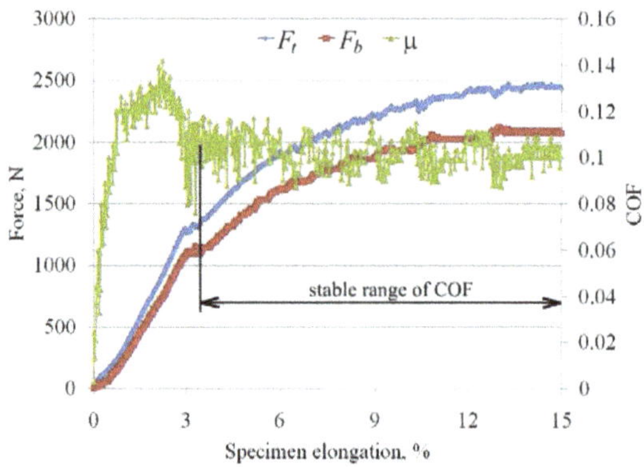

Figure 5. The effect of the specimen elongation on the values of front (F_t) and back (F_b) tension forces and coefficient of friction (COF); test conditions: $Ra = 0.32$ μm, heavy-drawing oil (HDO) lubrication.

The initial unstable range of changes in the coefficient of friction is related to the tension of the sample around roll surface with minimization of possible clearances. The increase of the friction coefficient at the beginning of the tribology test is usually attributed to the accommodation of contacting surfaces before reaching a stable stage meaning of linear friction conditions. In this respect, the coefficient of friction determined for the initial stage of the test by Equation (6) as the ratio of two process forces may not make any physical sense. The results were in agreement with the commonly known Amontons–Coulomb law that the COF does not depend on the value of the normal pressure and contact area. The unstable region of the changes of both front F_t and back F_b tension forces was not considered in the determination of COF. The average values of COFs presented in this section (Section 4.1) and Section 4.1.2 were determined in the stable range of COFs changes. Although both the front and back tensile forces increased during the test, which results in increasing of normal pressure on the surface of contact of workpiece with rounded countersample, the COF variation kept at a constant trend (Figure 5). This conclusion may be attributed to all conducted tests.

Another phenomenon that should be mentioned is the continuous change in surface topography of the sheet metal due to specimen elongation. This causes the real contact area to increase simultaneously with the normal pressure. The real contact area depends on, for instance, susceptibility to strain hardening of roughness asperities, roughness parameters of the tools and the sheet metal and the geometry of the contact surface [21].

Plots of variations of the COF as a function of friction conditions are shown in Figure 6. In the case of tests carried out with the use of rollers with a roughness $Ra = 1.25$ μm, the relatively highest values of the COF were observed for all lubrication conditions. At the same time, the lubricating oils used minimized the value of the COF.

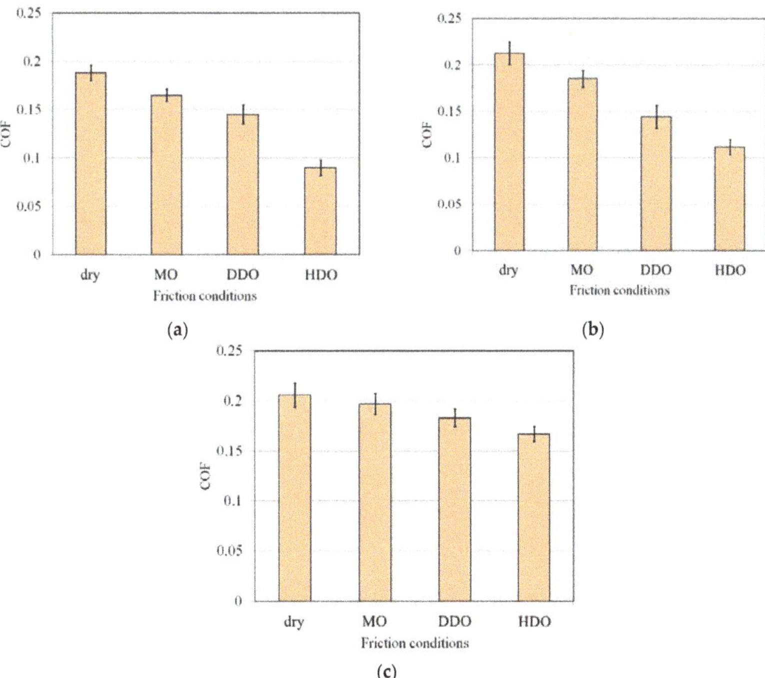

Figure 6. The effect of Ra of the roll on the value of COF: (**a**) $Ra = 0.32$ μm, (**b**) $Ra = 0.63$ μm and (**c**) $Ra = 1.25$ μm.

For dry friction conditions, the largest value of the COFs were recorded for rolls with a roughness $Ra = 0.63$ μm (Figure 6b). An increase or decrease in the value of roll roughness resulted in a decreased value of the COF. This is due to the balance of two mechanisms depending on the roughness of the tool: (1) adhesion of the workpiece surface to tool material, in the conditions of small roughness and (2) asperities flattening of the workpiece surface by high roughness of the tool. In the case of high surface roughness, the load pressure acts on the asperities, which results in a higher degree of surface flattening and increased frictional resistance [1]. The share between these phenomena in the total frictional resistance is particularly important in the conditions of sheet lubrication in SMF [2]. The lowest values of the coefficient of friction were recorded during tests using a roll with a roughness $Ra = 0.32$ μm (Figure 6a). Under these conditions, frictional resistances resulting from interactions between the peaks of cooperating asperities were the smallest. At the same time, the volume of "oil pockets" [2] between contacting surface asperities was large enough to provide an adequate reservoir of lubricant at the contact area.

4.1.2. Effectiveness of Lubrication

The lubricant used in the SMF should effectively separate friction surfaces, protect the tool against excessive wear, have high resistance to normal loads and exhibit easy flow in a tangential direction. The chemical composition of lubricants and viscosity are determined by manufacturers in such a way that they have the ability to transfer pressure and at the same time that the lubricant film is not easily broken. The lubricant viscosity plays a key role in mixed lubrication regime when applied load is partly carried out by fluid film and interacting asperities [36,37]. The mixed lubrication regime is the intermediate zone between the boundary lubrication regime and the elasto-hydrodynamic lubrication regime, where the applied load is partly carried by the interacting asperities and the remaining part by the fluid film. In these conditions, the suitable lubricant viscosity plays a key role [36].

One way to check how effectively the lubricant reduces the friction in specific load conditions is to determine the value of the factor of effectiveness of lubrication δ as a ratio of the coefficient of friction determined in dry conditions μ_d to the coefficient of friction determined in lubricated conditions μ_l, according to the following equation:

$$\delta = \frac{\mu_d - \mu_l}{\mu_l} \times 100\% \tag{12}$$

The highest lubrication efficiency was recorded for HDO (Figure 7). The used lubricants were able to reduce the value of the friction coefficient approximately by 3–52% in relation to the surface roughness of rolls. In the case of the HDO lubricant, the greater the roughness of the roll, the lower the degree of reduction of the friction resistance by the lubricant. The reduction in lubrication efficiency as the roughness increases is most evident in HDO. MO and DDO lubricants showed a local increase in efficiency in reducing frictional resistance for intermediate roll roughness $Ra = 0.63$ μm. Under these conditions, the amount of space between the asperities was sufficient to accumulate the appropriate volume of lubricant. For the largest tool roughness analyzed $Ra = 1.25$ μm, the share of the mechanism of ploughing of the workpiece surface by countersample surface was large enough that the grease did not have the proper conditions to reduce COF. An increase of the prestrain value causes an increase of the sheet surface roughness, and as a consequence, the frictional resistance increases due to intensification of the roughness flattening mechanism. In the case of high surface roughness of tool ($Ra = 1.25$ μm), the load pressure acts dominantly on the asperities, which results in a higher degree of surface flattening of sheet metal and increased frictional resistance. The hardness of the tool material made of cold-work tool steel was considerably higher than the hardness of sheet material. For proper operation, the lubricant requires sufficient pressure to work in conditions of the specific cushion separating the surface being in contact. When the surface roughness of the tool is high, the pressure is carried out by roughness asperities of the tool surface. In these conditions, even if the voids between surface asperities are high, the interaction of high asperities do guarantee the sufficient squeezing of the lubricant in the sheet–tool interface. High tool roughness facilitates the escape of the entrapped lubricant in the pockets. In this way, the lubricant may leak from the contact area. To identify the optimal surface roughness for the best lubrication effectiveness, three aspects need to be considered: surface roughness of tool (the surface of the sheet metal is determined in manufacturing process), lubricant type and contact pressure. The total frictional resistance should take into account appropriate balance between ploughing mechanism (when the surface roughness is too high) and a lack of sufficient lubricant pressure when the lubricant pockets are too small, and mechanical contact of asperities dominate. Due to many parameters and phenomena that influence the frictional resistance there is no universal efficient method to determine a priori appropriate conditions of the forming process. Still, the experimental testing, although time- and cost-consuming, is the best way to select optimal friction conditions that guarantee the best performance of lubricant.

The effectiveness of lubrication may be aided by the mechanism of strain hardening. The DC04 steel sheets are characterized by high susceptibility to work hardening. In the case of cold forming, this phenomenon consists of two main mechanisms, i.e., dislocation glide and twinning. These mechanisms strongly depend on the density of dislocations in the material. During elongation, the surface topography of the sheet metal leads to strong evolution (Figure 8). Scanning electron microscopy (SEM) micrographs revealed directional frontal deformation. These fronts may be in macroscale assignment to the motion of the dislocations.

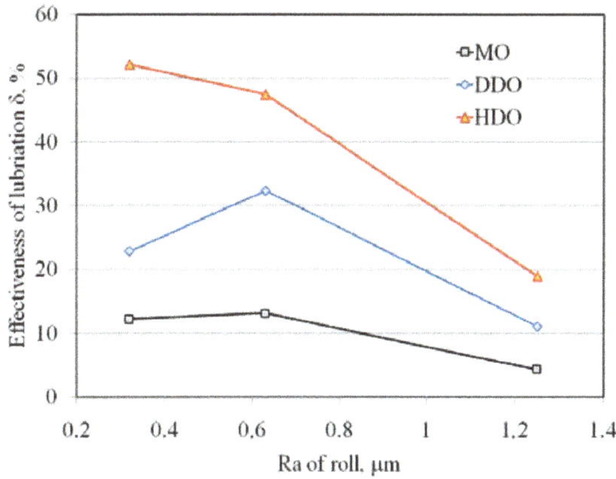

Figure 7. The effect of the oil type on the effectiveness of lubrication δ.

The rough topography assists the better adhesion of the lubricant to the workpiece surface. This influence is as expected, since the low viscous MO does not assure high effectiveness of lubrication, whereas the higher viscous DDO and the high viscous HDO may support microhydrodynamic lubrication in sheet metal processes as found by Sulaiman et al. [38]. They also concluded that low-viscous mineral oil does not promote micro-plasto-hydrodynamic lubrication in sheet metal forming. This can be revealed from Figure 7, where the performance of this oil in a reduction of COF did not exceed 12%. The highest effectiveness of lubrication was found for HDO at Ra = 0.32 μm. High viscosity lubricant reduces the occurring friction significantly. Oil with high viscosity is characterized by high sticking with contacting surfaces, so the oil film is difficult to break. It is noticed here that the tool roughness of Ra = 1.25 μm has reduced the effectiveness of lubrication as compared to the smoother tool surface (Ra = 0.32 μm and 0.63 μm), when testing with all viscosity oils. This leads to effective separation between a tool and a workpiece on the plateaus of the tool asperities. If a pocket of fluid layers is sheared, the individual fluid layers are displaced in the direction of the shearing force. Molecular forces create resistance to shearing and this resistance is given by the viscosity and the difference in velocity between two given fluid layers, related to the shear rate. The mixed lubrication regime, commonly existing in sheet metal forming to a large extent, based on the kinematic viscosity of lubricant to form a chemical or physical bond with the tool surfaces and steel sheet. When viscosity of lubricant is not able to fulfill the requirements for hydrodynamic lubrication, it will immediately lead to asperity peaks breaking through the lubricant film, provoking metallic contact [39]. A very fine polished tool surface lowers the COF as the lubricant is better retained. Higher viscosity of the lubricant reduces the COF, presumably due to the fact that lubricant escape is diminished.

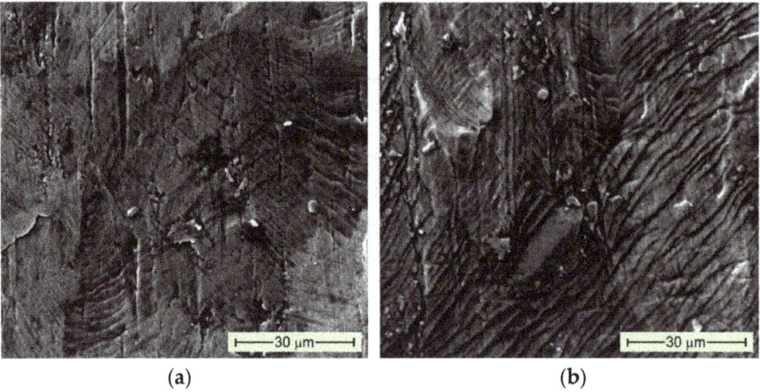

Figure 8. SEM micrographs of the surface topography of the DCO4 steel sheet subjected to the (**a**) 10% and (**b**) 15% elongation, magnification 1200×.

4.2. Results of Numerical Modeling

4.2.1. Distribution of Specimen Elongation

Numerical simulations of the BUT test allow one to better understand the way the sheet deforms and the phenomena that occur on the interface between the strip specimen and the counter-sample. Obtaining such results on an experimental basis would be very difficult. Numerical analyses were carried out on the numerical model corresponding to the friction of samples against the surface of rolls with a roughness Ra = 0.32 µm (Figure 6a); for these conditions the largest difference in COF values for dry friction and most effective HDO lubrication were observed. In the numerical simulations, the values of the COF determined in experimental studies were taken into account according to the schematic relationship in Figure 5. As this figure shows, the value of COF initial continued to increase with the elongation of the sample and it reached its stable range at a deformation of 3.5%. In the elongation range between 3.5% and 15%, a constant value of COF was obtained and this result was assumed in Figure 6a for the case of HDO.

Figure 9 shows the distribution of mean normal stresses on the contact surface for all the analyzed friction conditions and the various deformations of the specimen. In the case of the smallest analyzed sample deformation of 5%, the distribution and value of normal stress was generally similar for all friction conditions. For the HDO lubrication conditions, which was the most effective lubricant, the higher the deformation value, the more uniform the stress distribution in the contact zone (Figure 9d). Observation of the distribution of mean normal stresses for the largest sample deformation (Figure 9c,d) allowed us to draw a conclusion that the lubricant reduced the stress value on the sample surface in the contact area. Lower friction resistance makes the sample more easily move over the surface of the tool. The most loaded section was the vicinity of the place where the specimen was loaded by the front tensile force leaves contact with the roll. This is due to the accumulation of stresses in the material as a result of the braking frictional effect of the roll surface on the inner side (see Figure 3b) of the sample. The non-homogeneous distribution of stress across the sample width is associated with (i) anisotropy of the mechanical properties of the sheet and (ii) the occurring friction limiting the flow of the sample material over the width of the countersample.

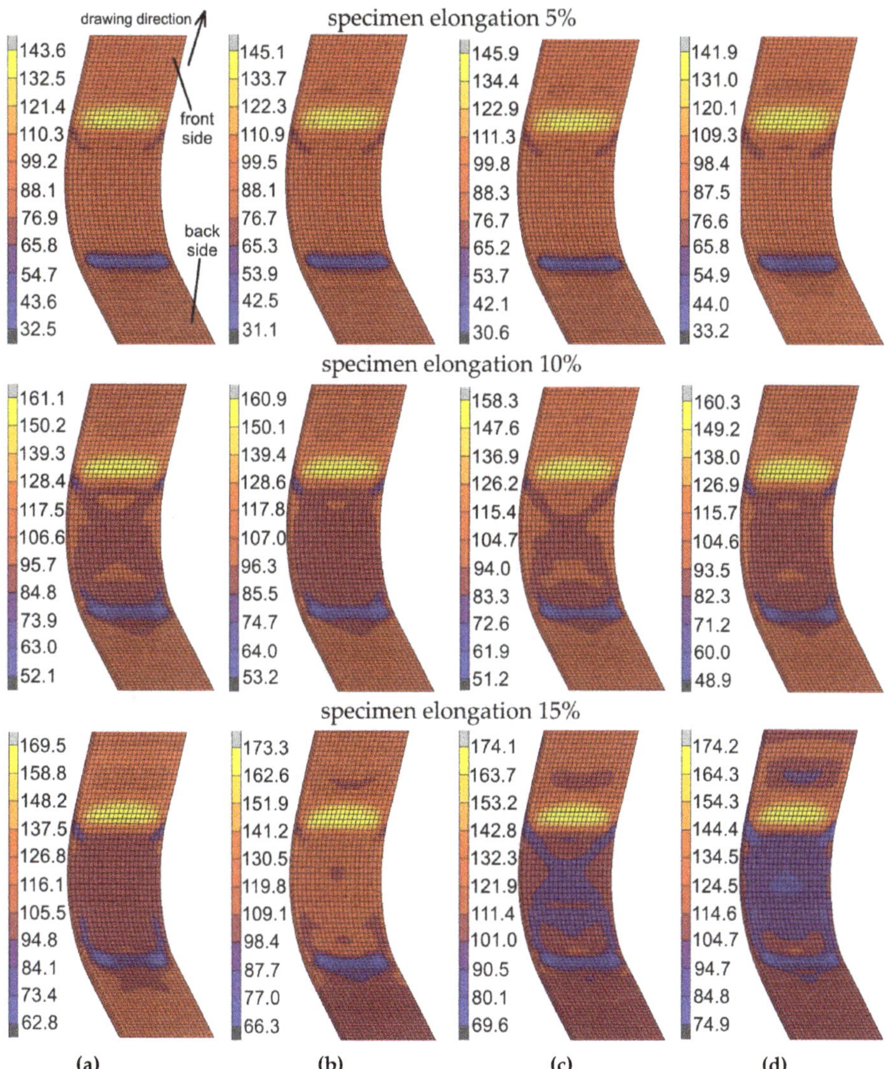

Figure 9. Effect of specimen elongation on the value of the mean normal stress (MPa) for the friction conditions: (**a**) dry, (**b**), machine oil (MO), (**c**) deep-drawing oil (DDO) and (**d**) HDO.

4.2.2. Flexuring of the Specimen

Existence of friction on one inner side of the sheet causes local flexure of the specimen. This phenomenon was observed at the front and back side of the specimen (Figure 10). Of course, the flexuring on the exit side was greater and resulted from different lengths of the free part of the samples (Figure 3b). The flexuring of the strip specimen affected the length of the contact area along the sheet-roll contact surface. The bending deformation of the sample was greater for greater values of the coefficient of friction. It also indicated that the non-uniform distribution of contact normal force along the specimen width and along the contact area (Figure 10) was a result of flexuring of the specimen in a plane perpendicular to the strip direction.

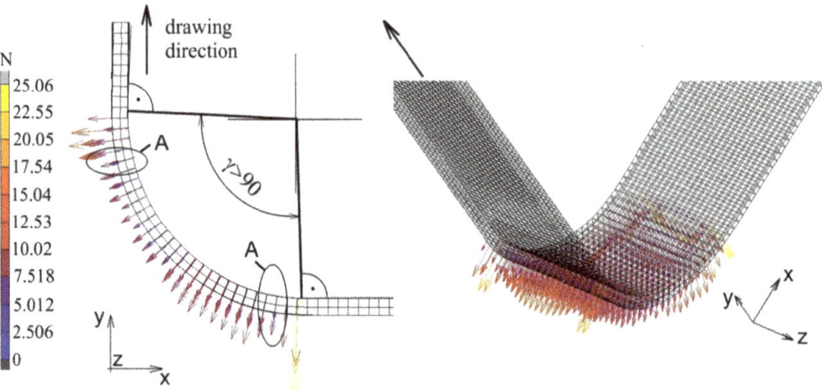

Figure 10. Distribution of contact normal force (in N) at the contact area of sheet metal and countersample surface; testing conditions: dry friction, roughness of countersample Ra = 0.32 μm.

4.2.3. Normal and Friction Forces

The distribution of the contact normal force and contact friction force is non-uniform along the contact area, as it is shown in Figure 11a,b, respectively. Local peaks of forces were observed at the start and the end of contact. This can be associated with the local flexuring of the strip sheet over the countersample surface (Figure 10). In this case, the flexure of the strip sheet locally unloads the material in contact (regions A in Figures 10 and 11). The distribution of normal and friction forces in dry as well as DDO and HDO lubricated conditions was approximately uniform in the range of 3–10 mm of contact length.

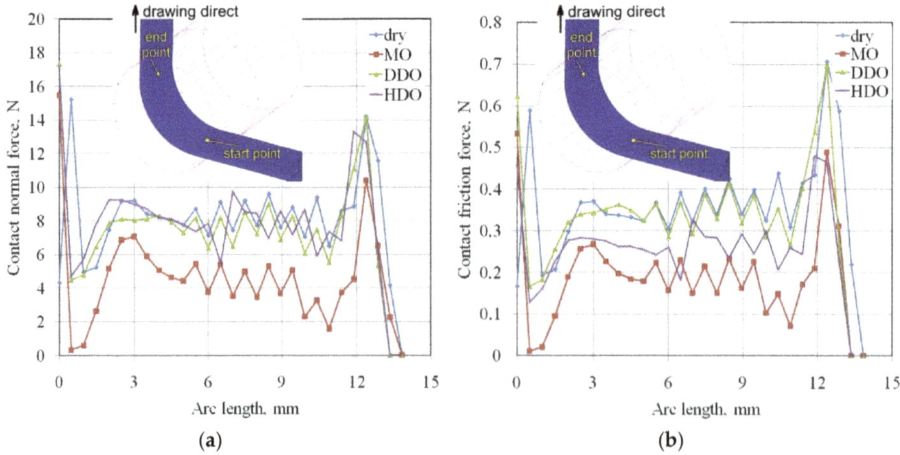

Figure 11. Distribution of the (**a**) contact normal force and (**b**) contact friction force on the contact surface of metal strip and countersample surface; analyzed roughness of the countersample Ra = 0.32 μm.

Triangular oscillations of the contact normal force value on the side, when the sample leaves contact with a countersample, might be the result of loss of sheet stability due to flexure of the sheet metal shown in Figure 10 (Left). In addition, the strip sheet in the vicinity of the contact with the countersample tended to bend across the width, which is shown in Figure 10 (Right), where the contact normal forces were not uniformly distributed across the specimen width. Due to the direction of specimen movement, the zone located at the bottom of the sample, immediately after coming into contact with the countersample, was most heavily loaded along the entire width of the sample. In this area, the distribution of values of contact normal forces between neighboring nodes was very stable, without oscillations.

The distribution of the mentioned parameters in this range shows that the values of contact and friction forces for MO lubrication were about 2 times lower than for DDO and HDO lubrication conditions. The lowest values of friction force along the path lying in the middle of contact area for MO lubricant did not indicate that this lubricant effectively reduced frictional resistances. Figure 11 allowed us to compare the forces generated by the different friction conditions at a specific cross-section. However, the decisive factor in the flowing resistance of the specimen along the roll surface is the distribution of the friction force values along the specimen width.

5. Conclusions

This study was devoted to the experimental investigation of the frictional phenomena at rounded edges of the die and punch in sheet metal forming operations. Numerical modeling allowed us to study the material flow over the rounded countersample. The change in different process parameters was considered in the investigations. The following conclusions were drawn from the research:

- The normal pressure in the BUT test continuously increased with increasing specimen elongation, and this was due to the strain hardening phenomenon. However, the COF was very stable during the tests realized in all friction conditions. This conclusion is in contrast to the recent investigations of authors [2,3] on the friction determination in a strip drawing test when the nonlinear relation between friction and normal force was found.
- The effectiveness of the lubrication depended on the balance between two mechanisms accompanied with friction: (1) adhesion of the surfaces in contact and (2) roughening of workpiece asperities by the tool surface. High surface roughness of tool released the dominant share of ploughing in total frictional resistance. In these conditions, all of the lubricants used were not able to decrease the COF in to sufficient extent.
- Lubricants destined for application in SMF operations were able to reduce a value of friction coefficient approximately by 3–52% in relation to the surface roughness of rolls.
- Friction in the sheet–tool interface caused flexuring of the strip during flowing the sheet through the BUT test. This effect results in the non-uniformity of the contact normal force and depended on the value of COF.
- The lubricated conditions, by making the sample to move over the tool surface more easily, reduced the mean normal stress value on the sample surface in the contact area. Moreover, in the case of the most effective lubricant, i.e., HDO, the higher the deformation value, the more uniform the stress distribution in the contact zone was observed.
- The distribution of contact friction force and contact normal forces was non-uniform along the width and length of the strip material being in contact with the roll surface. This could be associated with the local flexuring of the strip sheet over the countersample surface.

Author Contributions: Conceptualization and methodology, T.T.; investigation, H.G.L., T.T.; data curation, H.G.L., T.T.; software, T.T.; writing—original draft, T.T.; H.G.L. contributed in funding acquisition, project administration, validation, and writing—review and editing. All authors have read and agreed on the final version of the manuscript.

Funding: This research received no external funding.

Conflicts of Interest: The authors declare no conflict of interest.

References

1. Fratini, L.; Casto, S.L.; Valvo, E.L. A technical note on an experimental device to measure friction coefficient in sheet metal forming. *J. Mater. Process. Technol.* **2006**, *172*, 16–21. [CrossRef]
2. Trzepiecinski, T.; Fejkiel, R. On the influence of deformation of deep drawing quality steel sheet on surface topography and friction. *Tribol. Int.* **2017**, *115*, 78–88. [CrossRef]
3. Trzepiecinski, T.; Bochnowski, W.; Witek, L. Variation of surface roughness, micro-hardness and friction behaviour during sheet-metal forming. *Int. J. Surface Sci. Eng.* **2018**, *12*, 119–136. [CrossRef]
4. Trzepiecinski, T. A study of the coefficient of friction in steel sheets forming. *Metals* **2019**, *9*, 988. [CrossRef]
5. Van Der Heide, E.; Schipper, D.J. Friction and Wear in Lubricated Sheet Metal Forming Processes. In *Handbook of Lubrication and Tribology*; Taylor & Francis: Abingdon, UK, 2006; Volume 1, p. 28.
6. Weidel, S.; Engel, U. Surface characterisation in forming processes by functional 3D parameters. *Int. J. Adv. Manuf. Technol.* **2006**, *33*, 130–136. [CrossRef]
7. Dou, S.; Xia, J. Analysis of sheet metal forming (stamping process): A study of the variable friction coefficient on 5052 aluminum alloy. *Metals* **2019**, *9*, 853. [CrossRef]
8. Kirkhorn, L.; Bushlya, V.; Andersson, M.; Stahl, J.-E. The influence of tool steel microstructure on friction in sheet metal forming. *Wear* **2013**, *302*, 1268–1278. [CrossRef]
9. Trzepiecinski, T.; Lemu, H.G. Recent developments and trends in the friction testing for conventional sheet metal forming and incremental sheet forming. *Metals* **2019**, *10*, 47. [CrossRef]
10. Oliveira, M.C.; Fernandes, J.V. Modelling and simulation of sheet metal forming processes. *Metals* **2019**, *9*, 1356. [CrossRef]
11. Trzepiecinski, T.; Lemu, H.G.; Fejkiel, R. Numerical simulation of effect of friction directionality on forming of anisotropic sheets. *Int. J. Simul. Model.* **2016**, *16*, 590–602. [CrossRef]
12. Trzepiecinski, T. 3D elasto-plastic FEM analysis of the sheet drawing of anisotropic steel sheet. *Arch. Civ. Mech. Eng.* **2010**, *10*, 95–106. [CrossRef]
13. Sulaiman, M.H.; Farahana, R.; Bienk, K.; Nielsen, C.; Bay, N. Effects of DLC/TiAlN-coated die on friction and wear in sheet-metal forming under dry and oil-lubricated conditions: Experimental and numerical studies. *Wear* **2019**, 203040. [CrossRef]
14. Gonzalez-Pociño, A.; Alvarez-Antolin, F.; Asensio-Lozano, J. Improvement of adhesive wear behavior by variable heat treatment of a tool steel for sheet metal forming. *Materials* **2019**, *12*, 2831. [CrossRef]
15. Recklin, V.; Dietrich, F.; Groche, P. Influence of Test stand and contact size sensitivity on the friction coefficient in sheet metal forming. *Lubricants* **2018**, *6*, 41. [CrossRef]
16. Hol, J.; Meinders, V.T.; Geijselaers, H.; Boogaard, A.H.V.D. Multi-scale friction modeling for sheet metal forming: The mixed lubrication regime. *Tribol. Int.* **2015**, *85*, 10–25. [CrossRef]
17. Löfgren, H.B. A first order friction model for lubricated sheet metal forming. *Theor. Appl. Mech. Lett.* **2018**, *8*, 57–61. [CrossRef]
18. Littlewood, M.; Wallace, J.F. The effect of surface finish and lubrication on the fictional variation involved in the sheet-metal-forming process. *Sheet Met. Ind.* **1964**, *41*, 925–1930.
19. Nanayakkara, N.K.B.M.P.; Kelly, G.L.; Hodgson, P.D. Application of bending under tension test to determine the effect of toolradius and the contact pressure on the coefficient of fiction in sheet metal forming. *Mater. Forum* **2005**, *29*, 114–118.
20. Coubrough, G.; Alinger, M.; Van Tyne, C.; Van Tyne, C. Angle of contact between sheet and die during stretch–bend deformation as determined on the bending-under-tension friction test system. *J. Mater. Process. Technol.* **2002**, *130*, 69–75. [CrossRef]
21. Lemu, H.G.; Trzepieciński, T. Numerical and experimental study of the frictional behaviour in bending under tension test. *J. Mech. Eng.* **2013**, *59*, 41–49. [CrossRef]
22. Hoffmann, H.; Nürnberg, G.; Ersoy-Nürnberg, K.; Herrmann, G. A new approach to determine the wear coefficient for wear prediction of sheet metal forming tools. *Prod. Eng.* **2007**, *1*, 357–363. [CrossRef]
23. Berglund, J.; Brown, C.; Rosén, B.G.; Bay, N.O. Milled die steel surface roughness correlation with steel sheet friction. *CIRP Ann.* **2010**, *59*, 577–580. [CrossRef]
24. Pereira, M.; Yan, W.; Rolfe, B. Contact pressure evolution and its relation to wear in sheet metal forming. *Wear* **2008**, *265*, 1687–1699. [CrossRef]

25. Pereira, M.; Duncan, J.L.; Yan, W.; Rolfe, B.F. Contact pressure evolution at the die radius in sheet metal stamping. *J. Mater. Process. Technol.* **2009**, *209*, 3532–3541. [CrossRef]
26. Ceron, E.; Bay, N.O. Determination of friction in sheet metal forming by means of simulative tribo-tests. *Key Eng. Mater.* **2013**, *549*, 415–422. [CrossRef]
27. Trzepiecinski, T.; Fejkiel, R. A 3D FEM-Based Numerical Analysis of the Sheet Metal Strip Flowing Through Drawbead Simulator. *Metals* **2019**, *10*, 45. [CrossRef]
28. Trzepiecinski, T.; Lemu, H.G. Frictional conditions of AA5251 aluminium alloy sheets using drawbead simulator tests and numerical methods. *J. Mech. Eng.* **2014**, *60*, 51–60. [CrossRef]
29. ISO 6892-1. *Metallic Materials—Tensile Testing—Part 1: Method of Test at Room Temperature*; International Organisation for Standardization: Geneva, Switzerland, 2016.
30. *Msc.MARC 2010. Element Library*; MSC.Software Corporation: Santa Ana, CA, USA, 2010.
31. *Msc.MARC 2010. Theory Manual*; MSC.Software Corporation: Santa Ana, CA, USA, 2010.
32. Hill, R. A theory of the yielding and plastic flow of anisotropic metals. *Proc. R. Soc. London. Ser. A. Math. Phys. Sci.* **1948**, *193*, 281–297. [CrossRef]
33. Taherizadeh, A.; Green, D.E.; Yoon, J.W. A non-associated plasticity model with anisotropic and nonlinear kinematic hardening for simulation of sheet metal forming. *Int. J. Solids Struct.* **2015**, *69*, 370–382. [CrossRef]
34. Banabic, D.; Comsa, D.S.; Gawad, J. Plastic Behaviour of Sheet Metals. In *Multiscale Modelling in Sheet Metal Forming*; Springer Science and Business Media LLC: Berlin, Germany, 2016; pp. 1–46.
35. Farahnak, P.; Urbanek, M.; Džugan, J. Investigation study on determination of fracture strain and fracture forming limit curve using different experimental and numerical methods. *J. Phys. Conf. Ser.* **2017**, *896*, 12082. [CrossRef]
36. Lovell, M.R.; Khonsari, M.M.; Marangoni, R.D. The response of balls undergoing oscillatory motion: crossing from boundary to mixed lubrication regimes. *J. Tribol.* **1993**, *115*, 261–266. [CrossRef]
37. Karupannasamy, D.; Hol, J.; De Rooij, M.; Meinders, T.; Schipper, D. Modelling mixed lubrication for deep drawing processes. *Wear* **2012**, *294*, 296–304. [CrossRef]
38. Sulaiman, M.H.; Christiansen, P.; Bay, N.O. The influence of tool texture on friction and lubrication in strip reduction testing. *Lubricants* **2017**, *5*, 3. [CrossRef]
39. Bech, J.; Bay, N.O.; Eriksen, M. A study of mechanisms of liquid lubrication in metal forming. *CIRP Ann.* **1998**, *47*, 221–226. [CrossRef]

 *metals*

Article

Residual Stresses and Surface Roughness Analysis of Truncated Cones of Steel Sheet Made by Single Point Incremental Forming

Ján Slota [1], Bogdan Krasowski [2], Andrzej Kubit [3], Tomasz Trzepiecinski [4,*],
Wojciech Bochnowski [5], Kazimiera Dudek [6] and Miroslav Neslušan [7]

[1]   Institute of Technology and Material Engineering, Faculty of Mechanical Engineering, Technical University of Košice, Mäsiarska 74, 04001 Košice, Slovakia; jan.slota@tuke.sk

[2]   Department of Mechanical Engineering, State School of Higher Vocational Education, Rynek 1, 38-400 Krosno, Poland; b_krasowski@wp.pl

[3]   Department of Manufacturing and Production Engineering, Rzeszow University of Technology, al. Powst. Warszawy 8, 35-959 Rzeszów, Poland; akubit@prz.edu.pl

[4]   Department of Materials Forming and Processing, Rzeszow University of Technology, al. Powst. Warszawy 8, 35-959 Rzeszów, Poland

[5]   Centre for Innovative Technologies, University of Rzeszów, ul. Pigonia 1, 35-959 Rzeszów, Poland; wobochno@univ.rzeszow.pl

[6]   Institute of Technology, Faculty of Mathematics and Natural Sciences, University of Rzeszów, al. Rejtana 16c, 35-959 Rzeszow, Poland; kaziadudek@o2.pl

[7]   Department of Machining and Manufacturing Technology, University of Žilina, Univerzitna 1, 010-26 Žilina, Slovakia; Miroslav.Neslusan@fstroj.uniza.sk

*   Correspondence: tomtrz@prz.edu.pl; Tel.: +48-17-743-2527

Received: 31 December 2019; Accepted: 7 February 2020; Published: 10 February 2020

Abstract: The dimensional accuracy and mechanical properties of metal components formed by the Single Point Incremental Forming (SPIF) process are greatly affected by the prevailing state of residual stress. An X-ray diffraction method has been applied to achieve an understanding of the residual stress formation caused by the SPIF process of deep drawing a quality steel sheet drawpiece. The test object for an analysis of residual stress distribution was a conical truncated drawpiece with a slope angle of 71° and base diameter of the cone of 65 mm. The forming process has been carried out on a 3-axis HAAS TM1P milling machine. Uniaxial tensile tests have been carried out in the universal tensile testing machine to characterize the material tested. It was found that the inner surface of the drawpiece revealed small linear grooves as a result of the interaction of the tool tip with the workpiece. By contrast, the outer surface was free of grooves which are a source of premature cracking. The stress profile exhibits a nonlinear distribution due to different strengthening of the material along the generating line of the truncated conical drawpiece. The SPIF parts experienced a maximum residual stress value of about 84.5 MPa.

Keywords: mechanical engineering; truncated cone; incremental sheet forming; SPIF

---

1. Introduction

Single Point Incremental Forming (SPIF) is a sheet metal forming process which involves forming by local stretching of the sheet with a pin tool. The tool moves along a path controlled by a computer numerical control (CNC) machine. Incremental Sheet Forming (ISF) allows the forming of complex parts with a much higher degree of deformation than conventional deep drawing or stretch forming processes [1,2].

Springback is one of the main problems associated with incremental forming processes. The amount of springback is affected by the geometry of the drawpiece, the mechanical properties of the sheet material and the tension applied [3]. In SPIF, the state of residual stress can be influenced by adjusting the process parameters during the machining process [4]. A literature review of SPIF suggests that the geometrical accuracy of drawpieces may be increased by using a smaller tool diameter, a smaller tool step-down, and thicker sheets [5,6]. The large levels of deformation occurring in SPIF induce highly non-uniform residual stresses that affect the shape and dimensional accuracy of the formed parts.

Analysis of surface topography in ISF has gained research interest in the past few years [7,8]. According to Lasunon et al. [9] feed rate has little effect on the surface roughness, while the wall angle, depth increment, and its interaction play an important role. Hagan and Jeswiet [10] analyzed the effect of several forming variables, such as step-down size and spindle speed, on surface roughness. They concluded that the surface finish is a resultant of large-scale waviness created by the tool path and small-scale roughness induced by large surface strains. Attanasio et al. [11] investigated effect of variable step size and constant step size on surface roughness. It was found that varying step size with constant scallop height produced better surface finish. Powers et al. [12] found that the maximum height of the profile *Rz* is greater with rolling marks perpendicular to forming orientation. The surface roughness increases with the post-forming sheet strength, forming force, and residual stress, thereby showing that strain hardening has a direct influence on the roughness [13]. Furthermore, Al-Ghamdi and Hussain [13] found that the free-surface roughness and the contact-surface roughness are inversely related. Durante et al. [14] investigated the effect of spindle speed on the average surface of pyramid frusta. Roughness of surface was found to decrease when spindle speed passed from not rotating to rotating conditions. A detailed review of the current state-of-the-art of ISF processes in terms of its technological capabilities and discussions on the ISF process parameters and their effects on ISF processes may be found in the works of Gatea et al. [15] and Behera et al. [8]. Maqbool and Bambach [16] quantified the respective contribution of each forming mechanism (i.e., bending, membrane stretching, through-thickness shear) involved in the SPIF process and the dependence of geometrical accuracy on the dominant deformation mechanism. The practical consequence of the mode of deformation on the geometrical accuracy of pyramid-shaped parts has been demonstrated. It was found that a decrease in the dissipation energy in the bending mode leads to lower residual moments.

Residual stresses are defined as the internal stresses in a material that are present in the absence of external loading. In ISF the residual stresses are mainly caused due to local yielding of the material as a consequence of its plastic deformation. The lifetime and integrity of the components are greatly affected by the intensity of residual stresses [17]. The residual stresses also affect the fatigue life of the protective coatings which is especially important in the case of the low-carbon deep-drawing quality steel sheets used in the automotive industry. The combination of tensile residual stresses with service stresses adversely affects the fatigue life of the components, whereas compressive residual stresses are beneficial in improving the fatigue life of formed components [17–19]. An increase in the geometrical accuracy is achieved due to the decreased contribution of the bending deformation mode and lower residual moment [16]. A better understanding of the residual stresses is necessary to increase the geometrical accuracy of incrementally formed parts.

In the last decade, many papers have been devoted to residual stress analysis in incremental forming. There are various methods for measuring residual stresses such as hole drilling [20,21], slitting [22], the contour method [23], and numerical computations [4,17,20]. Maaß et al. [4] analyzed the significance of kinematic hardening effects on the state of residual stress based on numerical simulations. They found that the average deviation of the residual stress amplitudes in the clamped and unclamped parts may reach 18%. The results of numerical computations made by Tanaka et al. [24] revealed that the smaller the radius of the forming stylus, the larger the residual stresses become. Radu et al. [20] inspect, experimentally and by finite element-based simulation, the state of the residual stresses induced in incrementally formed double frustums of pyramids made from 1050 aluminium-based alloy. The hole drilling strain gauge method has been used to find the distribution of residual stresses in the sheet thickness. The results indicated compressive residual stresses with a magnitude that varies through the sheet thickness. In another paper, Radu et al. [25] investigated the effect of residual stresses on the accuracy of parts processed by SPIF using the strain-rosette method. The results of the investigations revealed that a favorable state of residual stresses, and implicitly a good accuracy of parts, can be obtained when small values of tool diameter and vertical step sizes are used. Shi et al. [21] investigated the residual stresses in incrementally formed Cu/steel bonded laminates using the hole-drilling method. They found that the most significant process parameters influenced by residual stresses were the tool diameter and wall angle, while the tool rotation was the least significant process parameter. Different unidirectional and bidirectional tool path strategies were used by Maaß et al. [26] to analyze the residual stress development in SPIF formed parts. The results suggest that the influence of the tool path strategy on the amplitude of the resulting residual stress is not decisive. Abdulrazaq et al. [27] used three types of tool shape in their work on the forming of pyramid like drawpieces from 1050 aluminium alloy which gave higher residual stresses than other types. It has been found that residual stresses increased with increasing step size and feed rate values. Al-Ghamdi and Hussain [28] experimentally analyzed the through-thickness stress gradient across the thickness of Cu/steel bonded laminates. It was found that ISF induces a gradient in compressive stress gradient, which can be much greater (about 18 times) than that the rolling process induces in the parent laminates while bonding. An experimental and numerical investigation of the influence of process parameters in ISF on residual stresses has been carried out by Maqboll and Bambach [17]. Residual stresses were determined in parallel and perpendicularly to the direction of tool motion at the center of a strip cut from the numerical model in the clamped and unclamped states. The results of the analyses reveal that the most significant parameter in the build-up of residual stress and the reduction of geometrical accuracy is the wall angle. Different strategies that analyze the intensity of the residual stresses and its effects on the accuracy of parts formed have been discussed in the literature [25,29].

Hajavifard et al. [30] used ISF to improve properties of conical annular discs by introduction residual stresses in material. The main mechanism of residual stress generation was the transformation of metastable austenite into martensite under the action of the forming tool. The residual stresses induced by the selective laser melting process on the fabricated AlSi10Mg metallic sheets, as well as those produced during their ISF operation were analyzed by López et al. [31].

To the best of authors' knowledge, measurement of the residual stresses by the X-ray diffraction (XRD) method in the SPIF process has not so far been reported. So, in the current study, residual stress distribution was investigated in a conical truncated drawpiece formed by the incremental forming process. The residual stresses were measured in the subsurface layer at a depth of about 0.005 mm. Therefore, verification will be carried out of the possible interaction between the value of the residual stress measured in different parts of the conical truncated drawpieces and the roughness parameters measured on the surface of the drawpieces. In this manuscript residual stresses were measured separately in the axial and tangential directions. The experimental investigations to form a conical truncated drawpiece were carried out on a 3-axis TM1P milling machine (HAAS, Oxnard, CA, USA). A Proto iXRD Combo diffractometer (Proto Manufacturing Inc., Taylor, MI, USA) using CrK_α radiation was used to determine the distribution of residual stresses.

2. Experimental Methodology

2.1. Material

The experimental forming of conical truncated drawpieces has been carried out for cold rolled DC04 deep drawing quality steel sheet (1.0338 acc. to EN 10130:2009) with a thickness of 0.8 mm supplied by Arcellor-Mittal (Dąbrowa Górnicza, Poland). The DC04 steel sheet is characterized by high ability to deform, weldability and susceptibility to springback. This material is intended for use in forming processes where good strength, rigidity and ductility are required. Typical applications are to be found in the automotive industry, metal furniture and the domestic appliance sector. The results of the element-analysis obtained by optical emission spectroscopy are presented in Table 1.

Table 1. The chemical composition (in weight%) of DC04 steel sheet.

C	Mn	Cu	Ni	Cr	Al	Ti	Mo	V	S	Si	Fe
0.016	0.188	0.06	0.042	0.021	0.035	0.022	0.003	0.020	0.004	0.003	Rest

The basic mechanical properties of the sheet tested (Table 2) were determined in the uniaxial tensile test according to EN ISO 6892-1:2016-09. The mechanical properties of the sheet metal were determined through tensile tests along three directions with respect to the rolling direction: 0°, 45° and 90°. Three specimens were tested for each cut direction and an average value of a specific parameter was determined. The strain hardening parameters (strength coefficient K and strain hardening exponent n) were determined by approximation of the true stress-true strain relationship using the Hollomon equation

$$\sigma_p = K \cdot \varepsilon^n \tag{1}$$

where σ_p is the yield stress, ε is the true strain.

Table 2. Basic mechanical properties of DC04 steel sheet.

Specimen Orientation	Yield Stress R_{p02}, MPa	Ultimate Tensile Stress R_m, MPa	Elongation A_{50}, %	Strengthening Coefficient K, MPa	Strain Hardening Exponent n
0°	184.5	303.9	23.0	490.4	0.205
45°	193.7	314.9	22.1	489.9	0.164
90°	176.1	296.0	22.8	465.7	0.169

2.2. Incremental Forming

The experimental investigations to form the conical truncated drawpiece (Figure 1a) were carried out on a 3-axis TM1P milling machine (HAAS, Oxnard, CA, USA). Blanks of size 120 mm × 120 mm were placed in the machine, and were clamped at the edges using screws (Figure 1b). The die supports the sheet, and its opening defines the working area of operation of the forming tool. The tool, which is attached in the head of the machine through the ER collet system, continuously indents into the sheet by step size and follows a spiral path for the desired part. The final height drawpiece was equal to 70 mm. The slope angle of the truncated cone and the diameter of the base of the cone were 71° and 65 mm, respectively. The tool was made of high-speed steel and had a rounded tip with a radius R = 3.5 mm and a diameter of 7 mm.

During machining, based on the recent investigations and experience the following processing conditions were applied:

– feed rate f = 1500 mm·min^{-1},

– tool rotational speed n = 87 rpm,

– incremental depth (step size) a_p = 0.3, 0.5 and 0.7 mm,

– lubricant: full synthetic 75W-85 (Castrol Ltd., Liverpool, UK) lubricant (viscosity 74.0 mm^2/s (at 40 °C), density 874 kg/m^3 (at 15 °C), pour point −45 °C).

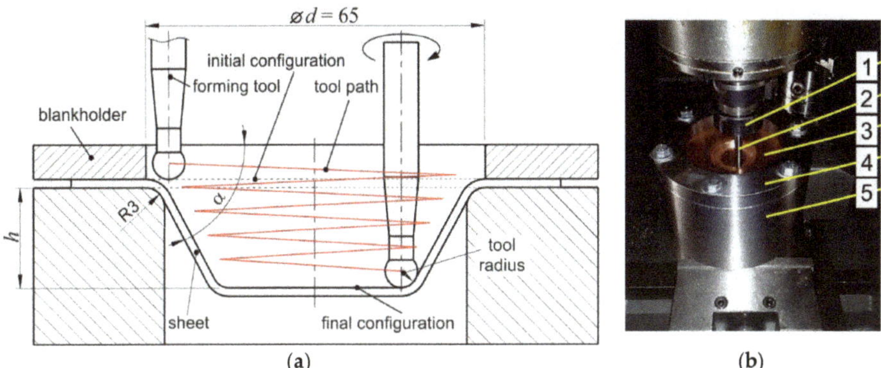

Figure 1. Geometry (**a**) and view (**b**) of the forming tool: 1—ER collet, 2—spindle, 3—workpiece, 4—holding ring, 5—body.

The tool path (Figure 2) was generated using the EDGECAM software (Hexagon, Reading, UK) based on the 3D model of a conical truncated drawpiece. After the control program was created, the forming device and tool were mounted on the milling machine. The control program, created in the simulation mode, was then validated in the test mode.

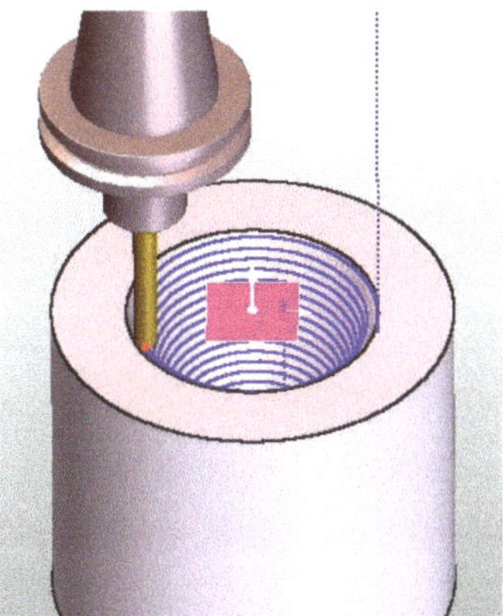

Figure 2. Tooling model with the tool path generated.

2.3. X-ray Diffraction Analysis

The XRD technique was used to measure the residual stress within the surface layer formed, as a result of interaction between the SPIF tool and the surface of the material being processed. The XRD method is based on the change of position of a diffracted plane which corresponds to a change of the material lattice parameter as a result of material deformation. Residual stresses after unloading were measured by the XRD technique carried out on a Proto iXRD Combo diffractometer (Proto Manufacturing Inc., Taylor, MI, USA) using CrK_α radiation (average effective penetration depth ~5 μm, scanning angle ±39° and Bragg angle 156.4°). The residual stresses were calculated from shifts of the 211 reflection. The Winholtz and Cohen method and X-ray elastic constants $\frac{1}{2}s_2 = 5.75$ TPa^{-1}, $s_1 = -1.25$ TPa^{-1} were used for residual stress determination. Residual stresses were measured in the axial as well as the tangential directions in the center of the sample (with respect to their width and length). The strain analysis is based on the interplanar spacing *d* according to Bragg's law:

$$n\lambda = 2d(hkl)sin\Theta \tag{2}$$

where λ is the wavelength, Θ is the Bragg angle, n is the diffraction order of the interference (hkl).

The interplanar lattice spacings were measured for different specimen orientations. The residual stress can then be calculated using the $\sin^2\psi$ method based on the Bragg symmetrical diffraction, and the appropriate elastic material constants. This method uses a Ψ–type goniometer (Proto Manufacturing Inc., Taylor, MI, USA), which allows one to obtain appropriate inclinations of the diffraction vector by angles Ψ_i (Figure 3) in the plane perpendicular to the diffraction plane. In XRD analyses a round collimator with a diameter of 2 mm was used. The exposure time during which the specimen was overexposed for each diffraction angle was 12 s. The reflection profiles for the were measured in the 2Θ angular range of $-39°$ to $39°$. Figure 4 shows the location of the measuring point in the element of the cone.

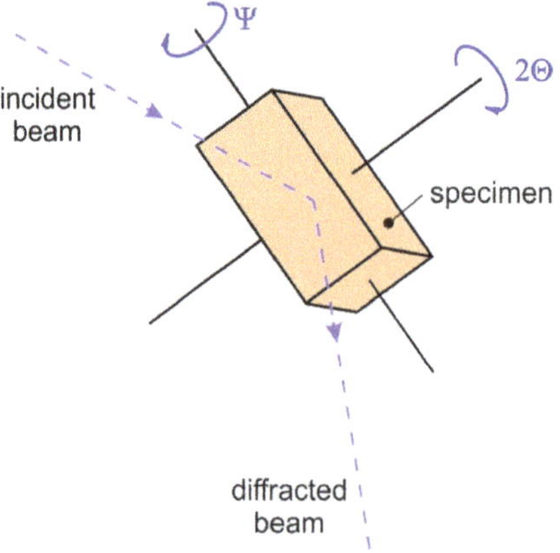

Figure 3. X-ray diffraction (XRD) methodology.

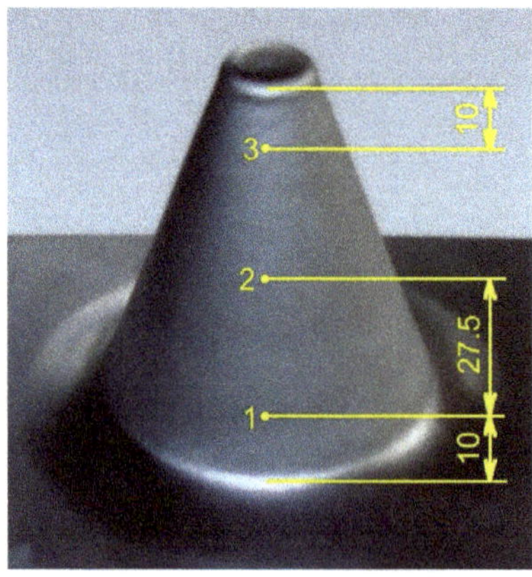

Figure 4. Location of the stress measurement points.

2.4. Surface Characterisation

Topographies of the as-received and deformed surfaces of truncated cones were characterized by 3D main surface roughness parameters. Measurement of the surface roughness parameters was carried out using a Contour GT 3D optical microscope (Bruker, MA, USA) with vision 64 software. Surface topography was measured at both internal and external surfaces of drawpieces. The main standard 3D parameters determined by this measurement are the average roughness Sa, the highest peak of the surface Sp, the root mean square roughness parameter Sq, and the maximum pit depth Sv. Selection of surface roughness parameters was based on the work of Ham et al. [32,33] which suggested that it was more useful to use areal surface texture parameters in evaluating surface texture and topography in single point incremental sheet forming.

The Sa parameter shows a strong correlation with the Sq parameter. Both the Sa and Sq parameters are used to represent the roughness of a two-dimensional surface topography. Sa describes surface roughness more comprehensively than the arithmetical mean height Ra, the arithmetical mean deviation of the profile that is assessed, which reflects the roughness of a one-dimensional contour [34]. In ISF, the Sa parameter is more suitable for analysis of surface topography than the Ra parameter [35,36].

3. Results and Discussion

3.1. Formability

During the experimental tests truncated cones were made with a slope angle $\alpha = 70$–$72°$ and with different step size a_p, and feed rates. The assumed total depth of the formed part was 75 mm. Two drawpieces were tested for each set of input parameters. The aim of the investigations was to obtain, for all configurations of the input parameters of the forming process, truncated cones with a height of 75 mm. It is noteworthy that a slope angle α in combination with step size a_p are responsible for the limit conditions for obtaining the successful deformation of drawpieces without cracks.

The step size a_p determines the amount of material that is subjected to deformation during one pass. According to the results of tests carried out, this parameter has little effect on the change of plasticization of the processed material. With an increase in step size a_p, a slight increase in the slope angle α was obtained at which the truncated conical drawpiece had a shape that was correct and did not have any defects. The reduction of the a_p value caused the drawpieces to crack at the initial stage of forming, in the range of a slope angle of 71–72° (Table 3). The heights of the drawpieces at fracture, shown in Table 3, are the averages of the two drawpieces that are formed. In addition, the reduction in a_p results in a longer duration of treatment, which is disadvantageous from the point of economic production. The only positive effect of decreasing the value of the step size is a clear improvement in the quality of the outer and inner surface of the drawpiece (Table 3).

Table 3. Results of experimental tests ($\sqrt{}$ means that drawpiece has been successfully formed)

Slope Angle α (°)	Step Size a_p (mm)		
	0.3	0.5	0.7
70	$\sqrt{}$	$\sqrt{}$	$\sqrt{}$
71	Fractured at $h = 15.6$ mm	$\sqrt{}$	$\sqrt{}$
71.5	Fractured at $h = 16.5$ mm	$\sqrt{}$	$\sqrt{}$
72	Fractured at $h = 16.5$ mm	Fractured at $h = 18.9$ mm	Fractured at $h = 20.4$ mm

The SPIF of the cones with larger wall angles was limited by their premature fracture (Table 3). It is well known [37], in fact, that at increasing the steepness of the walls, process conditions become heavier, so that plastic collapse occurs in the blank. Obviously, this result is strongly dependent on sheet thinning, which increases at increasing the steepness of the walls. All drawpieces with a slope angle of 72 were fractured at a depth much smaller than the assumed total depth. In contrast the truncated cones with the 70° wall angles were successfully formed until they reached the design depth without fracture. Since at a specific wall angle, a cone can only be successfully formed to a certain depth using single-stage SPIF, the maximum forming angle alone cannot properly represent the formability of the material. Hence, according to Shamsari et al. [38], the formability of a sheet metal in SPIF can be best illustrated by the maximum forming depth at various wall angles, as depicted in Table 3.

Figure 5 shows the deformed parts at fracture. In the case of all the fractured surfaces a fracture occurred as a continuous crack along the circumference of the component. The corner of the drawpiece is the most strained region in the sheet due to its complex stress state. The formability and morphology of crack initiation in the sheet (region of local sheet thinning) is a complex function of the tool path strategy [7,39,40], shape of the drawpiece [15,41], forming temperature [8,15], friction conditions [8,42], and vertical step down [15,38] used. The crack initiation was observed in the vicinity of the rib bottom and propagates along the edge of the truncated cone (Figure 5). The fracture region observed coincides with that seen in the results of the investigations of many authors [38,41]. The drawpieces were fractured due to excessive thinning of the sheet and sharp edges were produced from the crack surface.

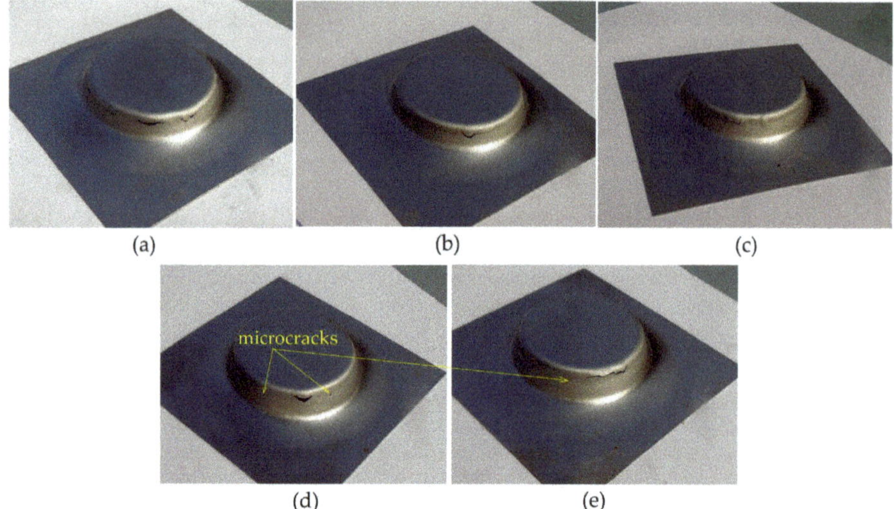

Figure 5. Morphology of fracture of drawpieces formed with the following parameters: (**a**) $\alpha = 71°$, $a_p = 0.3$ mm; (**b**) $\alpha = 71.5°$, $a_p = 0.3$ mm; (**c**) $\alpha = 72°$, $a_p = 0.3$ mm; (**d**) $\alpha = 72°$, $a_p = 0.5$ mm; (**e**) $\alpha = 72°$, $a_p = 0.7$ mm.

3.2. Surface Roughness

As mentioned above, a slope angle in combination with a step size a_p are responsible for obtaining a successful drawpieces without cracks. The step size influences the material deformation limit to a small degree. However, the step size had a very large impact on the surface finish of the surface of the part. Although the small linear scratch bands resulting from the interaction of the tool tip with the workpiece material have been revealed on the inner surface of the drawpiece, no clear grooves have been found on the outer surface. Small areas of galling with a wavy appearance are also visible on the internal surface (Figure 6). In SPIF, the surface finish on the inner surface can be mainly be characterized as a resultant of large-scale waviness created by the forming path [43]. The surface roughness of the outer surface is mainly caused by large surface strains, which usually leads to an orange peel phenomenon (Figure 6).

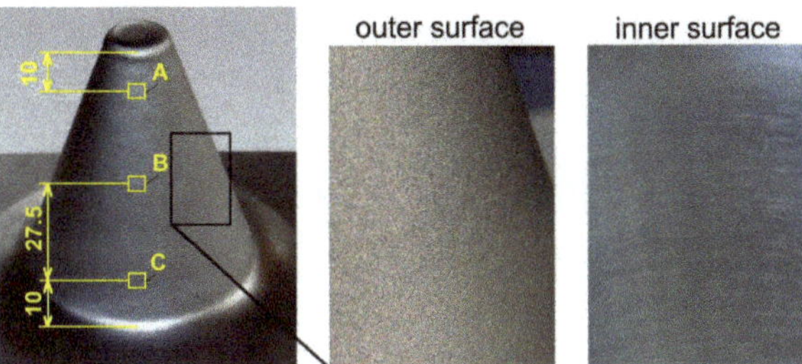

Figure 6. View of the outer and inner surface of the drawpiece.

Figures 7 and 8 show the surface topographies of the outer (Figure 7) and inner (Figure 8) surface of the drawpieces with a slope angle $\alpha = 71°$ formed at step size $a_p = 0.5$ mm. Surface topographies were measured in three locations A, B, and C along the generating line of the cone, according to the Figure 6. There is no clear change in the maximum height of the profile at outer surface of the drawpiece. In the outer surface there are clear peaks and valleys (scratch bands) resulted from the local contact of tool tip with sheet surface. The width between adjacent peaks equals the values of step size. The surface topography was measured at half a height of drawpieces (vicinity of point 2 in Figure 4). The distance between scratch bands do not increases with the increasing of the drawpiece heights (Figure 8). In means that the thickness of sheet metal which undergone the deformation due to interaction of tool tip did not change due to further increasing of drawpiece height.

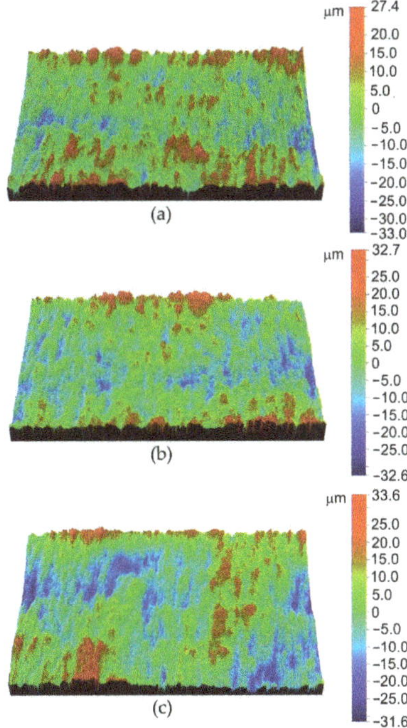

Figure 7. Surface topography of the outer surface of the drawpiece with a slope angle $\alpha = 71°$ formed at step size $a_p = 0.5$ mm; surface topographies (**a**), (**b**) and (**c**) are measured in locations of A, B and C indicated in Figure 6, respectively.

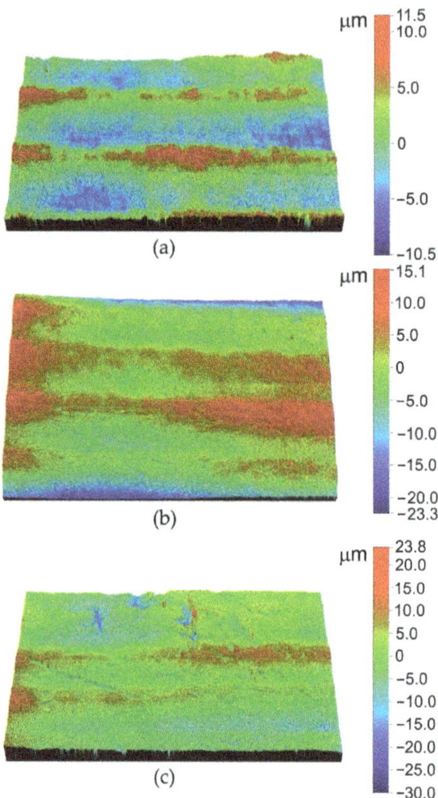

Figure 8. Surface topography of the inner surface of the drawpiece with a slope angle $\alpha = 71°$ formed at step size $a_p = 0.5$ mm; surface topographies (**a–c**) are measured in locations of A, B and C indicated in Figure 6, respectively.

Figures 9 and 10 show the effect of step size on the value of the basic roughness parameters on the outer and inner surfaces of the drawpiece, respectively. The parameters were measured in the vicinity of the points indicated in Figure 4. The trend of change in the values of Sv was found to be a decrease on the outer surface of the drawpiece (Figure 9). In the case of the drawpieces formed at $a_p = 0.3$ and $a_p = 0.5$ (Figure 9a,b) an increase of the Sa, Sp and Sq parameters along the height of drawpiece was found. Drawpieces formed at $a_p = 0.7$ (Figure 9c) are characterized by clear reduction of the highest peak of the surface Sp. The location of the roughness measurement at the outer surface do not effect significantly on the change in value of roughness Sv parameter (Figure 9).

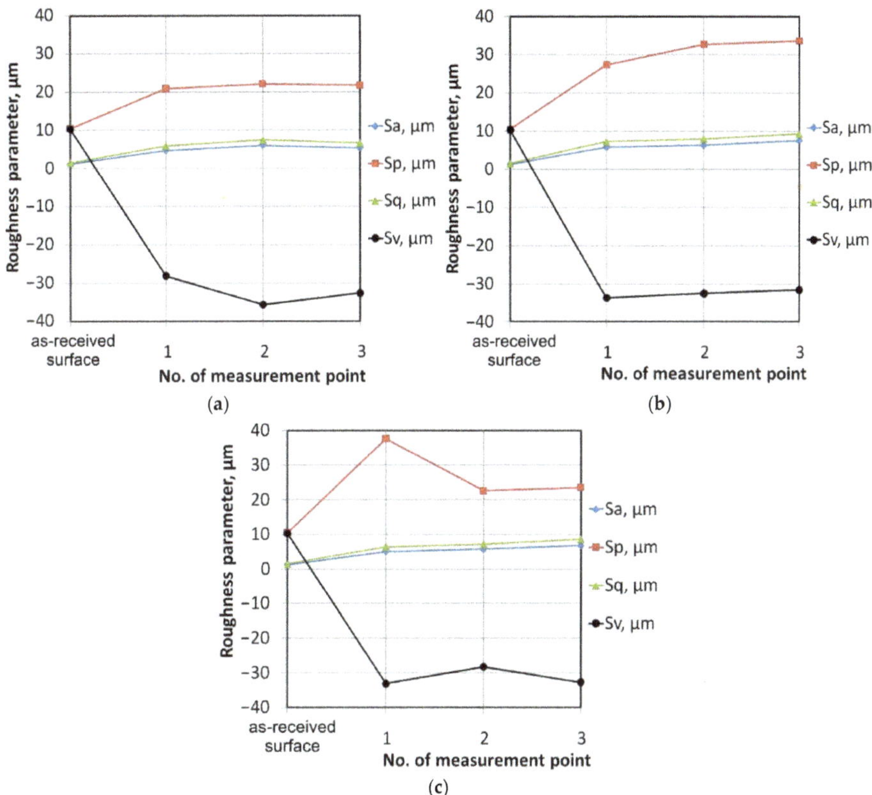

Figure 9. Variation of surface roughness parameters on the outer surface of a drawpiece with a slope angle $\alpha = 70°$ formed at step size (**a**) $a_p = 0.3$ mm, (**b**) $a_p = 0.5$ mm and (**c**) $a_p = 0.7$ mm.

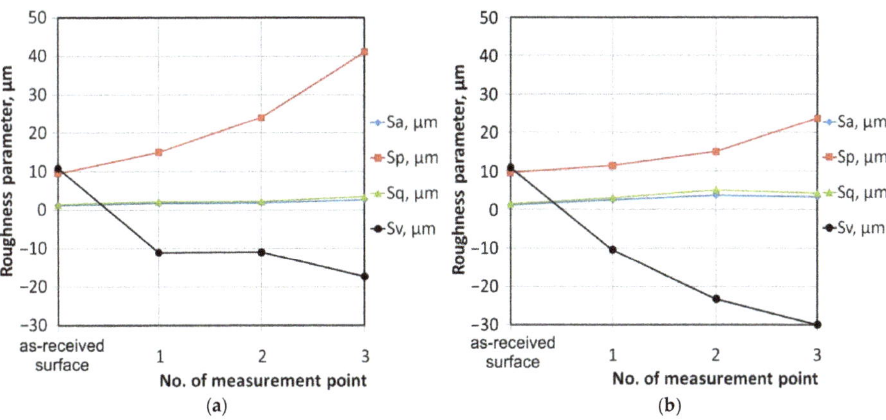

Figure 10. *Cont.*

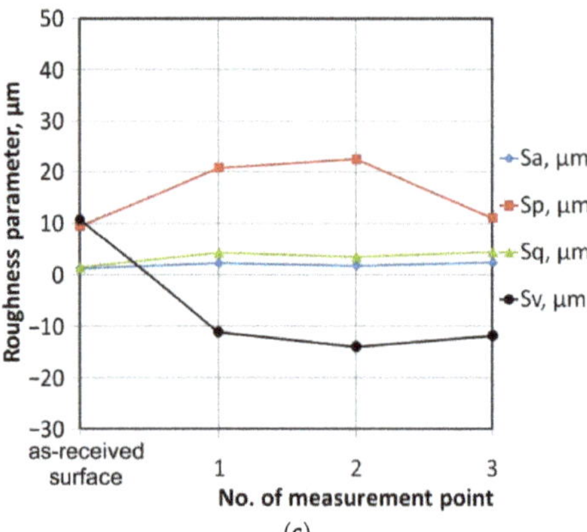

(c)

Figure 10. Variation of surface roughness parameters on the inner surface of a drawpiece with a slope angle α = 70° formed at step size (**a**) a_p = 0.3 mm, (**b**) a_p = 0.5 mm and (**c**) a_p = 0.7 mm.

It is very interesting to show the influence of ISF on the surface topography of the inner surface of drawpieces. In the case of the step sizes of 0.3 mm and 0.5 mm, as the surface roughness was measured at increasingly high locations along the drawpiece height, the value of Sp clearly increased (Figure 10b). In contrast, the value of the maximum pit depth Sv decreased. Surface roughness measurements of the drawpiece formed at a_p = 0.7 mm show the Sp parameter value is close to the one measured on the as-received surface.

The value of the two commonly used surface roughness parameters Sa and Sq did not change significantly with the increasing of the height of drawpiece. It can be concluded that the surface finish of outer surface of a drawpiece is a resultant of small-scale roughness induced by large surface strains. While the surface quality of inner surface of drawpiece is a result of large-scale waviness created by the tool path. This is in accordance with results of Hagan and Jeswiet et al. [10]. In general, observations of the surface roughness change are in accordance with the results of obtained by Liu et al. [43]. They concluded that surface roughness on the external noncontact surface is always higher than that of the internal tool-sheet contact surface.

Figure 11 shows the effect of slope angle on the main roughness parameters on the outer surface of the drawpieces. A clear increase in the average roughness Sa (Figure 11a) and Sq (Figure 11c) is observed for the slope angles α = 70° and α = 71°. In the same conditions, the value of these parameters decreases along the height of the drawpiece. An increase in the slope angle of the drawpiece causes more proportional stretching of the sheet in the circumferential direction and in the axial direction. In addition, ISF causes a decrease in the maximum pit depth Sv, a clear relationship between the location of place measurement and the value of this parameter was not revealed.

Final inner and outer average roughness was greater than on the initial surface of the blanks, whereas Oleksik et al. [44] concluded that the outer surface of parts was unaffected by the input parameters. In fact, in our investigations the average roughness Sa increases to a smaller extent on the inner surface than the outer surface.

A decrease in step size increases the processing time. However, the quality of the outer and inner surface characterized by the value of the average roughness was not significantly affected by step size (Figures 8 and 9). In the case of the outer surface, a small increase in the Sa parameter is observed along the generating line of the cone (Figure 9). Many researchers investigated the effect of process parameters on the resulting surface roughness at one selected location on the drawpiece surface. As found in this paper, the surface roughness changes with drawpiece height. So, to receive reasonable conclusions, the investigations must be focused on the evaluation of overall surface roughness.

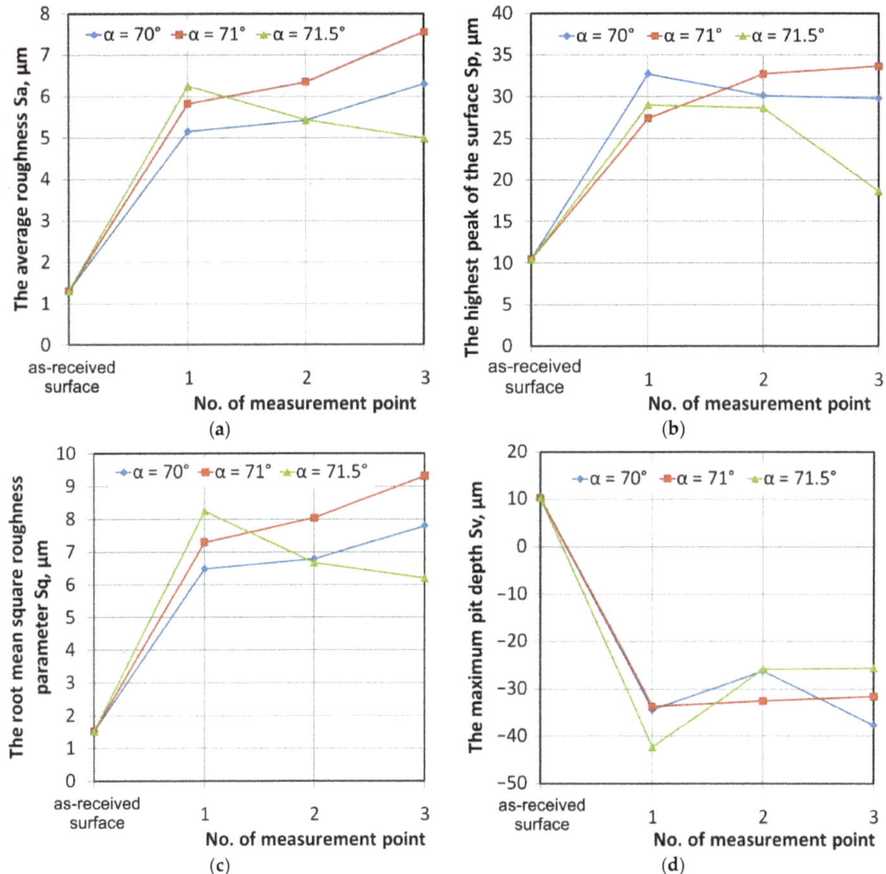

Figure 11. Variation of surface roughness parameters (**a**) *Sa*, (**b**) *Sp*, (**c**) *Sq* and (**d**) *Sv* on the outer surface of a drawpiece formed at step size a_p = 0.5 mm.

3.3. Residual Stresses

The outer surface of the drawpiece exhibits a compressive magnitude of residual stress (Figure 12). Radu [45], when studying DC04 steel with a thickness of 0.8, found that residual stresses in the material were of a compression type. As for the drawpieces formed at a_p = 0.5 mm, α = 70° and a_p = 0.5 mm, α = 71°, the residual stress starts to decrease along the surface until it reaches its maximum compressive magnitude value at point 3 (Figure 4). It was found that tangential residual stress was lower than axial stress. In the case of a drawpiece formed at a_p = 0.5 mm, α = 71.5° another relationship between the axial and tangential residual stresses than for drawpiece formed at a_p = 0.5 mm, α = 70° was revealed. The tangential stress increases between points 1 and 3, while the axial residual stresses decrease.

This variation of stress magnitudes between the measurement points 1 and 3 is an indication of flexural stresses on the conical truncated drawpiece formed from steel. The increase in the slope angle lead to increase of the circumference of the drawpiece at specific cross-sections parallel to the cone base. Base on the constancy-of-volume relationship, when the circumference of the drawpiece at specific cross-section increases, the sheet thickness decreases. In these conditions the stiffness of the drawpiece is smaller and sheet material in smaller extend accumulate the plastic deformation.

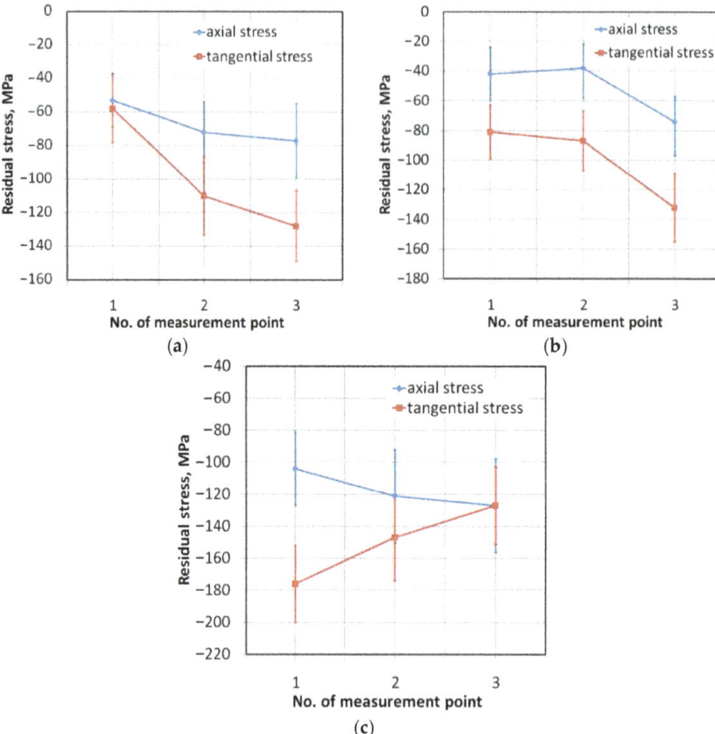

Figure 12. Variation of residual stress on the outer surface of a drawpiece with a slope angle (**a**) $\alpha = 70°$, (**b**) $\alpha = 71°$ and (**c**) $\alpha = 71.5°$, formed at step size $a_p = 0.5$ mm.

The surface roughness of drawpieces was measured at the same locations as the residual stresses (Figure 4). Considering that the stresses were measured in the subsurface layer at a depth of about 5–10 μm, this permitted them to investigate the possible effect of surface topography on the amount of residual stress. The effect of roughness parameters on the value of residual stress is presented in Figures 13–15. Only in the case of the Sa parameter, which plays an important role in the characterization of surface roughness in manufacturing processes, can a clear relation between slope angle, average roughness and residual stress be found. Residual stresses measured in a tangential direction decrease with the increasing value of the Sa parameter (Figure 13b). A similar trend is observed for the axial stresses, but only for the drawpieces with slope angles of 70° and 71°. Only in the case of a drawpiece with a slope angle of 71.5° does the increase of the Sa parameter lead to an increase in residual stresses (Figure 13a). Due to the similar trends of the changes of the Sa and Sq parameters (Figures 9 and 10), the above-mentioned conclusions are also valid for the effect of the Sq parameter on the residual stresses.

An increase in step size causes an increase in the axial (Figure 16a) and tangential (Figure 16b) residual stresses, as was also found by Radu et al. [25] and Jimenez et al. [46]. By increasing the tool step-down, the magnitude of the compressive residual stresses on the non-contact surface also increases when forming pyramidal drawpieces [17]. As for step size $a_p = 0.3$ mm, the value of the residual stresses measured in an axial and a tangential direction shows great proportionality (Figure 17a). When step size started to increase, this relationship led to a disturbance. Increase of step size to the maximal value considered, the value $a_p = 0.7$ mm, causes a reversal of the changes of the value of tangential residual stress, and at point 3 the value of the tangential stress is higher than the axial stress. The results allow us to draw the conclusion that the values of the residual stresses measured in the subsurface layer depend on the direction of measurement. The effect of surface roughness on the residual stresses is not entirely clear because the mechanical properties of the sheets change between points 1 and 3 due to the strengthening phenomenon.

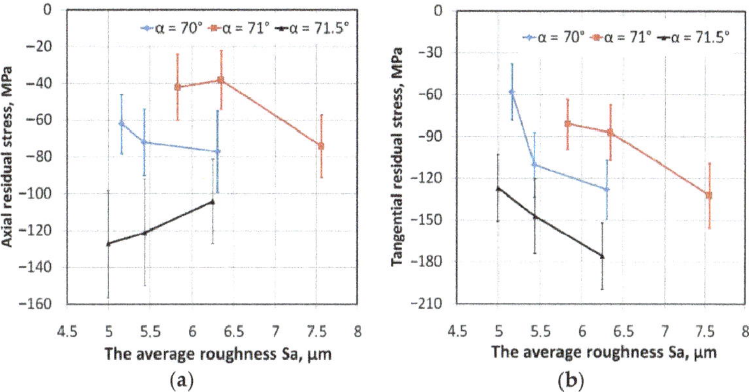

Figure 13. Effect of the average roughness *Sa* on the variation of (**a**) axial and (**b**) tangential residual stresses in the outer surface of a drawpiece formed at a step size $a_p = 0.5$ mm.

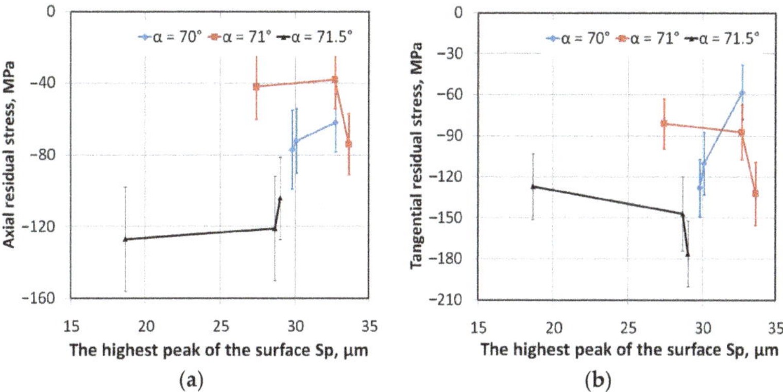

Figure 14. Effect of the highest peak of the surface *Sp* on the variation of (**a**) axial and (**b**) tangential residual stresses in the outer surface of a drawpiece formed at step size $a_p = 0.5$ mm.

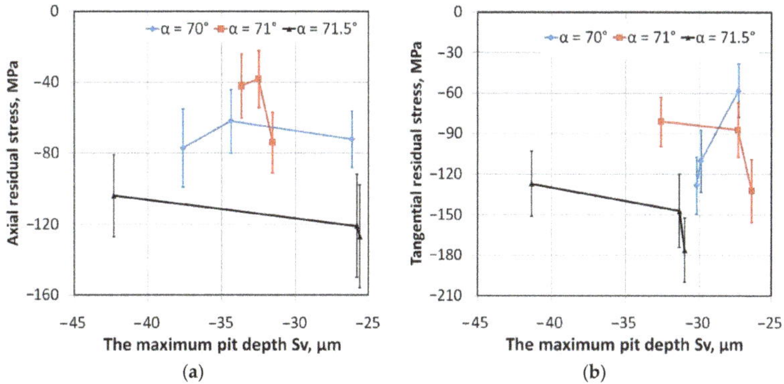

Figure 15. Effect of the maximum pit depth *Sv* on variation of (**a**) axial and (**b**) tangential residual stresses in the outer surface of a drawpiece formed at a step size $a_p = 0.5$ mm.

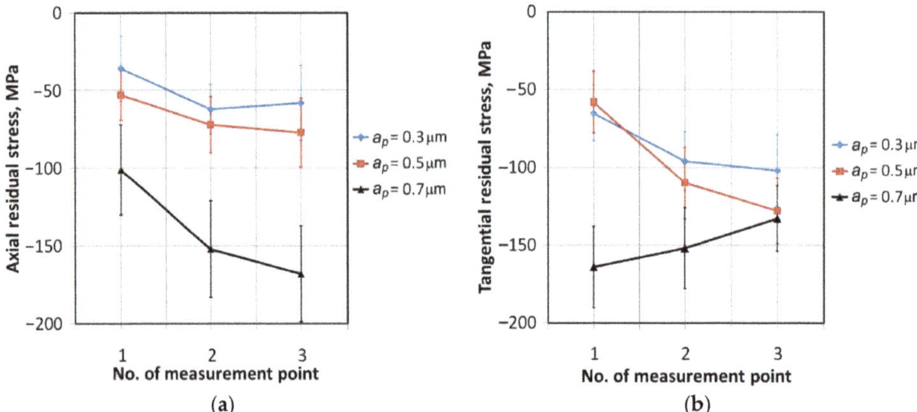

Figure 16. Effect of the step size a_p on the variation of (**a**) axial and (**b**) tangential residual stresses on the outer surface of a drawpiece with a slope angle $\alpha = 70°$.

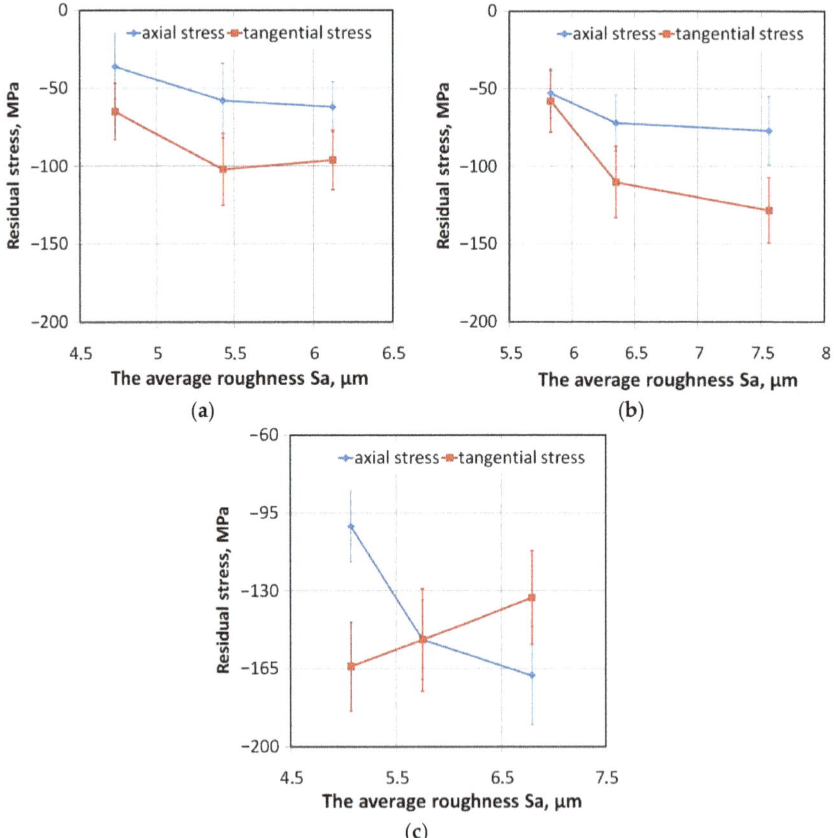

Figure 17. Effect of the average roughness Sa on the variation of residual stresses in the outer surface of a drawpiece with a slope angle $\alpha = 70°$ formed at a step size (**a**) $a_p = 0.3$ mm, (**b**) $a_p = 0.5$ mm and (**c**) $a_p = 0.7$ mm.

The change of residual stresses along the generating line of the cone may be attributed to the strain hardening phenomenon and material field stress. The mechanism of strain hardening is enhanced by the increase in density of the dislocations. On the other hand, dislocations are needed to produce the lattice rotation, and at the same time the remaining dislocation will increase the residual stresses [16]. The intensity and distribution of the residual stresses are controlled by the underlying plastic deformation mechanisms of the forming process, i.e., bending, stretching and through-thickness shear [16]. As Tanaka et al. [24] concluded, the higher the contribution of bending deformation, the more pronounced will be the geometric deviations. The non-compatible strain fields form the other source of residual stresses. The characteristics of the plastic zone are crucial for an understanding of the underlying physics of the deformation process and the generation of residual stress around a crack tip under fatigue loading.

4. Conclusions

The residual stress distribution was investigated along the outer surface of a DC04 steel sheet conical truncated drawpiece processed by SPIF. A XRD analysis has been carried out in an axial and tangential direction. The following conclusions are drawn from the research:

1. The inner surface of the drawpiece revealed small linear grooves as a result of the interaction of the tool tip with the workpiece. The surface finish of the outer surface of a drawpiece is the result of small-scale roughness induced by large surface strains which leads to an orange peel phenomenon.
2. On both the inner and outer surfaces of the drawpiece, an increase of the Sa, Sp, and Sq parameters was found along the generating line of the cone when compared to the as-received surface. Moreover, a clear reduction in the Sv parameter was revealed.
3. The inner and outer surfaces, characterized by their average roughness values, was not significantly affected by step size. In the case of the outer surface, only a small increase in the Sa parameter is observed along the generating line of the cone.
4. The outer surface of the drawpiece exhibits a compressive magnitude of residual stress in both the axial and tangential direction. The value of tangential residual stresses was lower than axial stresses.
5. Residual stresses measured in a tangential direction decrease with an increase in the value of the Sa parameter.
6. Step size had a very large impact on the value of residual stresses. An increase in step size causes an increase in the absolute values of axial and tangential residual stresses.
7. Residual stresses measured in a tangential direction decrease with the increasing value of the Sa and Sq parameters. A similar trend is observed for the axial stresses, but only for the drawpieces with slope angles of 70° and 71°. No clear relationship was found between the rest of roughness parameters and residual stress values.
8. The geometric deviations found in the incrementally formed drawpieces are a result of local springback behind the actual forming process as well as springback upon unclamping and upon trimming. To better understand the role of residual stress on the geometric accuracy of the SPIFed part, the effect of the process parameters on the residual stresses after different stages of drawpiece fabrication will be analyzed in future research.

Author Contributions: Conceptualization, B.K., A.K. and T.T., investigation, B.K., A.K., T.T., W.B., K.D. and M.N.; formal analysis, J.S.; validation, B.K., A.K., T.T., K.D. and M.N.; project administration, J.S.; funding acquisition, J.S.; writing–original draft, T.T.; writing–review and editing, T.T. All authors have read and agreed to the published version of the manuscript.

Funding: This research was funded by the Grant Agency of the Ministry of Education, Science, Research and Sport of the Slovak Republic, grant number VEGA 1/0259/19.

Conflicts of Interest: The authors declare no conflict of interest.

References

1. Duflou, J.R.; Habraken, A.M.; Cao, J.; Malhotra, R.; Bambach, M.; Adams, D.; Vanhove, H. Single point incremental forming: State-of-the-art and prospects. *Int. J. Mater. Form.* **2018**, *11*, 743–773. [CrossRef]
2. Martins, P.A.F.; Bay, N.; Skjoedt, M.; Silva, M.B. Theory of single point incremental forming. *Cirp Ann. Manuf. Technol.* **2008**, *57*, 247–252. [CrossRef]
3. Trzepiecinski, T.; Lemu, H.G. Effect of computational parameters on springback prediction by numerical simulation. *Metals* **2019**, *7*, 380. [CrossRef]
4. Maaß, F.; Gies, S.; Dobecki, M.; Brömmelhoff, K.; Tekkaya, A.E.; Reimers, W. Analysis of residual stress state in sheet metal parts processed by single point incremental forming. *AIP Conf. Proc.* **2018**, *1960*, 160017.
5. Micari, F.; Ambrogio, G.; Filice, L. Shape and dimensional accuracy in single point incremental forming: State of the art and future trends. *J. Mater. Process. Technol.* **2007**, *191*, 390–395. [CrossRef]
6. Ambrogio, G.; Cozza, V.; Filice, L.; Micari, F. An analytical model for improving precision in single point incremental forming. *J. Mater. Process. Technol.* **2007**, *191*, 92–95. [CrossRef]
7. Blaga, A.; Oleksik, V. A study on the influence of the forming strategy on the main strains, thickness reduction, and forces in a single point incremental forming process. *Adv. Mater. Sci. Eng.* **2013**, *2013*, 382635. [CrossRef]
8. Behera, A.K.; de Sousa, R.A.; Ingarao, G.; Oleksik, V. Single point incremental forming: An assessment of the progress and technology trends from 2005 to 2015. *J. Manuf. Process.* **2017**, *27*, 37–62. [CrossRef]
9. Lasunon, O.U. Surface roughness in incremental sheet metal forming of AA5052. *Adv. Mater. Res.* **2013**, *753-755*, 203–206. [CrossRef]
10. Hagan, E.; Jeswiet, J. Analysis of surface roughness for parts formed by computer numerical controlled incremental forming. *Proc. Inst. Mech. Eng. Part B J. Eng. Manuf.* **2004**, *218*, 1307–1312. [CrossRef]
11. Aldo, A.; Ceretti, E.; Giardini, C. Optimization of tool path in two points incremental forming. *J. Mater. Process. Technol.* **2006**, *177*, 409–412.
12. Powers, B.M.; Ham, M.; Wilkinson, M.G. Small data set analysis in surface metrology: An investigation using a single point incremental forming case study. *Scanning* **2010**, *32*, 199–211. [CrossRef]
13. Al-Ghamdi, K.A.; Hussain, G. On the free-surface roughness in incremental forming of a sheet metal: A study from the perspective of ISF strain, surface morphology, post-forming properties, and process conditions. *Metals* **2019**, *9*, 553. [CrossRef]
14. Durante, M.; Formisano, A.; Langella, A.; Minutolo, F.M.C. The influence of tool rotation on an incremental forming process. *J. Mater. Process. Technol.* **2009**, *209*, 4621–4626. [CrossRef]
15. Gatea, S.; Ou, H.; McCartney, G. Review on the influence of process parameters in icremental sheet forming. *Int. J. Adv. Manuf. Technol.* **2016**, *87*, 479–499. [CrossRef]
16. Maqbool, F.; Bambach, M. Dominant deformation mechanisms in single point incremental forming (SPIF) and their effect on geometrical accuracy. *Int. J. Mech. Sci.* **2018**, *136*, 279–292. [CrossRef]
17. Maqbool, F.; Bambach, M. Experimental and numerical investigation of the influence of process parameters in incremental sheet metal forming on residual stresses. *J. Manuf. Mater. Process.* **2019**, *3*, 31. [CrossRef]
18. Huber, N.; Heerens, J. On the effect of a general residual stress state on indentation and hardness testing. *Acta Mater.* **2008**, *56*, 6205–6213. [CrossRef]
19. Bambach, M.; Araghi, B.T.; Hirt, G. Strategies to improve the geometric accuracy in asymmetric single point incremental forming. *Prod. Eng.* **2009**, *3*, 145–156. [CrossRef]
20. Radu, C.; Herghelegiu, E.; Tampu, N.C.; Cristea, I. The residual stress state generated by single point incremental forming of aluminum metal sheets. *Appl. Mech. Mater.* **2013**, *371*, 148–152. [CrossRef]
21. Shi, X.; Hussain, G.; Butt, S.I.; Song, F.; Huang, D.; Liu, Y. The state of residual stresses in the Cu/Steel bonded laminates after ISF deformation: An experimental analysis. *J. Manuf. Process.* **2017**, *30*, 14–26. [CrossRef]
22. Kotobi, M.; Honarpisheh, M. Uncertainty analysis of residual stresses measured by slitting method in equal-channel angular rolled Al-1060 strips. *J. Strain Anal. Eng. Des.* **2017**, *52*, 83–92. [CrossRef]
23. Alinaghian, I.; Amini, S.; Honarpisheh, M. Residual stress, tensile strength, and macrostructure investigations on ultrasonic assisted friction stir welding of AA6061-T6. *J. Strain Anal. Eng. Des.* **2018**, *53*, 494–503. [CrossRef]
24. Tanaka, S.; Nakamura, T.; Hayakawa, K.; Nakamura, H.; Motomura, K. Residual stress in sheet metal parts made by incremental forming process. *AIP Conf. Proc.* **2017**, *908*, 775.

25. Radu, C.; Tampu, C.; Cristea, I.; Chirita, B. The effect of residual stresses on the accuracy of parts processed by SPIF. *J. Mater. Manuf. Process.* **2013**, *28*, 572–576. [CrossRef]

26. Maaß, F.; Hahn, M.; Dobecki, M.; Thannhauser, E.; Tekkaya, A.E.; Reimers, W. Influence of tool path strategies on the residual stress development in single point incremental forming. *Procedia Manuf.* **2019**, *29*, 53–58. [CrossRef]

27. Abdulrazaq, M.M.; Gazi, S.K.; Ibraheem, M.Q. Investigation the influence of SPIF parameters on residual stresses for angular surfaces based on ISO-planar tool path. *Al-Khwarizmi Eng. J.* **2019**, *15*, 50–59. [CrossRef]

28. Al-Ghamdi, K.A.; Hussain, G. Stress gradient due to incremental forming of bonded metallic laminates. *J. Mater. Manuf. Process.* **2017**, *32*, 1384–1390. [CrossRef]

29. Alinaghian, M.; Alinaghian, I.; Honarpisheh, M. Residual stress measurement of single point incremental formed Al/Cu bimetal using incremental hole-drilling method. *Int. J. Lightweight Mater. Manuf.* **2019**, *2*, 131–139. [CrossRef]

30. Hajavifard, R.; Maqbool, F.; Schmiedt-Kalenborn, A.; Buhl, J.; Bambach, M.; Walther, F. Integrated forming and surface engineering of disc springs by inducing residual stresses by incremental sheet forming. *Materials* **2019**, *12*, 1646. [CrossRef]

31. López, C.; Elías-Zúñiga, A.; Jiménez, I.; Martínez-Romero, O.R.; Siller, H.; Diabb, J.M. Experimental determination of residual stresses generated by single point incremental forming of AlSi10Mg sheets produced using SLM additive manufacturing process. *Materials* **2018**, *11*, 2542. [CrossRef]

32. Ham, M.; Powers, B.M.; Loiselle, J. Surface topography from single point incremental forming using an acetal tool. *Key Eng. Mater.* **2013**, *549*, 84–91. [CrossRef]

33. Ham, M.; Powers, B.M.; Loiselle, J. Multiscale analysis of surface topography from single point incremental forming using an acetal tool. *J. Phys. Conf. Ser.* **2014**, *483*, 012008. [CrossRef]

34. Li, X.; Liu, Z.; Liang, X. Tool wear, surface topography, and multi-objective optimization of cutting parameters during machining AISI 304 austenitic stainless steel flange. *Metals* **2019**, *9*, 972. [CrossRef]

35. Oraon, M. Surface roughness evaluation of AA 3003 alloy in single point incremental forming technique. *Int. J. Sci. Eng. Res.* **2018**, *9*, 28–34.

36. Behera, A.K.; Ou, H. Effect of stress relieving heat treatment on surface topography and dimensional accuracy of incrementally formed grade 1 titanium sheet parts. *Int. J. Adv. Manuf. Technol.* **2016**, *87*, 3233–3248. [CrossRef]

37. Ambrogio, G.; Filice, L.; Gagliardi, F.; Micari, F. Sheet thinning prediction in single point incremental forming. *Adv. Mater. Res.* **2005**, *6–8*, 479–486. [CrossRef]

38. Shamsari, M.; Mirnia, M.J.; Elyas, M.; Baseri, H. Formability imrpovement in singe point incremental forming of truncated cone using a two-stage hybrid deformation strategy. *Int. J. Adv. Manuf. Technol.* **2018**, *94*, 2357–2368. [CrossRef]

39. Liu, Z.; Li, Y.; Meehan, P.A. Tool path strategies and deformation analysis in multi-pass incremental sheet forming process. *Int. J. Adv. Manuf. Technol.* **2014**, *75*, 395–409. [CrossRef]

40. Said, L.B.; Mars, J.; Wali, M.; Dammak, F. Effects of the tool path strategies on incremental sheet metal forming process. *Mech. Ind.* **2016**, *17*, 411. [CrossRef]

41. Gatea, S.; Xy, D.; Ou, H.; McCartney, G. Evaluation of formability and fracture of pure titanium in incremental sheet forming. *Int. J. Adv. Manuf. Technol.* **2018**, *95*, 625–641. [CrossRef]

42. Kim, Y.H.; Park, J.J. Effect of process parameters on formability in incremental forming of sheet metal. *J. Mater. Process. Technol.* **2002**, *130–131*, 42–46. [CrossRef]

43. Liu, Z.; Liu, S.; Li, Y.; Meehan, P.A. Modeling and optimization of surface roughness in incremental sheet forming using a multi-objective function. *Mater. Manuf. Process.* **2014**, *29*, 808–818. [CrossRef]

44. Oleksik, V.; Pascu, A.; Deac, C.; Fleacă, R.; Bologa, O.; Racz, G. Experimental study on the surface quality of the medical implants obtained by single point incremental forming. *Int. J. Mater. Form.* **2010**, *3*, 935–938. [CrossRef]

45. Radu, C. Analysis of the correlation accuracy-distribution of residual stresses in the case of parts processed by SPIF. Mathematical models and methods in modern science. In Proceedings of the 14th WSEAS International Conference on Mathematical Methods, Computational Techniques and Intelligent Systems, Porto, Portugal, 1–3 July 2012; pp. 195–199.

46. Jiménez, I.; López, C.; Martinez-Romero, O.; Mares, P.; Siller, H.R.; Diabb, J.; Sandoval-Robles, J.A.; Elías-Zúñiga, A. Investigation of residual stress distribution in single point incremental forming of aluminum parts by X-ray diffraction technique. *Int. J. Adv. Manuf. Technol.* **2017**, *91*, 2571–2580. [CrossRef]

 *metals*

Article

Microstructure and Texture Evolution with Relation to Mechanical Properties of Compared Symmetrically and Asymmetrically Cold Rolled Aluminum Alloy

Jakob Kraner [1,2,*], Peter Fajfar [2], Heinz Palkowski [3], Goran Kugler [2], Matjaž Godec [1] and Irena Paulin [1]

[1] Institute of Metals and Technology, Lepi pot 11, SI-1000 Ljubljana, Slovenia; matjaz.godec@imt.si (M.G.); irena.paulin@imt.si (I.P.)
[2] Department of Materials and Metallurgy, Faculty of Natural Sciences and Engineering, University of Ljubljana, Aškerčeva cesta 12, SI-1000 Ljubljana, Slovenia; peter.fajfar@ntf.uni-lj.si (P.F.); goran.kugler@ntf.uni-lj.si (G.K.)
[3] Department of Metal Forming and Processing, Institute of Metallurgy, Faculty of Natural and Materials Science, Clausthal University of Technology, Robert-Koch-Straße 42, DE-38678 Clausthal-Zellerfeld, Germany; heinz.palkowski@tu-clausthal.de
* Correspondence: jakob.kraner@imt.si; Tel.: +386-1-4701-990

Received: 17 December 2019; Accepted: 17 January 2020; Published: 21 January 2020

Abstract: The impact of asymmetric cold rolling was quantitatively assessed for an industrial aluminum alloy AA 5454. The asymmetric rolling resulted in lower rolling forces and higher strains compared to conventional symmetric rolling. In order to demonstrate the positive effect on the mechanical properties with asymmetric rolling, tensile tests, plastic-strain-ratio tests and hardness measurements were conducted. The improvements to the microstructure and the texture were observed with a light and scanning electron microscope; the latter making use of electron-backscatter diffraction. The result of the asymmetric rolling was a much lower planar anisotropy and a more homogeneous metal sheet with finer grains after annealing to the soft condition. The increased isotropy of the deformed and annealed aluminum sheet is a product of the texture heterogeneity and reduced volume fractions of separate texture components.

Keywords: asymmetric rolling; aluminum alloy; planar anisotropy; mechanical properties; microstructures

---

1. Introduction

Rolling is one of the most common metal-forming processes, frequently used for steels and aluminum alloys [1–5]. With their lower density and higher corrosion resistance, aluminum alloys will be used for increasing numbers of components in the automotive industry [6–8]. Of particular interest is the AA 5xxx series, because its mechanical properties have the potential to be improved by asymmetric rolling, endowing it with mechanical properties that are closer to those of steels [9–11].

Asymmetry in the rolling process can be introduced in different ways. Kinetics, geometry and friction are the major parameters that can be influenced to create the asymmetry. This can involve rolling with different roller diameters, rolling with the uneven use of lubricants, rolling with single-drive roller and rolling with different rotation speeds of the rollers. The latter two are more appropriate for rolling mills in the industry [12–17]. Rolling with different rotation speeds, where the so-called "grabbing problems" are less significant than with a single-drive roller, have an impact on changes to the workpiece's thickness as a consequence of the uneven distribution of the longitudinal velocity. Higher contact and shear stresses as well as friction differences on the contact surfaces were introduced while rolling with a speed difference between the upper and lower rollers of 5 m·s^{-1}. A larger difference between the velocity of the upper and lower work roller contributes to a greater reduction of the rolling force [18–22]. At the same time, the speed difference results in the creation of a larger shear deformation area, which appears as a consequence of the changed position of the neutral points in the deformation zone [23,24]. On the other hand, a negative consequence of asymmetric rolling is the bending of the workpiece, more often known as the "ski effect" [25–27].

The asymmetric rolling of aluminum was in most cases investigated for thin sheets with small dimensions [28,29]. This means it would be interesting to study the impact of asymmetry in the rolling process on thicker sheets. In addition to achieving higher strains and creating a more homogeneous microstructure with smaller crystal grains, investigations of asymmetric rolling need to focus on the texture components. It would be particularly useful to know which of the rolling, shear and recrystallization texture components are the reason for the more effective heat treatments and the lower anisotropy [30–35]. For thinner sheets, the Erichsen test is normally used to assess the formability properties. In the case of thicker sheets, the plastic-strain-ratio test can be performed as an alternative. The calculated values of the so-called Lankford factor were in some cases, in previous investigations, much smaller for asymmetrically rolled samples than for symmetrically rolled samples, which means the formability is improved [36–39].

In our investigation the influences of asymmetric rolling in comparison to symmetric rolling were studied with respect to the technological, mechanical and metallographic perspectives. The interactions between the measured rolling forces, the achieved strains, the tensile and yield strength, the hardness, the indicators of planar anisotropy, the microstructures with an average size of grains and crystallographic textures with volume fractions of different rolling, shear and recrystallization texture components were investigated, with the aim to determine appropriate asymmetric rolling conditions for industrial applications.

2. Materials and Methods

2.1. Material and Geometry

The experimental materials were industrially produced via vertical direct chill casting with a subsequent homogenization of the slab, which was further hot rolled and coiled. A sheet with a thickness of 6.7 mm was cut into plates with an entry size of approximately 510 mm × 230 mm. The chemical composition of the AA 5454 aluminum alloy plates (Table 1) was controlled in the laboratory with X-ray fluorescence (XRF) and with inductively coupled plasma, optical emission spectrometry (ICP-OES).

Table 1. Chemical composition of AA 5454 aluminum alloy (wt.%).

Mg	Mn	Fe	Si	Cr	Al
2.43	0.61	0.25	0.18	0.09	Bal.

2.2. Rolling Parameters

The plates were rolled using two different roll gaps, i.e., 4.0 mm and 3.1 mm. With the roll gap of 4.0 mm the imposed strains were around 33%, and with the roll gap of 3.1 mm, around 44%. The rolling asymmetry results from the different rotation speeds of the two identical rollers of diameter 295 mm. In the laboratory experiments the rotation speed of the upper roller was kept constant at 10 rpm, while the rotation speed of the lower roller was set at 10 rpm, 15 rpm and 20 rpm. The corresponding asymmetry factors ($\omega_{lower}/\omega_{upper}$), as a consequence of the speed difference between the upper and lower rollers, were equal to 1.0, 1.5 and 2.0. The asymmetry factor 1.0 represents symmetric rolling, while the 1.5 and 2.0 factors of asymmetry represent two different types of asymmetric rolling. To recreate industrially relevant conditions a special lubricant, Somentor^{TM} 32, was used. Minimally and for each rolling type same quantity of lubricant was added on both rollers, with intention to avoid the adhesion, which is frequent for rolled aluminum. With lubrication at symmetric and asymmetric rolling the friction was reduced and consequently the factors of asymmetry were maintained. When lubrication applied, the factors of asymmetry are expected to be lower than the design value. On the other hand, the cold dry rolling is very rarely used and without lubrication the right response of material will not be reached and comparable between symmetric and asymmetric rolling with industry-similar conditions.

2.3. Mechanical Tests

The mechanical tests and the metallographic analyses were carried out on deformed and heat-treated samples. The heat treatment involved 400 °C for 1 h, which is common in the industry for this alloy. The mechanical properties from the tensile test, in accordance with the ASTM E8 M standard, and the plastic-strain-ratio test, in accordance with the ASTM E517 standard, were evaluated along the rolling direction (0°), the transverse direction (90°) and the diagonal direction (45°). The Brinell hardness was measured on the surfaces on the top and the bottom of the plates and in the centre position of the cross-section.

2.4. Microscopy

Prior to the light microscopy (LM) and the electron-backscatter diffraction (EBSD), the samples were mechanically grinded and polished. The last preparation step was polishing for 10 min with OPS and ion etching 50 min. For the metallographic analyses and the determination of the average size of the grains, in the top, bottom and centre positions of the specimen's cross-section, a ZEISS AXIO Imager M2m microscope (Carl Zeiss AG, Oberkochen, Germany) was used. A JEOL JSM-6500F field-emission scanning electron microscope (JEOL, Tokio, Japan) equipped with an HKL Nordlys II EBSD camera using Channel5 version 5.0.9.0 software (Oxford Instruments HKL, Hobro, Denmark) was used to detect the texture components in the symmetrically and asymmetrically rolled samples. Twelve major texture components in the face-centered cubic (fcc) material were searched for. Four rolling texture components (C, S, B and D), three shear texture components (H, E and F) and five recrystallized texture components (G, R, P, Q and Cube) were investigated and their volume fractions determined. The crystallographic planes and directions, as well as the angles in Euler space, are listed in Table 2. The instrument was operated at 15 kV and a 1.3 nA current for the EBSD analysis, with a tilting angle of 70°. Individual diffraction patterns were obtained together with mapping of the areas of interest. The detection was set to 5–7 bands, with 4×4 binning. For each sample, a map 1235 µm $\times$ 1005 µm, in the cross-section centre and bottom surfaces, which was in contact with the faster roller, was measured with a maximum step size of 5 µm. The volume fraction of the texture components was determined with a 10° deviation.

Table 2. Major texture components in face-centered cubic (fcc) material [30].

Designation	Miller Indices {hkl} <uvw>	Euler Angles (ϕ_1, Φ, ϕ_2)
B	{011} <211>	(35°, 45°, 0°)
S	{123} <634>	(59°, 34°, 65°)
C	{112} <111>	(90°, 35°, 45°)
D	{4411} <11118>	(90°, 27°, 45°)
H	{001} <110>	(0°, 0°, 45°)
E	{111} <110>	(60°, 54.7°, 45°)
F	{111} <112>	(90°, 54.7°, 45°)
Cube	{001} <100>	(0°, 0°, 0°)
G	{110} <001>	(0°, 45°, 0°)
R	{124} <211>	(53°, 36°, 60°)
P	{011} <112>	(65°, 45°, 0°)
Q	{013} <231>	(58°, 18°, 0°)

2.5. Numerical Simulation

The commercial finite element package ABAQUS CAE 2018 (Dassault Systémes, Vélizy-Vallacoublay, France) was used with a model modified from an explicit three-dimensional rolling model. For numerical simulations all initial dimensions of workpieces and rolling mill were the same as at experimental laboratory rolling. The density (2690 kg·m^{-3}), modulus of elasticity (70.5 × 10^9 Pa), Poisson's ratio (0.33) and stress-strain values of initial (hot rolled) material, with the yield stress 151.0 MPa, were set as mechanical properties of specific aluminum alloy. The set time period was 2.5 s and the used friction coefficient were 0.1, 0.15 or 0.2, what is in accordance with the velocity of rollers. Rotation speed of rollers (0.1667, 0.2500 and 0.3333 radians·time^{-1}) and roll gap set (4.0 mm and 3.1 mm) were the same as at laboratory rolling, that way the factors of asymmetry and strain were approximately equal.

3. Results and Discussion

The results are presented as a comparison between symmetric and asymmetric rolling. The technological, mechanical and metallographic findings were examined with respect to the strain and the samples' condition, i.e., deformed or annealed.

3.1. Rolling Force and Ski Effect

An undesirable ski effect only appeared with the asymmetric rolling. Stated is clearly shown with the comparison of numerical simulations for symmetric (Figure 1a) and asymmetric (Figure 1b) rolling. The same phenomena were observed also at laboratory rolling. Comparing simulations and experiments the same functions, effects and phenomena were observed. Asymmetric rolling provided thinner rolled plates at the same roll gap set and that made the rolling process faster than symmetric rolling. With a higher factor of asymmetry exit thicknesses of plates were closer to the set roll gap, what was noted at simulations and laboratory rolling as well.

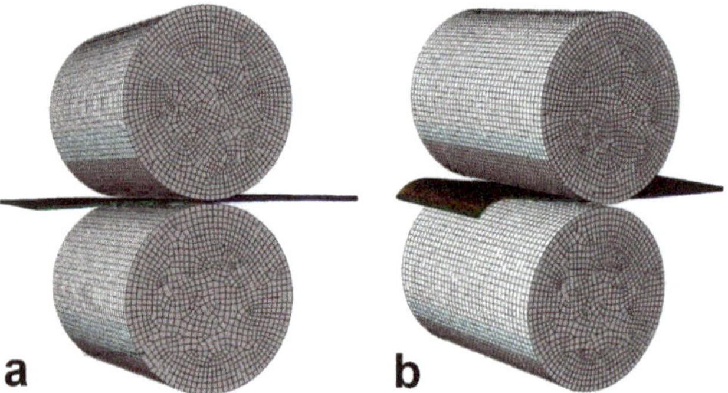

Figure 1. Numerical simulation of symmetric rolling (**a**) and asymmetric rolling (**b**).

Figure 2 shows the influence of the rolling-speed asymmetry factor on the strain of the plates and the rolling force for two set roll gaps of 4.0 mm and 3.1 mm. It is clear that with an increasing asymmetry factor the rolling forces were decreasing with the increasing strain. Explanation for that occurrence is in a much lower mean contact pressure at asymmetric rolling than in the case of symmetric rolling. That phenomena functioned by the denature shape of the deformation zone, which creates favorable circumstances for evolving longitudinal tensile stresses, whose action above the workpiece during the rolling match the previous and subsequent applied stress. Stated extensively reduces the mean pressure exerted on the contact surface of the rolls and as significance also rolling forces [22]. In the case of the symmetric rolling for a set roll gap of 4.0 mm, the rolling force was 1128 ± 3 kN at a 30.6 ± 0.6% strain. Compared to asymmetric rolling for the same set roll gap, the strain was higher and the rolling forces were lower. The obtained rolling force for a 1.5 factor of asymmetry was 1099 ± 2 kN, and for a factor of asymmetry 2.0 it was 1083 ± 7 kN. The differences between the strains were not significant. The achieved strain for the factor of asymmetry 1.5 was 31.4 ± 0.5% and for the factor of asymmetry 2.0 it was 31.5 ± 0.2%. Somewhat smaller differences in the strains for the asymmetric rolling were obtained for the set roll gap of 3.1 mm. In both cases the strain was up to 44.9 ± 0.5%. In spite of the similar strain, a rolling force of 1341 ± 2 kN for the lower factor of asymmetry and 1309 ± 3 kN for the higher factor of asymmetry were measured. Compared to the symmetric rolling, a rolling force of 1377 ± 2 kN and a strain of 42.7 ± 0.1% were obtained. The increase in the strain and the decrease of the rolling force with asymmetric rolling was the case for both set roll gaps. For the higher set roll gap, the difference in the achieved strain between the symmetric and the asymmetric rolling process was around 2.5%. At a lower set roll gap the difference was 5%. According to the presented strains, the decrease of the rolling force in the case of a factor of asymmetry equal to 1.5 was 2.5% and for a factor of asymmetry equal to 2.0 it was 5%. In general, the increases of the achieved strains and the decreases of the rolling forces were not significant, but in the case of more passes, with asymmetric rolling, the desired thickness would be achieved with a smaller number of passes, which has an important economic impact for the rolling mill's function and wear.

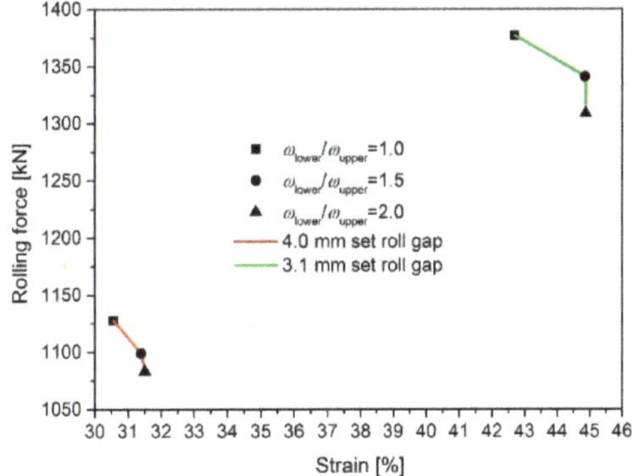

Figure 2. Measured rolling forces and calculated strains according to the two different roll gaps.

The ski effect only appeared on the asymmetrically rolled plates, and it was described using the length and the angle. The length of the ski effect was measured from the bended edge to the place where the plate became straight. The angle of the ski effect was defined as the highest angle of the plates' curvature. Ski effect on rolled plates is presented in Figure 3a. In Figure 3b the schematic presentation of the ski effect length and angle for separate rolling type is shown. Plates with exit dimensions from 684 mm to 785 mm had average values of the bended length from 57.0 to 63.6 mm (±3 mm). The differences between the measured angles of the bending area were small. At the same time, it should be noted that the factor of asymmetry had a higher impact on the formation of greater ski effect than the strain. Larger angles of 20.3 ± 0.5° and 21 ± 0°, as well as longer lengths of 60.3 ± 0.6 mm and 63.6 ± 1 mm, were measured on the plates rolled with a factor of asymmetry equal to 2.0. Shorter lengths of the ski effect equal to 57.0 ± 3 mm and 57.6 ± 0.6 mm were measured at a factor of asymmetry equal to 1.5. For this factor of asymmetry, the observed angles were between 18.3 ± 0.6° and 19 ± 0°. Asymmetric rolling with a faster lower roller results in a down-orientated ski effect. This can result in damage to the rolling table at the exit of the laboratory rolling mill or the transport rollers on industrial rolling mills, if in such cases asymmetric rolling is even possible. The ski effect represents a bent, useless area of rolled workpiece and must be cut off or made flat with mechanical intervention. The length of the ski effect, depending on selected asymmetric rolling type, will be significantly influenced by the vertical space between the deformation zone and the rolling table. The smaller the height between the workpiece on the exit and rolling table, the faster the bending process will finish and the workpiece with a limited ski effect will slide forward on the rolling table. In the cases of asymmetric rolling from 7 ± 1% to 9 ± 1% of the ski effect according to the whole rolled plate were calculated. Comparing the symmetric and asymmetric plates, where, with the asymmetric rolling, between 2% and 5% more useless material was produced. On the other hand, it is necessary to emphasize that besides a larger percentage of a bent workpiece, the plates were longer because higher strains were reached with asymmetric rolling and the material will have better properties

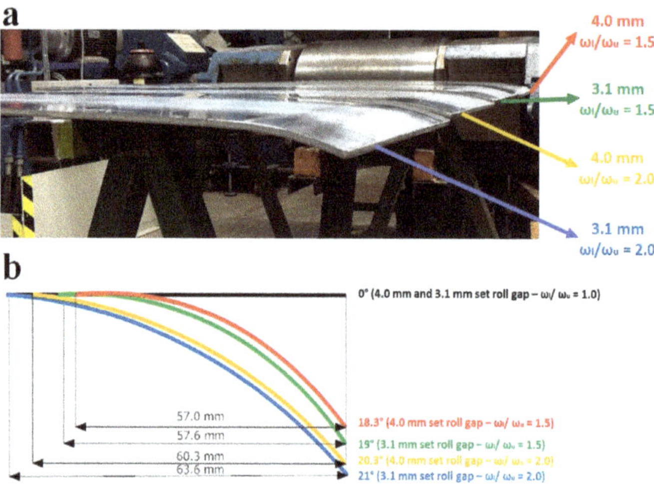

Figure 3. Ski effect on rolled plates (**a**) and schematic presentation of ski effect with measured lengths and angles (**b**).

3.2. Tensile and Yield Strength

With both set roll gaps the highest tensile-strength values were in the transverse direction. At the same time the tensile-strength values in the rolling direction with both set roll gaps were the lowest. For the 4.0 mm set roll gap the tensile-strength values were 279 ± 5 MPa for an asymmetry factor of 1.0 and 280 ± 5 MPa for asymmetry factors of 1.5 and 2.0. At a 3.1 mm set roll gap the average tensile strength values were higher. A factor of asymmetry of 1.0 brought us a tensile strength of 290 ± 6 MPa, with a factor of asymmetry of 1.5 the average value was 293 ± 6 MPa and for a factor of asymmetry 2.0 the tensile strength was 291 ± 7 MPa. Regarding the heat-treated samples for both set roll gaps, the highest values of the tensile strength were in the rolling direction. At the same time the lowest tensile-strength values after the heat treatment were observed in the diagonal direction. The average tensile-strength values after the heat treatment were similar for both set roll gaps and all three directions. Measured tensile strength was around 220 ± 3 MPa. In both cases of the roll-gap settings the highest values of the yield strength were in the rolling direction. The values for the factor of asymmetry equal to 2.0 were for both set roll gaps the highest and reached 235 ± 8 MPa with a set roll gap of 4.0 mm and 253 ± 1 MPa with a set roll gap of 3.1 mm. All the other yield-strength values for the deformed samples were for a higher set roll gap between 216 ± 6 MPa and 234 ± 5MPa and for a lower set roll gap between 244 ± 7 MPa and 251 ± 8 MPa. The highest yield-strength values were for a set roll gap of 3.1 mm with the heat-treated samples in the diagonal direction. All the values of the yield strength for heat-treated samples and all rolling types were lower and relatively even for both set roll gaps, with small deviations between 86 ± 0 MPa and 91 ± 1 MPa. The symmetrically and asymmetrically achieved strengths before and after heat treatment were very similar, with just smaller increases in both properties with asymmetric rolling. On the other hand, the tensile and yield strengths with asymmetric rolling stayed in the range of created strengths with symmetric rolling.

3.3. Anisotropy

For thicker sheets, the Erichsen test for deep-drawing properties could not be performed. For that reason the alternative plastic-strain-ratio test was conducted. The planar anisotropy indicator Δr was calculated from the Lankford factor r. The values of Δr could be positive or negative, depending on the relationship between the elongation and width increase of the tested sample. More important is that the Δr values were close to 0 [38]. For all the rolling types and for both set heights of the roll gap, the Δr values were negative (Table 3). The deformed sample's Δr values for both set roll gaps were, in the case of asymmetric rolling, higher in the negative section than with symmetric rolling. Higher negative values of Δr with the asymmetric rolling also mean a higher planar anisotropy, because the total isotropy of the material appears when Δr is 0. Comparing the Δr values of the deformed and heat-treated samples there was a strong contrast. The Δr values for the deformed samples were decreasing with an increasing factor of asymmetry, and the values of Δr for the heat-treated samples were increasing with an increasing factor of asymmetry. After the heat treatment Δr was -0.237 ± 0 for a factor of asymmetry equal to 1.0 at a 4.0 mm set roll gap. This was a more negative Δr value than for the same sample in the just-deformed condition. The same phenomenon was observed with a factor of asymmetry equal to 1.5 and a larger set roll gap, where the Δr after heat treatment was -0.206 ± 0.01. For a factor of asymmetry equal to 2.0 Δr was -0.169 ± 0.02, which was less negative than the Δr of the same, just-deformed sample. A more significant negative decrease in the values was visible with a smaller set roll gap. Of particular note is that for factors of asymmetry equal to 1.5 and 2.0 the Δr values were -0.052 ± 0 and -0.050 ± 0. This is a very good indicator of the very low anisotropy for the asymmetric deformed samples after heat treatment, knowing that the Δr after heat treatment with the same set roll gap and factor of asymmetry equal to 1.0 was -0.112 ± 0.02. Regarding the higher Δr values of the asymmetrically rolled samples in the deformed condition, after the heat treatment the asymmetrically rolled samples in all cases had a lower Δr than the symmetrically rolled samples. This phenomenon could be attributed to a more appropriate texture for the successful heat treatment [29].

Table 3. Planar anisotropy indicator Δr for deformed and heat-treated samples.

Roll Gap Set	$\omega_{lower}/\omega_{upper}$	Δr Deformed Condition	Δr Heat-Treated Condition
	1.0	-0.145 ± 0.05	-0.237 ± 0
4.0 mm	1.5	-0.197 ± 0.05	-0.206 ± 0.01
	2.0	-0.276 ± 0.02	-0.169 ± 0.02
	1.0	-0.134 ± 0.01	-0.112 ± 0.02
3.1 mm	1.5	-0.157 ± 0.02	-0.052 ± 0
	2.0	-0.155 ± 0.01	-0.050 ± 0

3.4. Hardness

Hardness values for all rolling types and both conditions are presented in Table 4. The highest hardness values were obtained in the centre for all the rolling types. The values for the deformed condition with the larger set roll gap and factor of asymmetry equal to 1.0 were somehow higher than in the cases with a factor of asymmetry equal to 1.5 and 2.0. What is more important is that the difference between the surfaces hardness for the asymmetric types was smaller. The same phenomenon was observed with deformed samples with a set roll gap of 3.1 mm. The differences in the surface's hardness decreased in the case of the set roll gap with a higher factor of asymmetry. All the measured hardness values at the top, bottom and centre for heat-treated samples were more evenly spread than with the deformed samples. The differences in the surface hardness were smaller for the heat-treated samples than for the deformed samples. However, at the same time, it is clear that the hardness differences were smaller after the heat treatment of the asymmetrically rolled samples than for the symmetrically rolled samples.

Table 4. Hardness for all rolling types in deformed and heat-treated condition.

Roll gap Set (mm; Factor of Asymmetry)	Measuring Position	Deformed Conditio (HB)	Heat-Treated Condition (HB)
4.0 (1.0)	Top	86 ± 0.5	57 ± 2
	Centre	90 ± 2	56 ± 3
	Bottom	83 ± 1	57 ± 2
4.0 (1.5)	Top	83 ± 0.5	57 ± 1
	Centre	88 ± 1	57 ± 1
	Bottom	82 ± 3	56 ± 1
4.0 (2.0)	Top	83 ± 1	57 ± 0.5
	Centre	88 ± 2	56 ± 0.5
	Bottom	80 ± 3	58 ± 1
3.1 (1.0)	Top	84 ± 1	56 ± 2
	Centre	92 ± 0.5	58 ± 1
	Bottom	85 ± 2	57 ± 0.5
3.1 (1.5)	Top	87 ± 2	58 ± 3
	Centre	92 ± 1	56 ± 1
	Bottom	86 ± 2	58 ± 3
3.1 (2.0)	Top	86 ± 1	56 ± 1
	Centre	92 ± 1	56 ± 1
	Bottom	86 ± 0.5	56 ± 0.5

3.5. Microstructure

Comparing the microstructures of the deformed samples (Figure 4a–c) with the microstructures of the heat-treated samples (Figure 4d–f), some differences were observed. Since higher strains were achieved with the asymmetric rolling, the average size of the grains, as a consequence, was smaller. The centre band of the longitudinally deformed grains was for the asymmetrically rolled microstructure smaller and it was decreased with a higher factor of asymmetry. Moreover, it is important that the grains in the asymmetrically deformed samples are, in all three measured sections, more evenly distributed than for the symmetrically rolled samples. The results for the higher and lower set roll gaps are presented in Figure 5a,b. In the deformed samples the grain sizes are, for a set roll gap of 4.0 mm, between 26.4 ± 0.7 μm and 17.7 ± 0.5 μm, and for a set roll gap of 3.1 mm, between 23.3 ± 0.3 μm and 17.5 ± 0.6 μm. The grains in the centre position of the cross-section are, for all three different factors of asymmetry, larger than those on the top or bottom position. The difference between the grain size on the top and bottom positions is very small, but at the same time the smaller grains were obtained at the bottom position, which can be the impact of the higher velocity of the lower roller. This is valid for the deformed samples as well as for heat-treated samples. After the heat treatment, there is no major difference in the distribution of the microstructural components for all three sections. The grains in the microstructure are very even, especially with the asymmetrically rolled samples. Some impacts of the asymmetry remained after the treatment to the soft condition. For the heat-treated samples the grain sizes were between 33.9 ± 1 μm and 25.3 ± 0.7 μm for the set roll gap of 4.0 mm and from 25.0 ± 0.5 μm to 19.3 ± 0.6 μm for the set roll gap of 3.1 mm at the top, centre and bottom positions of the samples' cross sections. The presented average grain sizes for the heat-treated samples are shown in Figure 5c,d. A major advantage of the asymmetric rolling is the creation of more homogeneous material, in our case, confirmed with the average size of the grains and the hardness throughout the cross-section. More homogeneous material, i.e., the material with the more evenly distributed properties throughout the cross-section, is better for further forming processes, heat treatment and quality of the products [30,31].

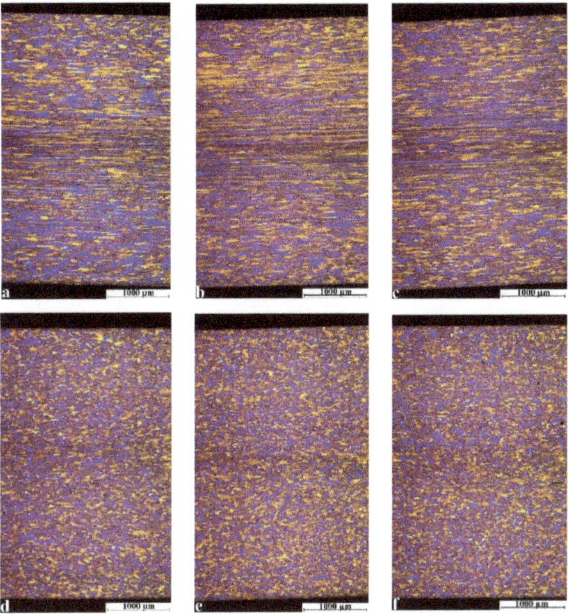

Figure 4. Microstructure in samples' cross-sections for symmetrically rolled 1.0 (**a**), asymmetrically rolled 1.5 (**b**), asymmetrically rolled 2.0 (**c**), symmetrically rolled 1.0 and heat treated (**d**), asymmetrically rolled 1. 5 and heat treated (**e**) and asymmetrically rolled 2.0 and heat treated (**f**).

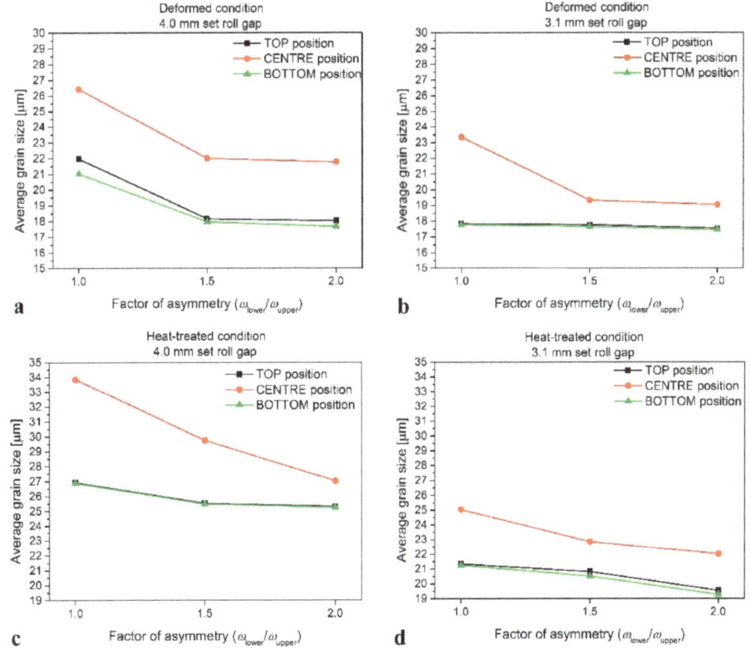

Figure 5. Average grain size for deformed condition with 4.0 mm set roll gap (**a**), deformed condition with 3.1 mm set roll gap (**b**), heat treated with 4.0 mm set roll gap (**c**) and heat treated with 3.1 mm set roll gap (**d**).

3.6. Texture

The texture components were detected with EBSD for the deformed and heat-treated samples. At the same position and for the same asymmetric rolling type, the deformed condition, with its strongly pronounced texture pattern and higher intensity of the pole figure, and heat-treated condition, with an even distribution of all the texture components and a very low intensity of the pole figure were observed. The EBSD mapping confirms the refinement of the crystal grains with asymmetric rolling, were with the higher factor of asymmetry the smaller grains were produced. The associated pole figures in Figure 6a–c present a lower intensity and that way also the higher heterogeneity of texture created with the asymmetric rolling. Observing the central position of the lower deformed and heat-treated sample, the texture component E at a factor of asymmetry equal to 1.0 and a texture component R at a factor of asymmetry equal to 2.0 stand out. All the other observed texture components for all three factors of asymmetry were from 0.0 to 2.5 vol.%. These results are presented in Figure 7a. The same figure shows that the volume fraction of all the rolling texture components in the case of asymmetric rolling is higher than for the symmetric rolling type. At the same time, a smaller difference in the volume fractions was observed between the separate texture components in the asymmetric rolling textures. The same phenomenon appears in Figure 7b, where the centre positions of the more deformed samples are presented. In the sample that was deformed with a factor of asymmetry equal to 1.0 and heat-treated higher volume fractions for all the rolling texture components were obtained in comparison to the asymmetric rolling types with factors of asymmetry equal to 1.5 and 2.0. Moreover, in the symmetrically rolled sample, the separate texture components were more obvious. This indicates that the volume fractions of all the texture components with asymmetric rolling were more even than for the symmetric rolling. In general, the volume fractions of all the observed texture components were, for the asymmetrically rolled and heat-treated samples with a set roll gap of 3.1 mm, between 0.1 and 1.8 vol.%. Closer to the faster roller, i.e., the bottom position of the sample, were with a set roll gap of 4.0 mm in a symmetrically rolled sample, there was the volume fraction of texture component E, with 6.3 vol.% and texture component G, with 3.9 vol.%. The highest volume fraction of the rolling texture components was, in Figure 7c, observed for a factor of asymmetry equal to 2.0. In Figure 7d, where the bottom section of the more symmetrically deformed and heat-treated sample was presented, the highest volume fraction of 3.1 vol.% was observed for the texture component R. The higher volume fractions of the rolling and shear texture components were obtained in this sample with a factor of asymmetry equal to 1.5. All the compared textures had a very even distribution of texture components, but at the same time, because the symmetrically rolled sample had more stood out recrystallization texture components, and the volume fractions of all the textures were, for the asymmetrically rolled samples with a factor of asymmetry equal to 1.5 and 2.0, significantly equivalent. For comparison of the observed texture components and the texture evolution the deformation history and the initial textures of the material are important [35]. The fibers as texture elements connected to different texture components must be exposed as a possible preference for successful heat treatment [31]. Nevertheless, a comparison between asymmetrically and symmetrically rolled textures is useful and can explain some changes in the mechanical properties. This is in addition to the detection of the C, S and B texture components, which could indicate the formation of β fibers as major texture element for highly deformed rolling materials with fcc.

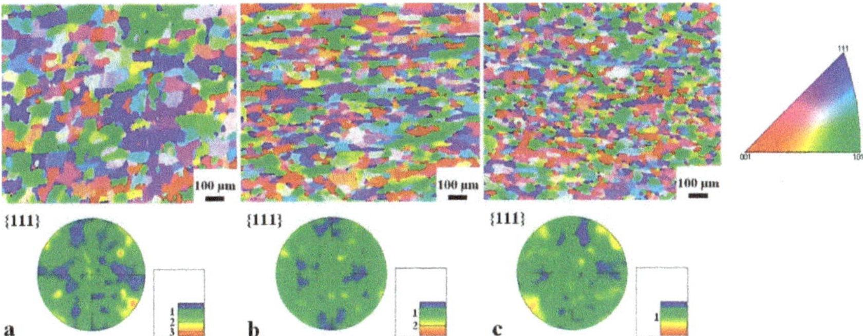

Figure 6. EBSD mapping (IPF Z) and pole figures for rolled and heat-treated samples in centre position for symmetrically rolled 1.0 (**a**), asymmetrically rolled 1.5 (**b**) and asymmetrically rolled 2.0 (**c**).

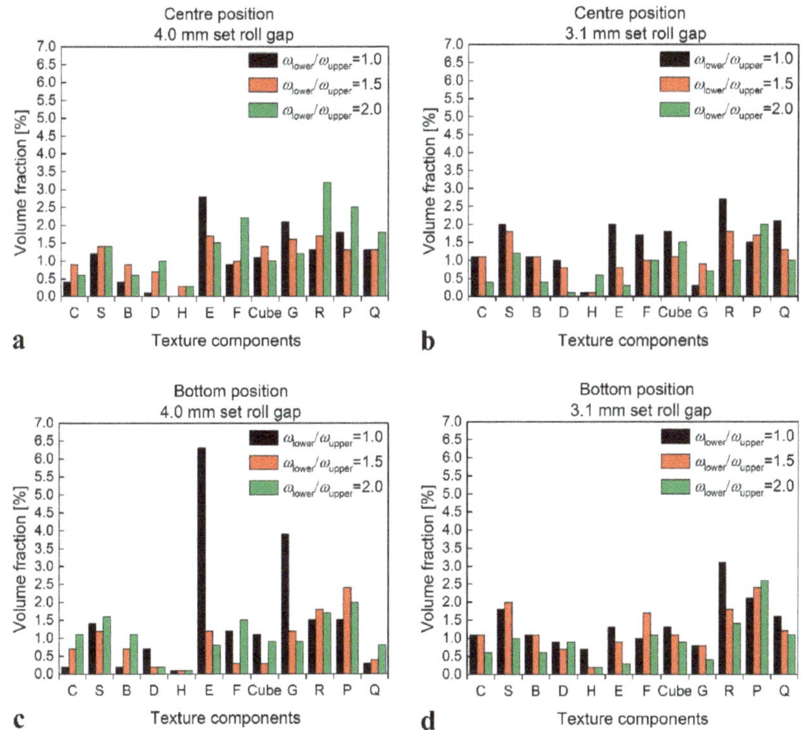

Figure 7. Volume fraction of texture components in heat-treated samples for 4.0 mm set roll gap and centre position (**a**), 3.1 mm set roll gap and centre position (**b**), 4.0 mm set roll gap and bottom position (**c**) and 3.1 mm set roll gap and centre position (**d**).

4. Conclusions

The asymmetric cold rolling of AA 5454 aluminum alloy plates was compared with symmetric rolling. Due to the uneven distribution of the upper and lower roller pressures and the workpiece velocity, the mechanisms in the deformation zone were more suitable for producing smaller crystal grains, which has an impact on improving most mechanical properties. The main advantage of asymmetric rolling is the presence of significant shear strain, leading to a gradient structure. At the same time, the asymmetric rolling enables the creation of more shear texture components, in addition to the rolling and recrystallized texture components. The texture heterogeneity presented with more different texture components and less stands out volume fraction of a separate texture component will have lower planar anisotropy.

The asymmetric conditions were created by using different roller speeds. The general aim was to determine the impact of the asymmetry on the rolling process for a specific alloy. At the same time, metallographic analyses were made to explain some changes to the technological and mechanical properties. For asymmetric rolling, a lower rolling force was needed to create higher strains at the same set roll gap. Thickness reductions as well as strains increased and the rolling forces were decreased with a higher factor of asymmetry. The lengths and the angles of the ski effects were very similar, nevertheless, it was clear that the ski effect depends much more on the factor of asymmetry than the strain. The highest values of the tensile and yield strengths were not present in the deformed and heat-treated condition in the same direction. The elongations after the heat treatment were quite similar, which indicates an effective heat treatment.

The planar anisotropy in the deformed condition of the asymmetrically rolled samples was higher than with the symmetrically rolled samples. The elimination of the planar anisotropy after the heat treatment was larger in the asymmetrically rolled samples. The production of a more homogeneous material is visible in the samples with the more evenly distributed hardness throughout the cross-section, with smaller deviations in the case of the asymmetrically rolled samples than with the symmetrically rolled samples. The impact of the asymmetry is through the hardness of the material, also shown after the heat treatment. In accordance with the homogeneity of the hardness is the average size of the grains, where with a higher factor of asymmetry grains decreased. The grain size deviations throughout the cross-section were also lower with the asymmetrically rolled samples. Different crystallographic textures were detected at the central position, as well as at the bottom position of the same samples. The volume fractions of the detected texture components were, for the asymmetrically rolled samples, more evenly distributed than for the symmetrically rolled samples, where more separate texture components stand out.

Author Contributions: Conceptualization, J.K. and P.F.; methodology, J.K., P.F. and H.P.; software, J.K., M.G. and I.P.; validation, J.K., P.F., H.P., M.G. and I.P.; formal analysis, J.K., H.P., M.G. and I.P.; investigation, J.K., P.F. and H.P.; data curation, G.K.; writing—original draft preparation, J.K., P.F. and I.P.; writing—review and editing, H.P., M.G. and G.K.; supervision, P.F., H.P., M.G. and I.P.; funding acquisition, H.P., G.K. and M.G. All authors have read and agreed to the published version of the manuscript.

Funding: The authors acknowledge the financial support from the Slovenian Research Agency, research corefundings No. P2-0132 and No. P2-0344. This work was also funded by the program MARTINA, supported by Slovenian Ministry of education, science and sport and European Regional Development Fund.

Conflicts of Interest: The authors declare no conflict of interest.

References

1. Minton, J.J.; Cawthorn, C.J.; Brambley, E.J. Asymptotic analysis of asymmetric thin sheet rolling. *Int. J. Mech. Sci.* **2016**, *113*, 36–48. [CrossRef]

2. Xie, H.B.; Manabe, K.; Jiang, Z.Y. A novel approach to investigate surface roughness evolution in asymmetric rolling based on three dimensional real surface. *Finite Elem. Anal. Des.* **2013**, *74*, 1–8. [CrossRef]

3. Dutta, S.; Kaiser, M.S. Effect of asymmetric rolling on formability of pure aluminium. *J. Mech. Eng.* **2014**, *44*, 94–99. [CrossRef]

4. Mekhtiev, A.D.; Azbanbayev, E.M.; Isagulov, A.Z.; Karipbayeva, A.R.; Kvon, S.S.; Zakariya, N.B.; Yermaganbetov, N.Z. Effect of asymmetric rolling with cone-shaped rolls on microstructure of low-carbon steel. *Metalurgija* **2015**, *54*, 623–626.

5. Nilsson, A.; Salvator, I.; Putz, P.D. *Using Asymmetrical Rolling for Increased Production and Improved Material Properties*; Report of Research Programme of the Research Fund for Coal and Steel; European Union: Brussel, Belgium, 2012.

6. Pesin, A.; Pustovoytov, D. Heat transfer modeling in asymmetrical sheet rolling of aluminium alloys with ultra high shear strain. In Proceedings of the MATEC Web of Conferences, Troyes, France, 4–7 July 2016; Volume 80, p. 4005.

7. Bintu, A.; Vincze, G.; Picu, R.C.; Lopes, A.B. Effect of symmetric and asymmetric rolling on the mechanical properties of AA5182. *Mater. Des.* **2016**, *100*, 151–156. [CrossRef]

8. Shore, D.; Kestens, L.A.I.; Sidor, J.; Van Houtte, P.; Van Bael, A. Process parameter influence on texture heterogeneity in asymmetric rolling of aluminium sheet alloys. *Int. J. Mater. Form.* **2018**, *11*, 297–309. [CrossRef]

9. Kawałek, A. The analysis of the asymmetric plate rolling process. *J. Arch. Mater. Manuf. Eng.* **2007**, *23*, 63–66.

10. Inoue, H. Texture Control of Aluminum and Magnesium Alloys by the Symmetric/Asymmetric Combination Rolling Process. *Mater. Sci. Forum* **2012**, *702*, 68–75. [CrossRef]

11. Tamimi, S.; Correia, J.P.; Lopes, A.B.; Ahzi, S.; Barlat, F.; Gracio, J.J. Asymmetric rolling of thin AA-5182 sheets: Modelling and experiments. *Mater. Sci. Eng. A* **2014**, *603*, 150–159. [CrossRef]

12. Hwang, Y.; Tzou, G. Analytical and Experimental Study on Asymmetrical Sheet Rolling. *Int. J. Mech. Sci.* **1997**, *39*, 289–303. [CrossRef]

13. Nikolaev, V.A.; Vasilyev, A.A. Analysis of strip asymmetrical cold rolling parameters. *Metall. Min. Ind.* **2010**, *2*, 405–412.

14. Kasai, D.; Komori, A.; Ishii, A.; Yamada, K.; Ogawa, S. Strip Warpage Behavior and Mechanism in Single Roll Driven Rolling. *ISIJ Int.* **2016**, *56*, 1815–1824. [CrossRef]

15. Polkowski, W. Differential Speed Rolling: A New Method for a Fabrication of Metallic Sheets with Enhanced Mechanical Properties. In *Progress in Metallic Alloys*; INTECH: London, UK, 2016.

16. Pesin, A.; Pustovoytov, D.; Sverdlik, M. Influence of Different Asymmetric Rolling Processes on Shear Strain. *Int. J. Chem. Nucl. Metall. Mater. Eng.* **2014**, *8*, 422–424.

17. Ko, Y.G.; Hamad, K. Structural features and mechanical properties of AZ31 Mg alloy warm-deformed by differential speed rolling. *J. Alloys Compd.* **2018**, *744*, 96–103. [CrossRef]

18. Liu, J.; Kawalla, R. Influence of asymmetric hot rolling on microstructure and rolling force with austenitic steel. *Trans. Nonferrous Met. Soc. China (Engl. Ed.)* **2012**, *22*, s504–s511. [CrossRef]

19. Richelsen, A.B. Elastic—Plastic analysis of the stress and strain distributions in asymmetric rolling. *Int. J. Mech. Sci.* **1997**, *39*, 1199–1211. [CrossRef]

20. Bakshi, S.D.; Leiro, A.; Prakash, B.; Bhadeshia, H. Dry rolling/sliding wear of nanostructured bainite. *Wear* **2014**, *316*, 70–78. [CrossRef]

21. Dyja, H.; Korczak, P.; Pilarczyk, J.W.; Grzybowski, J. Theoretical and experimental analysis of plates asymmetric rolling. *J. Mater. Process. Technol.* **1994**, *45*, 167–172. [CrossRef]

22. Alexa, V.; Rațiu, S.; Kiss, I. Metal rolling-Asymmetrical rolling process. In Proceedings of the IOP Conference Series, Materials Science and Engineering, Jinan, China, 30–31 August 2016.

23. Zhang, S.H.; Zhao, D.W.; Gao, C.R.; Wang, G.D. Analysis of asymmetrical sheet rolling by slab method. *Int. J. Mech. Sci.* **2012**, *65*, 168–176. [CrossRef]

24. Gudur, P.P.; Salunkhe, M.A.; Dixit, U.S. A theoretical study on the application of asymmetric rolling for the estimation of friction. *Int. J. Mech. Sci.* **2008**, *50*, 315–327. [CrossRef]

25. Wei, G.; Pang, Y.H.; Liu, C.R.; Yu, H.F.; Bin, L.U.; Zhang, M.M. Effect of Asymmetric Friction on Front End Curvature in Plate and Sheet Rolling Process. *J. Iron Steel Res. Int.* **2010**, *17*, 22–26.

26. Philipp, M.; Schwenzfeier, W.; Fischer, F.D.; Wödlinger, R.; Fischer, C. Front end bending in plate rolling influenced by circumferential speed mismatch and geometry. *J. Mater. Process. Technol.* **2007**, *184*, 224–232. [CrossRef]

27. Anders, D.; Münker, T.; Artel, J.; Weinberg, K. A dimensional analysis of front-end bending in plate rolling applications. *J. Mater. Process. Technol.* **2012**, *212*, 1387–1398. [CrossRef]

28. Aljabri, A.; Jiang, Z.Y.; Wei, D.B. Analysis of thin strip profile during asymmetrical cold rolling with roll crossing and shifting mill. *Adv. Mater. Res.* **2014**, *894*, 212–216.

29. Liu, X.; Liu, X.H.; Song, M.; Sun, X.K.; Liu, L.Z. Theoretical analysis of minimum metal foil thickness achievable by asymmetric rolling with fixed identical roll diameters. *Trans. Nonferrous Met. Soc. China (Engl. Ed.)* **2016**, *26*, 501–507. [CrossRef]

30. Sidor, J.; Miroux, A.; Petrov, R.; Kestens, L. Microstructural and crystallographic aspects of conventional and asymmetric rolling processes. *Acta Mater.* **2008**, *56*, 2495–2507. [CrossRef]

31. Wronski, S.; Bacroix, B. Microstructure evolution and grain refinement in asymmetrically rolled aluminium. *Acta Mater.* **2014**, *76*, 404–412. [CrossRef]

32. Hu, H. Texture of metals. *Text. Stress Microstruct.* **1974**, *1*, 233–258. [CrossRef]

33. Kang, S.B.; Min, B.K.; Kim, H.W.; Wilkinson, D.S.; Kang, J. Effect of asymmetric rolling on the texture and mechanical properties of AA6111-aluminum sheet. *Metall. Mater. Trans. A Phys. Metall. Mater. Sci.* **2005**, *36*, 3141–3149. [CrossRef]

34. Han, J.H.; Suh, J.Y.; Oh, K.H.; Lee, J.C. Effects of the deformation history and the initial textures on the texture evolution in an al alloy strip during the shear deforming process. *Acta Mater.* **2004**, *52*, 4907–4918. [CrossRef]

35. Goli, F.; Jamaati, R. Asymmetric cross rolling (ACR): A novel technique for enhancement of Goss/Brass texture ratio in Al-Cu-Mg alloy. *Mater. Charact.* **2018**, *142*, 352–364. [CrossRef]

36. Kim, K.H.; Lee, D.N.; Choi, C.H. The deformation textures and Lankford values of asymmetrically rolled aluminum alloy sheets. In Proceedings of the ICOTOM, Montreal, QC, Canada, 9–13 August 1999; Volume 12, pp. 267–272.

37. Cheon, B.H.; Kim, H.W.; Lee, J.C. Asymmetric rolling of strip-cast Al-5.5Mg-0.3Cu alloy sheet: Effects on the formability and mechanical properties. *Mater. Sci. Eng. A* **2011**, *528*, 5223–5227. [CrossRef]

38. Jin, H.; Lloyd, D.J. The reduction of planar anisotropy by texture modification through asymmetric rolling and annealing in AA5754. *Mater. Sci. Eng. A* **2005**, *399*, 358–367. [CrossRef]

39. Lee, J.K.; Lee, D.N. Texture control and grain refinement of AA1050 Al alloy sheets by asymmetric rolling. *Int. J. Mech. Sci.* **2008**, *50*, 869–887. [CrossRef]

 *metals*

Article

A 3D FEM-Based Numerical Analysis of the Sheet Metal Strip Flowing Through Drawbead Simulator

Tomasz Trzepiecinski [1,*] and Romuald Fejkiel [2]

[1] Department of Materials Forming and Processing, Rzeszow University of Technology, al. Powst. Warszawy 8, 35-959 Rzeszow, Poland

[2] State School of Higher Vocational Education, Rynek 1, 38-400 Krosno, Poland; rfejkiel@wp.pl

[*] Correspondence: tomtrz@prz.edu.pl; Tel.: +48-17-743-2527

Received: 29 November 2019; Accepted: 23 December 2019; Published: 25 December 2019

Abstract: Drawbeads are elements of the stamping die and they are used to compensate material flow resistance around the perimeter of the drawpiece or to change the stress state in specific regions of the drawpiece. This paper presents the results of experimental and numerical analyses of tests of sheet metal flowing through a drawbead. The tests have been carried out using a special tribological simulator of the drawbead. Experimental tests to determine the coefficient of friction (COF) have been carried out for three widths of sheet metal strip and two drawbead heights. The three-dimensional (3D) elastic-plastic numerical computations were performed using the MSC. Marc program. Special attention was given to the effect of material flow through the drawbead on the distribution of the normal stress on the tool-sheet interface. The mesh sensitivity analysis based on the value of the drawing force of the specimen being pulled through the drawbead allowed an optimal mesh size to be determined. The errors between the numerically predicted values of the COF and the values experimentally determined ranged from about 0.95% to 7.1% in the cases analysed. In the case of a drawbead height of 12 mm, the numerical model overestimated the value of the COF for all specimen widths analysed. By contrast, in the case of a drawbead height of 18 mm, all experimentally determined friction coefficients are underestimated by Finite Element Method (FEM). This was explained by the different character of sheet deformation under friction and frictionless conditions. An increase in the drawbead height, with the same sheet width, increases the value of the COF.

Keywords: drawbead; FEM; friction; material properties; numerical modeling; mechanical engineering; sheet metal forming; stamping process

---

1. Introduction

Drawbeads are used to compensate material flow resistance around the perimeter of the drawpiece or to change the stress state in specific regions of the drawpiece [1,2]. Typically, drawbeads are located between the die and blankholder surfaces. The quality of deep drawn parts largely depends on controlling the movement of the material into the die hole. For this purpose, a blankholder or drawbeads are used. When using the blankholder, the increased resistance of the material flow in the flange of the drawpiece is achieved by friction. When a greater flow resistance is required, especially locally in certain areas of the flange of the drawpiece, drawbeads are used. Then, the increased resistance in the flange results both from friction and from cyclic plastic deformation of the sheet metal [3,4] and drawbead geometry imposes a change in the curvature of the sheet. The most widely used geometrical form of a drawbead cross section is circular (Figure 1).

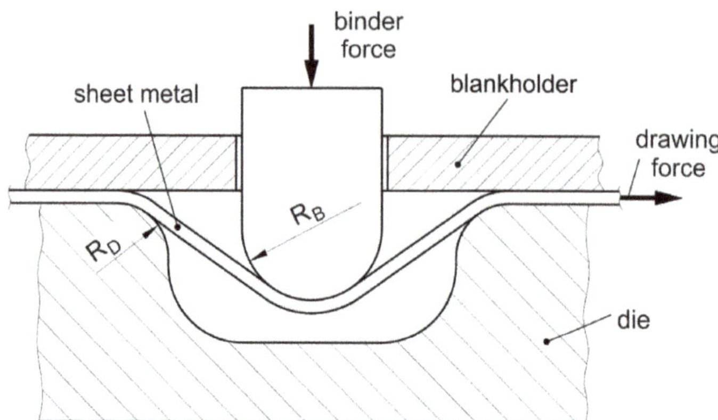

Figure 1. Cross-section of a typical drawbead with circular form.

The tribological and mechanical properties of the sheet metal and friction processes, i.e., the lubrication conditions and the dynamics of contact, play a decisive role in determining the quality of the parts formed [5–8]. To investigate the behaviour of the sheet while flowing through the drawbead, the drawbead test (DBT) is commonly attributed to friction modelling at the drawbead in sheet metal forming (SMF). A method of friction determination using the DBT has been developed by Nine [9]. Nine's drawbead simulator mainly tests the friction condition between the blankholder and the die, which fails to reflect the change of coefficient of friction (COF) in each region caused by different stress modes. Several authors have conducted research to better understand the frictional phenomena in sheet metal forming [10–13] and mechanics of sheet deformation at the drawbeads [14–16] and over the years, analytical, semi-empirical and numerical approaches have been developed.

Lee et al. [17] developed numerical procedures to predict the drawbead restraining force (DBRF) using a semi-analytical method. The effect of process parameters and material properties including the shape of the yield surface is investigated. Xie [18] studied the effect of the geometric parameters influencing the semi-circular drawbead force. The inverse model of the drawbead geometric parameters has been established based on the neural network and genetic algorithm. This enabled a nonlinear relationship between drawbead geometry and forming quality to be obtained. Zhang et al. [19] proposed a new method for automatic optimisation of drawbead geometry in SMF based on the iterative learning control model. The process improves the level of automation and the efficiency of drawbead design and decreases the dependency on engineering experience. Halkaci et al. [20] used a shallow drawbead to reduce the wrinkling tendency in hydromechanical deep drawing. A shallow drawbead added to the blankholder increased the limiting drawing ratio of aluminium alloy sheet.

Experimental testing of the material flowing through the drawbead using friction simulators permits not only the determination of the DBRF but also the evaluation of the value of the COF. Due to a strong effect of the type of material, coating, lubricant and surface topography of both the workpiece and the tool, etc. on the value of frictional resistances determined using DBS, there are no universal models of friction. Thus, the experimental testing of draw beads is still necessary. Bassoli et al. [21] proposed a versatile draw bead simulator to measure the restraining force exerted on sheet metal by the draw bead during deep drawing. It was found that the restraining force was slightly augmented when the sheet metal thickness was increased. As a consequence of cyclic bending and unbending of the sheet metal, the material experienced work hardening and wall thinning. Nanayakkara and Hodgson [22] proposed a method to determine the draw bead contact conditions as a function of bead penetration which was based on the drawbead simulator. The results give a useful relationship between the rates of change of the contact angle with increasing bead penetration.

Many studies on drawbeads have emerged, particularly encouraged by the rapid progress of computational methods. Numerically, there are still numerous difficulties in accurately modelling and describing drawbead geometries and actions [23]. Duarte et al. [24] proposed a hybrid approach for estimating the DBRF throughout the explicit dynamic finite element (FE) method. A comparison of the numerical results with the experimental data provided by Nine [9] shows that the average of absolute error with respect to experimental data bases was around 6%. Moon et al. [25] employed drawbead forces computed by equivalent drawbead models in a finite element modelling (FEM) of the stamping process. Joshi et al. [26] conducted the experimental investigations on different semi-circular drawbead geometries and compared them to those obtained from Stoughton's analytical drawbead model [27]. This model significantly overpredicted the DBRF, whereas the numerical two-dimensional (2D) plane strain drawbead model accurately predicted the DBRF. Bramhakshatriya et al. [28] conducted the numerical computations of hemispherical cup forming using a circular drawbead. The numerical results were in good agreement when compared with the experimental ones.

Knowledge of the frictional resistances that have arisen in the drawbead region of the stamping die is necessary in order to prepare an appropriate design of die tool and to forecast the material flow during the forming of drawpieces, especially those with complex shapes [7,29,30]. Most of the research is focused on the estimation of the drawbead restraining force in sheet metal forming. Academic investigations on the determination of the COF during sheet metal flow through the drawbead are limited. The majority of FEM-based computations are based on the analysis of a 2D plane-strain state model of the physical drawbead which does not accurately reflect the material deformation in the drawbead. The aim of this paper is experimental study and three-dimensional (3D) elastic-plastic FEM-based numerical modelling of the frictional phenomena and material deformation during the flow of the sheet metal through the drawbead. Different set-up conditions, including drawbead height and specimen widths, have been considered.

2. Experimental

2.1. Material

DC04 cold rolled steel sheet for drawing with a thickness of 0.8 mm was used as a test material. Due to good weldability and excellent formability, DC04 steel sheet is well-suited to the manufacturing of structural elements, especially in the transport and household appliance industries. Examples of typical applications for this steel are side panels, floor panels and car bodies. The chemical composition of this steel (Table 1) is restricted by the requirements of EN 10130:2009 [31].

Table 1. Chemical composition of DC04 steel sheet (in wt.%).

C	Mn	P	S	Fe
≤0.08	≤0.4	≤0.03	≤0.03	remainder

The mechanical properties of sheet metals (Table 2) were determined in a uniaxial tensile test in machine Z100 (Zwick/Roell, Ulm, Germany) according to EN ISO 6892-1:2016 [32]. The plastic properties are characterised by the strain hardening coefficient K and strain hardening exponent n. These parameters were determined by approximation of the true stress–true strain relation by the Hollomon equation:

$$\sigma = K \times \varepsilon^n \tag{1}$$

where σ is stress and ε is plastic strain. Three samples were tested for each direction of the specimen: At an angle of 0°, 45° and 90° with respect to the rolling direction of the sheet metal.

The values of the basic surface roughness parameters determined using a Talysurf CCI Lite 3D optical profiler (Taylor Hobson, Leicester, UK) are as follows: Average roughness $Sa = 1.31$ μm, root mean square roughness $Sq = 1.53$ μm, maximum pit depth $Sv = 6.98$ μm, total height $St = 11.39$ μm, 10-point peak-valley surface roughness $Sz = 9.41$ μm and the highest peak of the surface $Sp = 4.41$ μm.

Table 2. Selected mechanical properties of DC04 steel sheet.

Specimen Orientation	Yield Stress $R_{p0.2}$ (MPa)	Ultimate Tensile Stress R_m (MPa)	Elongation A_{50} (%)	Strengthening Coefficient K (MPa)	Strain Hardening Exponent n
0°	179	311	22	471	0.19
45°	189	305	21	488	0.19
90°	183	289	23	479	0.17

2.2. Method

The frictional resistances arising in the drawbead region of the stamping die have been determined using a special tribological drawbead simulator (DBS). The model of the simulator is shown in Figure 2. The device is designed to allow the separation of the deformation resistance of the sheet and the frictional resistance from the total resistance of the sheet metal deformation at the drawbead. The frame (1) of the device is mounted in the bottom grip of the universal tensile testing machine Z100 (Zwick/Roell, Ulm, Germany). The specimen (3) is mounted in the upper grip of this machine. During the tests the unsupported end of the specimen has a tendency to bend. The stability of the contact area of the sheet and the working rollers (4) was achieved by a supporting roller (5). During tests the values of the drawing (friction) and clamping (normal) force were measured by using two load cells (6 and 7) placed at surface of the upper tension member (2) and horizontal tension member (8), respectively. When the test was carried out with fixed rollers, the rotation of the working rollers (4) was blocked by using pins (9). Adjustment of the value of the drawbead height was realised by a wing nut (10). A personal computer was used to register the values of both the drawing and clamping forces. The measuring signals from both tension members (2 and 8) were transmitted to a 4-channel C series strain/bridge input module NI-9237 (National Instruments, Austin, TX, USA) into the LabView program (National Instruments, Austin, TX, USA). The values of the forces were registered with a frequency of 50 Hz.

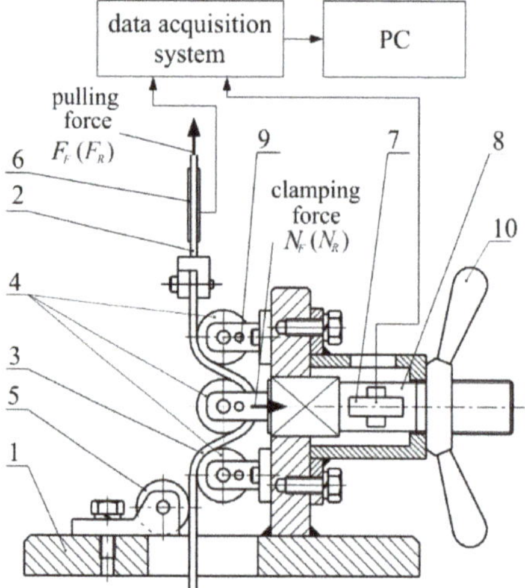

Figure 2. Cross-section of a typical drawbead with circular form: 1—frame, 2—upper tension member, 3—specimen, 4—working rollers, 5—support roller, 6 and 7—load cells, 8—horizontal tension member, 9—pin, 10—wing nut.

The method of determining the COF requires two tests of drawing a sheet metal strip over rotating (Figure 3a) and fixed (Figure 3b) rollers. Drawing the specimen over a set of rotating rollers allows one to minimise the frictional resistance. The drawing force in this case is mainly associated with overcoming the deformation resistance of the sheet metal strip. A set of fixed rollers represents the total resistance of drawing the specimen through the drawbead.

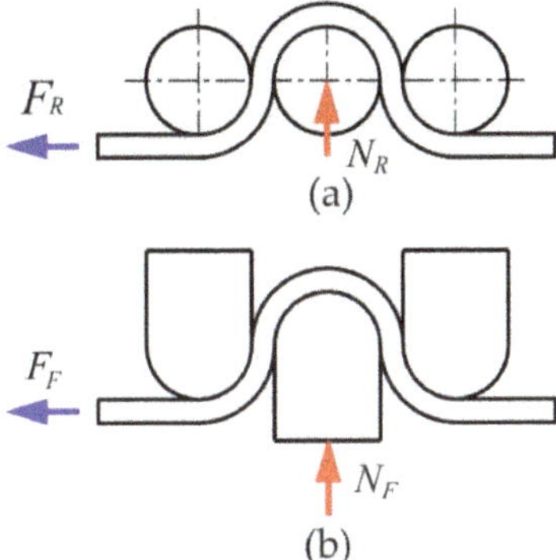

Figure 3. The concept behind the determination of frictional resistance: (**a**) rotatable rollers, (**b**) fixed rollers.

The value of the COF is determined from [33]:

$$\mu = \frac{sin\alpha}{2a} \times \frac{F_F - F_R}{N_F} \tag{2}$$

where α is the quarter contact angle of actual engagement of the strip over the middle roller, F_R is the pulling force obtained with the freely rotating rollers, F_F is the pulling force obtained with fixed rollers and N_F is the normal force obtained with fixed beads.

To avoid breaking the specimen at large wrap angles and when testing mild sheets due to excessive drawing resistance, it is necessary to ensure clearance c (Figure 4) between the rollers. The clearance value should be appropriately set. Too small an amount of clearance may lead to blocking the specimen, especially when testing the drawing specimen with fixed beads. On the other hand, excessive clearance may cause an unfavourable change in the character of the specimen deformation [34]. In this paper, the clearance is set according to the paper of Nanayakkara et al. [33], which suggests setting a clearance of 1.55 mm for a sheet thickness of 0.8 mm.

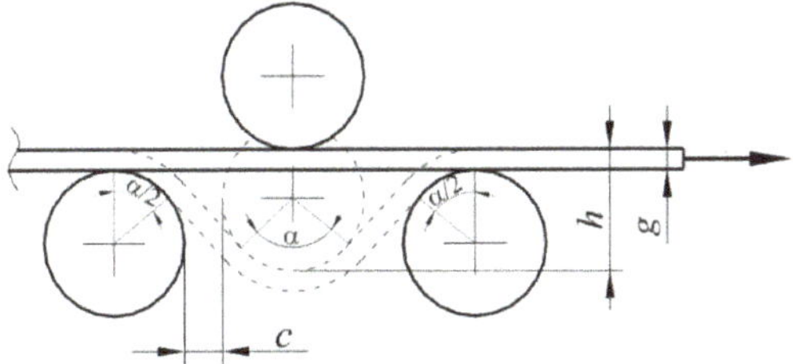

Figure 4. Parameters of a drawbead: c—clearance, h—drawbead height, g—sheet thickness, α—quarter contact angle of actual engagement of the strip over the middle roller.

A change of friction conditions in DBS may be obtained by changing the wrap angle of the counter-sample, the lubrication conditions, the shape and dimensions of the drawbead model, and the drawing speed of the sheet. The results of a wide range of experimental investigations prepared using a simulator that had been developed (Figure 2) were presented in a recent paper of the author [1]. In this paper, the aim of experimental research is to provide experimental data for the preparation and validation of the numerical model.

The research was carried out on sheet metal strips with dimensions 400 mm × w mm × 0.8 mm (length × width × thickness). The specimens were cut along the direction of sheet rolling. Three specimen widths, i.e., 7 mm, 14 mm and 20 mm, were considered. Moreover, two heights h of the drawbead (Figure 4), i.e., 12 and 18 mm, were investigated. The friction tests were conducted under dry friction conditions. The velocity of drawing the specimens was 10 mm·s^{-1}. The set of rollers that was used had a surface roughness characterised by an arithmetical mean deviation of the profile Ra = 0.63 μm.

3. Numerical Modeling

3.1. Description of the FE Model

Numerical modelling of the drawing of sheet metal through the drawbead model was carried out in the finite element-based MSC.Marc program (MSC Software, Newport Beach, CA, USA). The rollers were assumed to be non-deformable, which only allowed the inclusion of the external surfaces of the rollers in the numerical model. Displacement is applied to the first end of the specimen. In contrast, the second end of the specimen was free. The simulations took place in two stages. In the first stage, one of the ends of the sheet metal strip was blocked. Then, the displacement was imposed on the middle roller to reach the desired height of the drawbead h (Figure 5a). After that, the displacement was applied to one end of the specimen (Figure 5b).

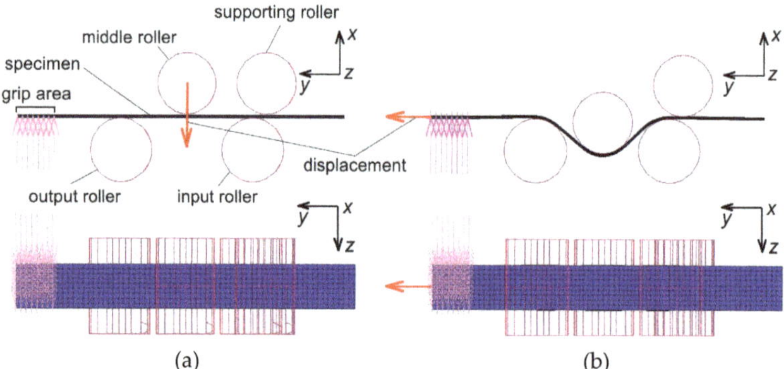

Figure 5. Geometry and boundary conditions of the Finite Element Method (FEM) model of the drawbead simulator: (**a**) first and (**b**) second stage of simulation.

The material model was assumed to be an elastic-plastic. The numerical computations have been performed with material behaviour described by the von Mises [35] yield criterion, including the strain hardening phenomenon according to the Hollomon equation. The average values of K and n constants in Equation (1), were determined based on Table 2. The remaining material parameters were as follows: Young's modulus E = 186,891 MPa, Poisson's ratio v = 0.3, material density ρ = 7860 kg·m^{-3}.

The contact conditions between the sheet and the surfaces of the rollers were described by Coulomb's law:

$$\sigma_t \leq -\mu\sigma_n t \tag{3}$$

where σ_t is the shear stress, σ_n is the normal stress, μ is COF, t is the tangential vector $\left(t = \frac{v_r}{\|v_r\|}\right)$ and v_r is the relative sliding velocity.

Alternatively, Equation (3) can be rewritten as a relationship between tangential and normal forces:

$$F_T \leq -\mu F_N t \tag{4}$$

where F_T is the tangential (friction) force and F_N is the normal force.

Experimentally determined variations of COF when pulling a sheet metal strip through a drawbead simulator are shown in Figure 6. The variation in COF can be divided into two regions (Figure 7):

(1) The first region (0–2.5 s) with an increase in the COF due to the positioning the middle roller to the desired value.

(2) The second region (2.5–17 s) with the changes in the COF stabilised during the stage of pulling the sheet strip through the DBS.

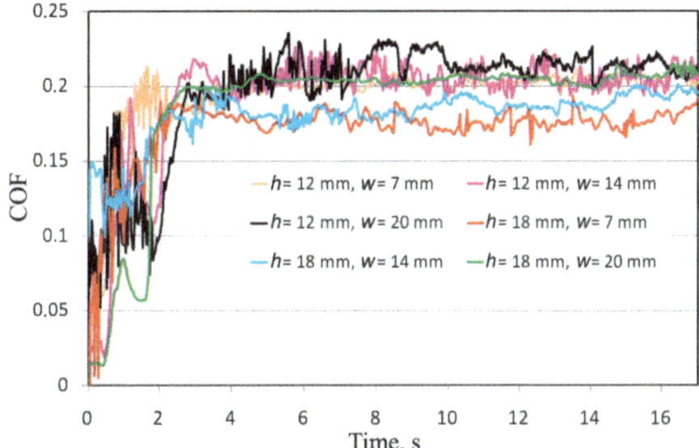

Figure 6. Experimentally determined variation of the coefficient of friction during a drawbead test.

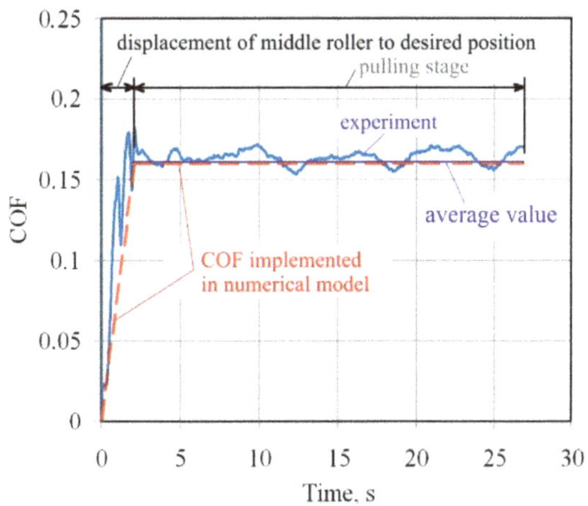

Figure 7. Experimentally determined variation of the COF during a DBS test for height of drawbead $h = 18$ mm and specimen width $w = 20$ mm.

Two simulations are required for the determination of the numerically predicted COF:

(1) In the model with blocked pins, the COF was assumed based on the experimental investigations. For the stage of setting the position of the middle roller, a linear increase in the COF was assumed. In the case of the drawing, a constant value of the COF was assumed, which corresponded to the average value of the experimentally determined COF (Table 3).

(2) In the model with freely rotatable rollers, the COF was assumed to be 0.

Table 3. Experimentally determined values of coefficient of friction.

Drawbead Height h, mm	Specimen Width w, mm	COF
12	7	0.202
	14	0.205
	20	0.221
18	7	0.176
	14	0.191
	20	0.206

3.2. Mesh Sensitivity Analysis

To discretise the sheet metal, the spatial 8-node isoparametric finite elements hex-8 were used with a formulation assumed strain that improves the behaviour of these elements during bending [36]. The hex-8 element has been successfully used by Fejkiel [37] to study the flowing of a sheet specimen through a drawbead. An appropriate element size is crucial in the analysis of the sheet deformation in metal forming operations. In the aforementioned investigations of Fejkiel [37] it was found that five layers of spatial hex-8 elements are appropriate to get accurate results. In this paper, a mesh sensitivity analysis (MSA) was carried out for five elements through the sheet thickness and different sizes of mesh $s_e \times s_e$ at the sheet plane: 3.334×3.334 mm^2, 1.667×1.667 mm^2, 0.833×0.833 mm^2 and 0.417×0.417 mm^2. MSA was carried out for the highest sheet width considered $w = 20$ mm and the highest drawbead height considered $h = 18$ mm. The aforementioned sizes of finite elements in the sheet metal plane correspond to dividing the sheet with 6, 12, 24 and 48 elements (Figure 8). More and more dense meshes of finite elements were determined by division of the superordinated mesh into four equal elements. In order to reduce computation time in MSA, the distance of sheet drawing was reduced to 70 mm. The constant value of COF in MSAs was assumed to be 0.2.

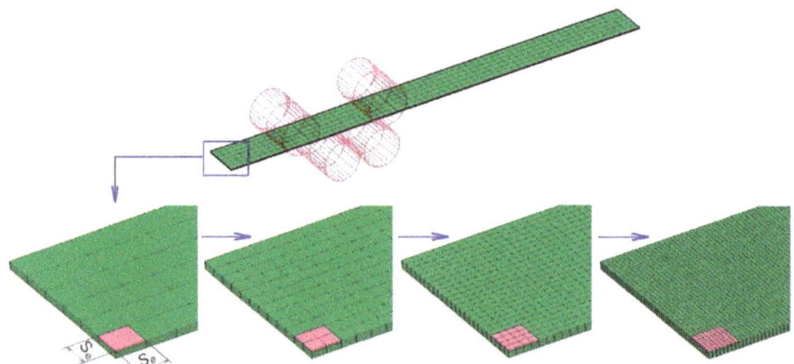

Figure 8. View of the mesh in the mesh sensitivity analysis (MSA).

4. Results and Discussion

The average drawing force of the metal strip in fixed rollers was adopted as the parameter constituting the basis for the selection of mesh size. Due to the extremely disproportional ratio of height in the thickness direction to length of the elements with a size $s_e = 3.334$ mm, the numerical calculations of a model including elements of such sizes failed (Figure 9).

In the case of the remaining mesh sizes, the computations were successfully completed. For the numerical model containing finite elements of size $s_e = 1.667$ mm, the value of the average drawing force was 1041.6 N with a computation time of about 102 min. Twice reducing the dimensions of the finite elements from $s_e = 1.667$ mm to $s_e = 0.833$ mm caused an increase in the average value of the drawing force by about 1%, while the central processing unit (CPU) time increased almost 10 times. Further reduction of the mesh size was not rational due to the long CPU time, i.e., over 105 h (Table 4).

From the point of view of the effect of mesh size on CPU time and the change in the value of the drawing force, the model with elements s_e = 1.667 mm is the most advantageous.

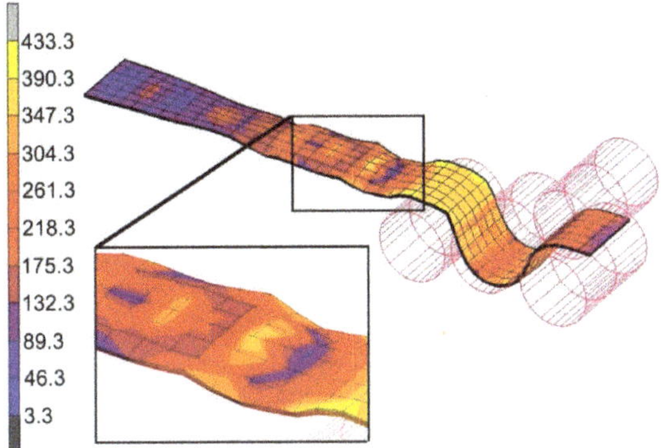

Figure 9. Distribution of von Mises equivalent plastic strain in the specimen discretised with elements with a size of s_e = 3.334 mm.

Table 4. Parameters and results of MSA.

Element Size $s_e \times s_e$ mm^2	Number of Elements	Number of Nodes	Average Drawing Force, N	Computation Time, s
3.334 × 3.334	1620	2695	-	1240
1.667 × 1.667	6480	8502	1041.6	6115
0.833 × 0.833	35,920	32,550	1050.1	56,829
0.417 × 0.417	103,680	127,302	1061.0	381,197

In the FEM numerical simulation, it is not possible to take the roughness of the rollers and friction mechanisms in the form of adhesive wear, ploughing, and the interaction of the roughness asperities of the sheet metal surface and tools into account. The value of the COF determined experimentally and introduced into the FEM model is a result of complex contact phenomena defined by the value of the COF introduced. Therefore, it is possible to determine the response of the numerical model in the form of values of both the drawing and clamping forces, and the determination of the value of the COF according to the modified version of Equation (2):

$$\mu = \frac{sin\alpha}{2\alpha} \times \frac{F_F^{FE} - F_R^{FE}}{N_F^{FE}} \tag{5}$$

where F_F^{FE} is the numerically predicted pulling force obtained with the fixed rollers, F_R^{FE} is the numerically predicted pulling force obtained with freely rotating rollers and N_F^{FE} is the numerically predicted normal force obtained with fixed beads.

Variations in the drawing and clamping forces of the middle roller under the conditions of height of drawbead h = 12 mm and specimen width w = 20 mm and under the conditions of freely rotatable and fixed rollers are presented in Figure 10a,b, respectively. Based on such results, the instantaneous values of process forces and the variations in the value of COF were determined from Equation (5).

The results of the variation of COF for the height of drawbead h = 18 mm are presented in Figure 11. The average value of COF was calculated for the stage of drawing the specimen (Figure 11). A comparison of the experimentally determined and numerically predicted values of COF for the drawbead heights of h = 12 mm and h = 18 mm are presented in Figure 12a,b, respectively. The errors between the predicted values of the COF and those experimentally determined in the cases analysed ranged from about 0.95% to 7.1%. For the drawbead height h = 18 mm smaller error values of COF were observed (Figure 12b) compared to the drawbead height h = 12 mm (Figure 12a). The next observation is that the numerical model overestimated the value of COF for all specimen widths analysed. In the numerical model increasing the drawbead height, at the same sheet width, increases the value of COF. This conclusion has been confirmed for the two sample widths analysed, 12 and 18 mm. The difference in the COF between the largest and smallest width of the samples was about 8.8%.

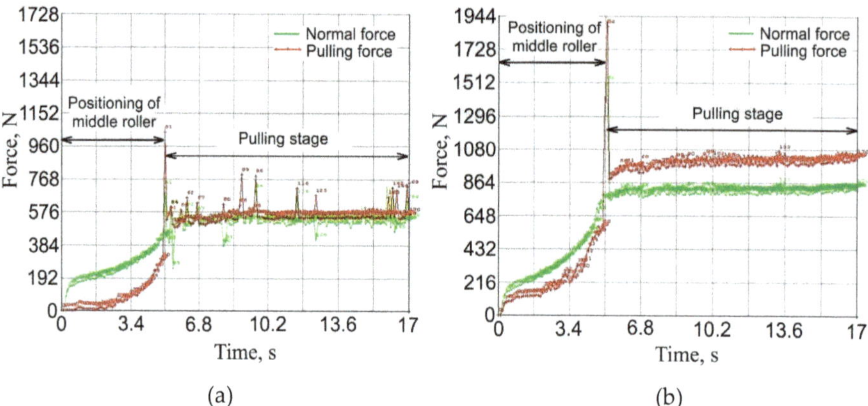

Figure 10. Numerically predicted variation of the drawing and normal force for a specimen pulled to flow by a system of (**a**) rotatable and (**b**) fixed rollers for height of drawbead h = 12 mm and specimen width w = 20 mm.

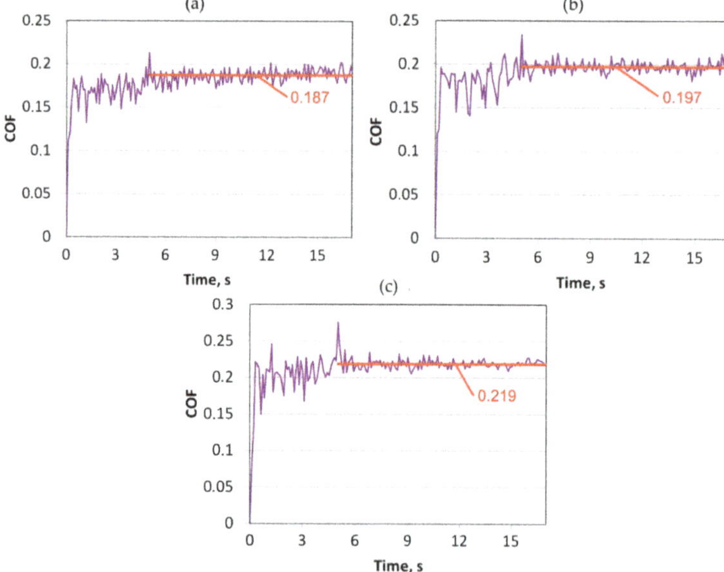

Figure 11. Numerically predicted variation of the COF for height of drawbead h = 18 mm and different specimen widths: (**a**) 20 mm, (**b**) 14 mm and (**c**) 7 mm.

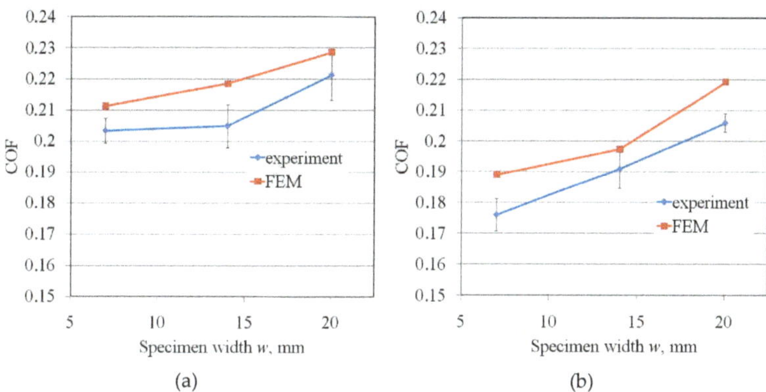

Figure 12. Comparison of average values of COF determined experimentally and based on FEM for drawbead heights of (**a**) $h = 12$ mm and (**b**) $h = 18$ mm.

The differences between the average values of COF determined in two ways can be partially explained by the deviation of the specimen profile from the ideal case. During the friction test in both the friction and frictionless cases, one end of the specimen was mounted in the grip of the tensile machine. The second end of the specimen was free (Figure 13a). As a result of the lack of tension force in the specimen material located opposite to the grip of the testing machine, the wrap angles of the specimen around the input roller α_1 and the output roller α_2 are not equal (Figure 13b,c). This also translates into a lack of symmetrical wrapping of the middle roller. The values of the wrapping angles of the rollers operating under friction and frictionless conditions α and β are different. The experimental method of determination of the COF assumes that the wrap angles of both the middle and output rollers are the same and result from the geometry of the drawbead. Thus, the COF determined on the basis of Equation (2) may be disturbed. Future work will be focused on the verification of this hypothesis for different specimen widths and thicknesses.

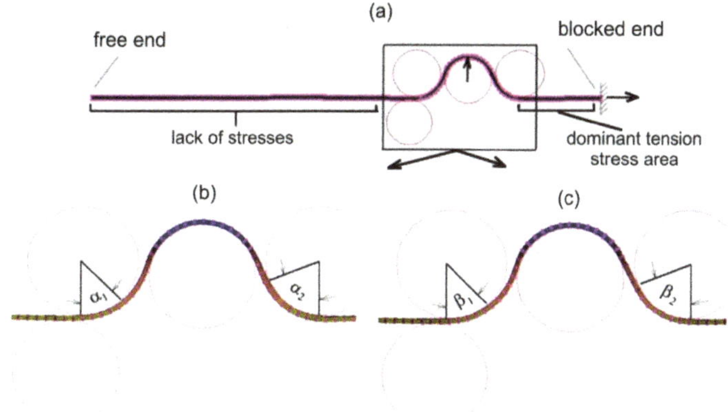

Figure 13. The character of the specimen deformation around the working rollers (**a**) for the drawbead height of $h = 18$ mm, and specimen width $w = 20$ mm for (**b**) the friction and (**c**) frictionless conditions.

The sheet strip between the middle roller and the output roller dominates the tensile stresses resulting from counteracting the pulling force by frictional forces on the surfaces of both the middle and input rollers. In the contact zone between the sheet metal strip and the input roller, the sheet metal is bent for the first time, thus, the elastic deformation of the sheet exerts a significant impact on the character of the material flow. As the sheet drawing progresses, the sheet material is bent and straightened, intensifying the effect of the strain hardening phenomenon on the resistance of specimen drawing. The varied character of sheet deformation during flow through the drawbead causes the wrap angle of the rollers to be different from the analytical value resulting from the geometry of the drawbead.

In addition to the difference in the wrap angles of individual rollers, another effect that has been noticed is the non-uniform distribution of normal stress in the sheet metal material (Figure 14). While the distribution of stresses is similar when sheet metal is drawn under friction and frictionless conditions, there is a significant difference in the maximum values of normal stresses. The value of maximum normal stresses in the contact area between the sheet metal and rollers under friction conditions is about 33% higher compared to the frictionless conditions. A similar observation was noticed in the case of the process of drawing the specimen through the drawbead with a height of $h = 12$ mm.

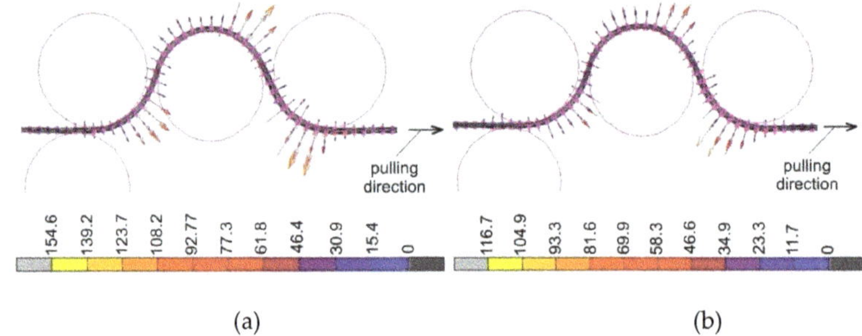

Figure 14. Distribution of normal stress in the specimen material for a specimen width $w = 20$ mm, and drawbead height of $h = 18$ mm under (**a**) friction and (**b**) frictionless conditions.

The next stage of the research determined the distribution of equivalent plastic strains (EPSs) along the width of the specimen. These results were obtained for two planes lying in the middle part of the specimen and at its edge on the internal (Figure 15a) and external (Figure 15b) surfaces of the specimen. The distributions thus correspond to the stage of passing the specimen through the drawbead.

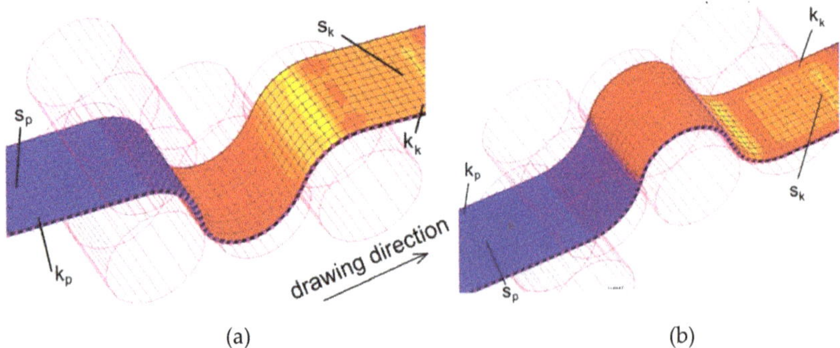

Figure 15. Paths of measuring the equivalent plastic strain on (**a**) the internal and (**b**) external surfaces of specimen: s_p and k_p—start points, s_k and k_k—end points.

Three zones with different levels of EPS can be distinguished both on the external (Figure 16a) and the internal (Figure 16b) surfaces of the sheet metal. The individual levels correspond to the locations where the specimen material contacts the surface of the rollers. In these locations, the material flow associated with the frictional contact between the rollers and the sheet metal strip is inhibited.

Therefore, rapid changes in the value of EPS occur in the zones that are between the rollers. The region where a large amount of EPS occurs corresponds to the region where the highest normal stresses occur on the surface of the output roller (Figure 14). The character of change of EPS on both surfaces of the specimen is similar (Figure 16a,b). The locations of measurement of the EPS over the specimen width and different friction conditions do not have a significant effect on the value and distribution of the EPS.

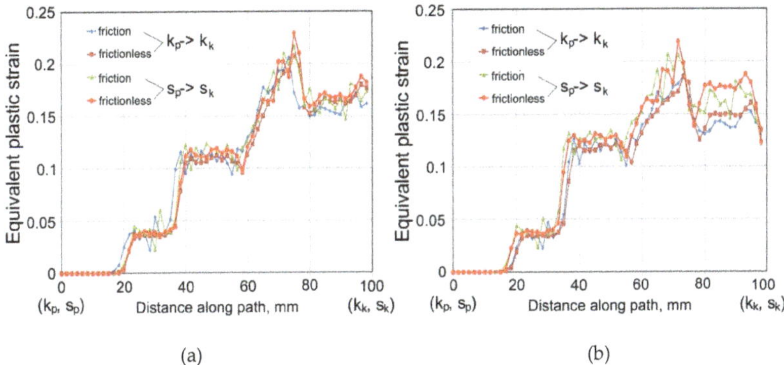

Figure 16. Distribution of EPS on (**a**) the internal and (**b**) external surface of specimen.

5. Conclusions

This article presents the results of experimental and numerical analyses of the sheet metal flowing through a drawbead simulator. Tribological tests have been carried out on DC04 steel sheets commonly used in the automotive industry using a special tribological drawbead simulator. The following conclusions are drawn from the research:

(1) From the point of view of the effect of mesh size on CPU time and the change in the value of the drawing force, the model element length $s_e = 1.667$ mm was the most advantageous.

(2) The errors between the experimentally determined and numerically predicted values of the COF in the cases analysed ranged from about 0.95% to 7.1%.

(3) In the case of drawbead height of 12 mm, the numerical model overestimated the value of COF for all the specimen widths analysed; in contrast, in the case of drawbead height $h = 18$ mm, all the experimentally determined friction coefficients are underestimated by FEM; this can be explained by the different character of the sheet deformation around input and output rollers in the experiment, which translates into a lack of symmetrical wrapping of the middle roller, and input and output rollers.

(4) In the numerical model increasing the drawbead height, at the same sheet width, increases the value of the COF; the difference in the COF between the largest and smallest widths of the samples was about 8.8%.

(5) The value of maximum normal stresses in the contact area between sheet metal and rollers under friction conditions was about 33% higher compared to frictionless conditions.

Author Contributions: Conceptualization, T.T.; Data curation, R.F.; Investigation, R.F.; Methodology, T.T.; Software, T.T.; Validation, R.F.; Writing—Original draft, T.T.; Writing—Review and editing, T.T. All authors have read and agreed to the published version of the manuscript.

Funding: This research received no external funding.

Conflicts of Interest: The authors declare no conflict of interest.

References

1. Trzepieciński, T.; Bazan, A.; Lemu, H.G. Frictional characteristics of steel sheets used in automotive industry. *Int. J. Automot. Technol.* **2015**, *16*, 849–863. [CrossRef]

2. Olsson, D.D.; Bay, N.; Andreasen, J.L. Direct Friction Measurement in Draw Bead Testing. In Proceedings of the 8th International Conference on Technology of Plasticity, Verona, Italy, 9–13 October 2005.

3. Trzepieciński, T.; Lemu, H.G. Effect of computational parameters on springback prediction by numerical simulation. *Metals* **2017**, *7*, 380. [CrossRef]

4. Zhang, R.; Shao, Z.; Lin, J. A review on modelling techniques for formability prediction of sheet metal forming. *Int. J. Lightweight Mater. Manuf.* **2018**, *1*, 115–125. [CrossRef]

5. Figueiredo, L.; Ramalho, A.; Oliveira, M.C.; Menezes, L.F. Experimental study of friction in sheet metal forming. *Wear* **2011**, *271*, 1651–1657. [CrossRef]

6. Wang, C.; Ma, R.; Zhao, J.; Zhao, J. Calculation method and experimental study of coulomb friction coefficient in sheet metal forming. *J. Manuf. Process.* **2017**, *27*, 126–137. [CrossRef]

7. Trzepieciński, T.; Gelgele, H.L. Investigation of anisotropy problems in sheet metal forming using finite element method. *Int. J. Mater. Form.* **2011**, *4*, 357–369. [CrossRef]

8. Dou, S.; Xia, J. Analysis of sheet metal forming (stamping process): A study of the variable friction coefficient on 5052 aluminium alloy. *Metals* **2019**, *9*, 853. [CrossRef]

9. Nine, H.D. Drawbead forces in sheet metal forming. In *Mechanics of Sheet Metal Forming*, 1st ed.; Koistinen, D.P., Wang, N.M., Eds.; Springer: Boston, MA, USA, 1978; pp. 179–211.

10. Jadhav, S.; Schoiswoh, M.; Buchmayr, B. Tribology test methods and simulations of the effect of friction on the formability of automotive steel sheets. *Metall. Ital.* **2018**, *9*, 56–63.

11. Sigvant, M.; Pilthammar, J.; Hol, J.; Wiebenga, J.H.; Chezan, T.; Carleer, B.; van den Boogard, T. Friction in sheet metal forming: Influence of surface roughness and strain rate on sheet metal forming simulation results. *Procedia Manuf.* **2019**, *29*, 512–519. [CrossRef]

12. Recklin, V.; Dietrich, F.; Groche, P. Influence of test stand and contact size sensitivity on the friction coefficient in sheet metal forming. *Lubricants* **2018**, *6*, 41. [CrossRef]

13. Li, G.; Long, X.; Yang, P.; Liang, Z. Advance on friction of stamping forming. *Int. J. Adv. Manuf. Technol.* **2018**, *96*, 21–38. [CrossRef]

14. Murali, A.; Gopal, B.; Rajadurai, C. Analysis of influence od draw bead location and profile in hemispherical cup forming. *Int. J. Eng. Technol.* **2010**, *2*, 356–360. [CrossRef]

15. Wang, Z.; Zhang, Q.; Liu, Y.; Zhang, Z. A robust and accurate geometric model for automated design of drawbeads in sheet metal forming. *Comput. Aided Des.* **2017**, *92*, 42–57. [CrossRef]

16. Larsson, M. Computational characterization of drawbeads: A basic modeling method for data generation. *J. Mater. Process. Technol.* **2009**, *209*, 376–386. [CrossRef]

17. Lee, M.G.; Chung, K.; Wagoner, R.H.; Keum, Y.T. A numerical method for rapid estimation of drawbead restraining force based on non-linear, anisotropic constitutive equations. *Int. J. Solids Struct.* **2008**, *45*, 3375–3391. [CrossRef]

18. Xie, Y. Geometric parameter inverse model for drawbeads based on grey relational analysis and GA-BP. *AIP Conf. Proc.* **2013**, *1567*, 1044–1047.

19. Zhang, Q.; Liu, Y.; Zhang, Z. A new method for automatic optimization of drawbead geometry in the sheet metal forming process based on an iterative learning control mode. *Int. J. Adv. Manuf. Technol.* **2017**, *88*, 1845–1861. [CrossRef]

20. Halkaci, H.S.; Turkoz, M.; Dilmec, M. Enhancing formability in hydromechanical deep drawing process adding a shallow drawbead to the blank holder. *J. Mater. Process. Technol.* **2014**, *214*, 1638–1646. [CrossRef]

21. Bassoli, E.; Sola, A.; Denti, L.; Gatto, A. Experimental approach to measure the restraining force in deep drawing by means of a versatile draw bead simulator. *Mater. Manuf. Process.* **2019**, *34*, 1286–1295. [CrossRef]

22. Nanayakkara, N.K.B.M.P.; Hodgson, P. Determination of drawbead contacts with variable bead penetration. *Comput. Methods Mater. Sci.* **2006**, *6*, 188–194.

23. Alves, J.L.; Oliveira, M.C.; Menezes, L.F. Modeling Drawbeads in Deep Drawing Simulations. In *III European Conference on Computational Mechanics*; Motasoares, C.A., Ed.; Springer: Dordrecht, The Netherlands, 2006; p. 533.

24. Duarte, E.N.; Oliveira, S.A.G.; Weyler, R.; Neamtu, L. A hybrid approach for estimating the drawbead restraining force in sheet metal forming. *J. Braz. Soc. Mech. Sci. Eng.* **2010**, *32*, 283–291. [CrossRef]

25. Moon, S.J.; Lee, M.G.; Lee, S.H.; Keum, Y.T. Equivalent drawbead models for sheet forming simulation. *Met. Mater. Int.* **2010**, *16*, 595–603. [CrossRef]

26. Joshi, Y.; Christiansen, P.; Masters, I.; Bay, N.O.; Dashwood, R. Numerical modelling of drawbeads for forming of aluminium alloys. *J. Phys. Conf. Ser.* **2016**, *734*, 032082. [CrossRef]

27. Stoughton, T.B. Model of Drawbead Forces in Sheet Mmetal Forming. In Proceedings of the 15th International Deep Drawing Research Group, Dearbon, MI, USA, 16–18 May 1988; p. 205.

28. Bramhakshatriya, G.M.; Sharma, S.K.; Patel, B.C. Analysis on Effect of Die Fillet Radius and Draw Bead on Hemispherical Cup Forming. In Proceedings of the 2012 Students Conference on Engineering and Systems, Allahabad, Uttar Pradesh, India, 16–18 March 2012; pp. 1–6.

29. Trzepieciński, T. 3D elasto-plastic FEM analysis of the sheet drawing of anisotropic steel sheet. *Arch. Civ. Mech. Eng.* **2010**, *10*, 95–106. [CrossRef]

30. Tisza, M.; Lukács, Z.; Kovács, P.; Budai, D. Some recent developments in sheet metal forming for production of lighteight automotive parts. *J. Phys. Conf. Ser.* **2017**, *896*, 012087. [CrossRef]

31. EN 10130 2009. *Cold Rolled Low Carbon Steel Flat Products for Cold Forming—Technical Delivery Conditions*; European Committee for Standardization: Brussels, Belgium, 2009.

32. ISO 6892-1:2016. *Metallic Materials—Tensile Testing—Part 1: Method of Test at Room Temperature*; International Organization for Standardization: Geneva, Switzerland, 2016.

33. Nanayakkara, N.K.; Kelly, G.L.; Hodgson, P.D. Determination of the coefficient of friction in partially penetrated draw beads. *Steel Gr.* **2004**, *2*, 677–680.

34. Meinders, V.T. Developments in Numerical Simulations of the Real-Life Deep Drawing Process. Ph.D. Thesis, University of Twenty, Enschede, Holland, 11 February 2000.

35. Von-Mises, R. Mechanik der festen Körper im plastisch—Deformablen Zustand. *Nachrichten von der Gesellschaft der Wissenschaften zu Göttingen. Mathematisch. Physikalische Klasse* **1913**, *1913*, 582–592.

36. *Marc® Volume B: Element Library*; MSC Software Corporation: Santa Ana, CA, USA, 2010.

37. Fejkiel, R. An Analysis of Frictional Resistance during Sheet Metal Passing Through the Draw Bead in the Sheet Metal Forming Process. Ph.D. Thesis, Rzeszow University of Technology, Rzeszów, Poland, 4 July 2019.

Article

Deformation Behavior and Experiments on a Light Alloy Seamless Tube via a Tandem Skew Rolling Process

Feilong Mao [1,*], Fujie Wang [1], Yuanhua Shuang [2], Jianhua Hu [2] and Jianxun Chen [2]

[1] Department of Mechanical and Electronic, Yuncheng University, Yuncheng 044000, China; yonghuikeji@126.com

[2] The Coordinative Innovation Center of Taiyuan Heavy Machinery Equipment, Taiyuan University of Science and Technology, Taiyuan 030024, China; yhshuang@tyust.edu.cn (Y.S.); clhjh@sina.com (J.H.); kdzgcjx@vip.163.com (J.C.)

* Correspondence: maofeilong@163.com; Tel.: +86-359-859-4433

Received: 29 November 2019; Accepted: 25 December 2019; Published: 29 December 2019

Abstract: As a process for producing seamless tubes, the tandem skew rolling (TSR) process was proposed. In order to study deformation characteristics and mechanism on tubes obtained by the TSR process, a numerical simulation of the process was analyzed using Deform-3D software. Simulation results demonstrated the distribution of stress, strain, velocity, and temperature of a seamless tube in the stable stage during the TSR process. Actual experiments of carbon steel 1045, high strength steel 42CrMo, and magnesium alloy AZ31 were carried out in a TSR testing mill. The results demonstrated that the TSR process is qualified for producing tubes of high quality, with an accuracy of ±0.2 mm in wall thickness and ±0.35 mm in diameter. This process is suitable for manufacturing seamless tubes that are difficult to deform or that have been deformed in a narrow range of temperature.

Keywords: tandem skew rolling; seamless tube; magnesium alloy; deformation behavior; high strength steel

---

1. Introduction

The steel industry is now in a post-industrial era of steel, whose theme is "green manufacture, make green." The short process is one way to meet the requirement for industrial development. The tandem skew rolling (TSR) process is an ingenious metal-formed technique for producing seamless steel tubes that can achieve continuous skew rolling by combining the piercing section with the rolling section [1,2]. Compared with the conventional process of manufacturing seamless steel tubes, a three-step forming process of piercing, rolling, and reducing, the TSR process is a shorter, two-step forming process.

The TSR process has been investigated for many years. In order to implement the TSR process, a TSR testing mill was designed and developed for experimental research. The mechanical structures of the mill and a method of adjusting process parameters were introduced and described [3,4]. A theoretic model of the TSR process was then established and derived in terms of dual stream functions, which could obtain the kinematically admissible velocity field of the billet during the rolling process [5]. Based on the times spent on the deformation process of the two sections, the velocity coordination of the rollers in both sections was researched, and a proper velocity match was obtained [6]. Moreover, to study the interstand tension in the tandem process, a tension testing device was developed and used with the testing mill, which reflected the changes in the stress of billet [7]. Further, how the process parameters affected the rolling force and the dimensional precision of the tubes was investigated [8]. Finally, aiming to improve the dynamic performance of the control system in the TSR process, a dynamic matrix control algorithm together with a particle swarm optimization algorithm was used to optimize the parameters in the process [9,10].

Due to the application of a finite element model (FEM), it was possible and economical to precisely analyze the changes in the shape of workpieces during metal deformation. Wang et al. [11] used the finite element method to analyze the three-roll skew piercing process. Bogatov et al. [12] studied the rotary piercing process using Deform-3D software and obtained the trajectories of points with different radial coordinates. Pater et al. [13] studied skew rolling with three tapered rolls for producing long main shafts of steel using the commercial FEM simulation software suite Simufact Forming v.12. They also studied the tube forming process in Diescher's mill [14], and the thermo-mechanical piercing process using the commercial software MSC SuperForm 2005 [15]. Stefanik et al. [16] studied the three-high skew process for producing aluminum bars using the Forge 2011® software. Romantsev et al. [17] studied the piercing process for a new four-high screw rolling mill without guides, which was designed by comparing existing piercing processes in screw rolling mills. Toporov et al. [18] used finite element modeling to improve the piercing operation on a piercing mill. Murillo-Marrodán et al. investigated the friction conditions between the rolls and the workpiece of a hot skew roll piercing process with three friction laws via the FORGE software [19].

Aiming at the precise representation of the deformation characteristics of billets during the TSR process, numerical simulation of the TSR process using the software Deform-3D was employed to study the distributions of stress and strain. Based on numerical analysis, experiments were carried out with carbon steel 1045, high-strength steel (HSS) 42CrMo, and magnesium alloy AZ31 on the TSR testing mill to verify the feasibility of the process. The TSR process was originally developed to manufacture seamless tubes that were difficult to deform (e.g., high strength steel 42CrMo) or could only be deformed with a narrow temperature range (e.g., AZ31 alloy) because the temperature drop was slower in this process.

2. Materials and Methods

By the TSR process of producing seamless tubes, the heated cylindrical billet was pierced and rolled continuously in the mill, in which the piercing and rolling processes were completed at once. The schema of the TSR process is presented in Figure 1. The process involves three Mannesmann barrel-shaped rollers in the piercing section, three Assel tapered rollers in the rolling section, and a plug together with a mandrel and a billet.

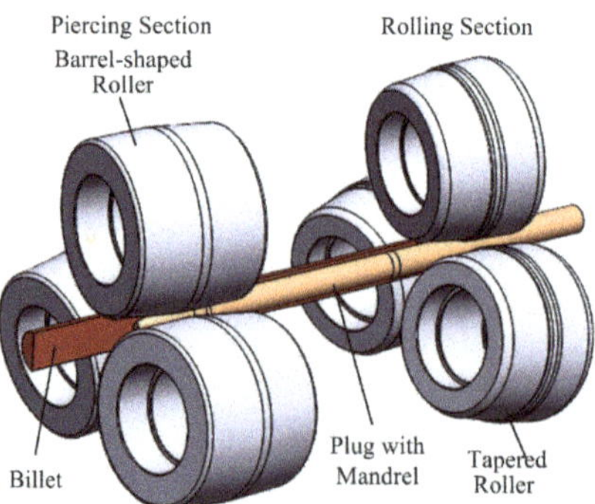

Figure 1. Schema of the tandem skew rolling (TSR) process.

2.1. Experimental Test

Through untiring efforts made over many years, the TSR testing mill was developed for experimental studies of seamless tubes, as shown in Figure 2. This mill consists of the driving part of the piercing section, the main mill (the piercing section combined with the rolling section), the mandrel carriage in which the plug is joined with the mandrel by a screw thread, and the driving part of the rolling section.

Figure 2. TSR testing mill.

In order to extend the range of the TSR process, in the experimental study, the cylindrical billets of carbon steel 1045, HSS 42CrMo, and magnesium alloy AZ31 were used with dimensions $\varphi 40 \times 500$ mm. The rollers in the piercing section were the same sizes with a maximum diameter of 180 mm and a length of 140 mm, while the rollers in the rolling section were the same sizes with a diameter of 180 mm at the shoulder and a length of 120 mm. The plug was used together with the mandrel with a diameter of 30 mm. The rollers in the piercing section, driven by A.C. motors, rotated in the same direction with the same invariable rotary speed of 169 rpm. The rollers in the rolling section, driven by D.C. motors, rotated in the same direction with the same variable rotary speed, which should match the speed of the rollers in the piercing section under different process parameters. The reasonable ranges of process parameters were obtained through previous multiple experiments. The rolling temperature was very important for hot plastic deformation, especially the alloy material. Therefore, the initial rolling temperatures for steel 1045, HSS 42CrMo, and AZ31 alloy were set at 1200, 1250, and 400 °C, respectively, which were selected to make full use of their plastic deformation capacity under corresponding high temperatures. The process parameters used in experiments are shown in Table 1.

Table 1. Process parameters of the TSR testing mill.

Item	Piercing Section				Rolling Section			
Process Parameter	Feed Angle (deg)	Entrance Face Angle (deg)	Roll Gap (mm)	Plug Advance (mm)	Feed Angle (deg)	Entrance Face Angle (deg)	Roll Gap (mm)	Roll Rotational Speed/rpm
Adjustable rang	0–15	0–9	30–48	0–35	0–15	0–9	35–48	0–200
Reasonable rang	7–9	0	34–36	20–25	7–9	4	37–39	174–190
Experimental values	8	0	35	23	9	4	39	186

2.2. Finite Element Model

For the analysis of the metal flow and deformation in the TSR process, the commercial software Deform-3D was applied. Due to the complexity, the analysis of the metal flow and deformation should be simulated only in a 3D state of strain conditions. In Figure 3, the worked-out FEM of the TSR process according to the schema in Figure 1 is presented. This model consists of three barrel-shaped rollers, three tapered rollers, a pusher, a plug together with a mandrel, and a billet. The billet is defined as a plastic body, and the others are modeled as rigid bodies. Additionally, the pusher was used to place the billet into the area between rollers. The process parameters (e.g., feed angle, entrance face angle, roll gap, and speed) used in the FEM were the same as those used in the experiment and are shown in Table 1. However, a short distance of 200 mm between the piercing section and the rolling section was used in the simulation in order to save computation time, which was 630 mm in the TSR testing mill.

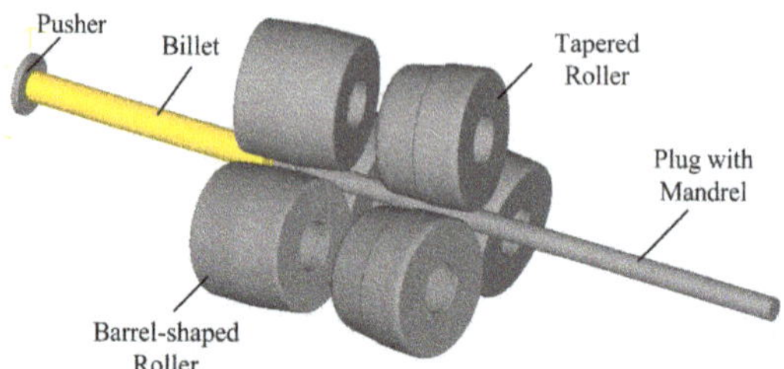

Figure 3. Finite element model (FEM) of the TSR process.

Aiming to reflect the metal flow and deformation rule, only the 1045 steel was selected in simulation. A rigid plastic material model, which is the 1045 steel of the software's own model, was employed and assigned to the billet with dimensions $\varphi 40 \times 200$ mm, whose flow curves are shown in Figure 4. The billet was meshed and approximately divided into 30,000 tetrahedral elements. It was assumed that the billet was heated to 1200 °C, and the temperatures of the rigid tools, the rollers, and the pusher were constant at 150 °C during the whole forming process. In the case of the plug and the mandrel, it was assumed that the temperature was the same as the environment temperature—20 °C. It has been stated that the coefficient of the heat exchange between the tools and formed material is 11 N/(s·mm·C). However, the value of this coefficient determining the heat exchange between the billet and the environment was 0.02 N/(s·mm·C). Friction is an important factor for a smooth rolling process. The greater the friction, the easier it is to roll smoothly, and the less time spent in simulation. Due to the purposeful roughness of the rollers, it was assumed that the friction factor for these rollers had a limiting value of 1.0. However, because the plug was made with a follow-up rotational movement with the billet during piercing, the friction of the surface of contact between the plug and the billet was very low. Therefore, the friction was set to 0.3 in the FEM.

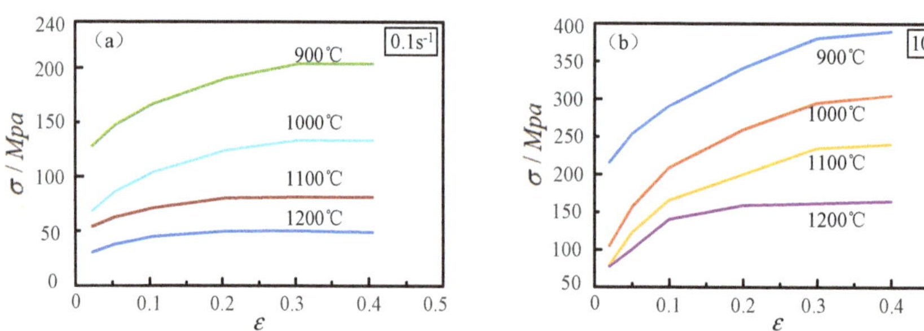

Figure 4. Flow curves of 1045 steel at different strain rates: (**a**) 0.1 s^{-1} and (**b**) 100 s^{-1}.

3. Results and Discussion

3.1. Numerical Modeling Results of the Steel 1045 during the TSR Process

The changes in the billet shape during the TSR process are presented in Figure 5, from which two rollers are omitted to improve readability.

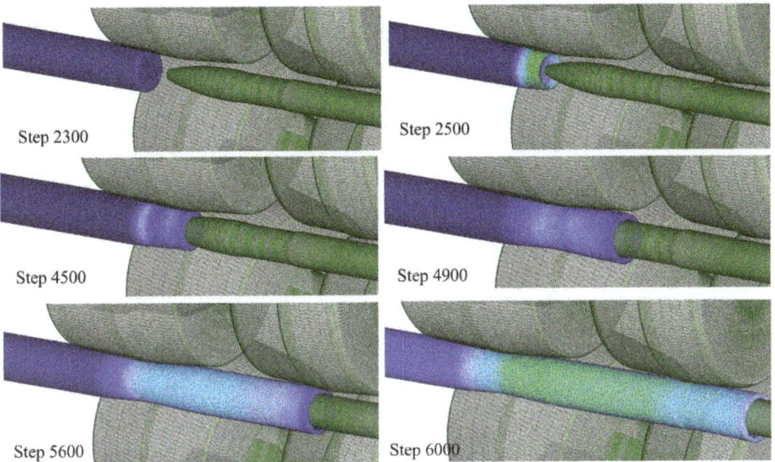

Figure 5. Change in billet shape during the TSR process.

Based on Figure 5, at the beginning, the billet is clamped by rollers in the piercing section, which puts it into a rotary motion and later transmits it in the axial direction. A process of rotational compression is then started, and this lasts until the head of the billet has made contact with the plug. Next, the forming process of the hollow shell internal hole begins, and its size is determined by the applied plug dimension. At approximately Step 4900, the head of the hollow shell moves straight out from the rollers in the piercing section. Further, with continuous rolling, the pierced hollow shell makes contact with the rollers in the rolling section at approximately Step 5600 and is rolled into the roll gap. Under the combination of the roller in the piercing section and the rolling section, the TSR process reaches a stable phase after Step 6000.

The distribution of stress determined for the stable stage in the longitudinal and cross-section planes is presented in Figure 6, and the strain is presented in Figure 7. The data in Figure 6 shows that higher stress intensity magnitudes occur at locations outside where the surface contacts the rollers and are equal to approximately 160 MPa in both sections, and the highest values are concentrated at the shoulder of the rollers in the rolling section. Meanwhile, the changes in the different cross sections indicate that it makes the sharpest contact only at the moment when the rollers bite the billet from three directions. As the billet moves along the deformation region, the magnitude of the stress intensity, while it deepens from outside to inside, increases with an increase in the contact surface. The same change regularity is presented in the rolling section.

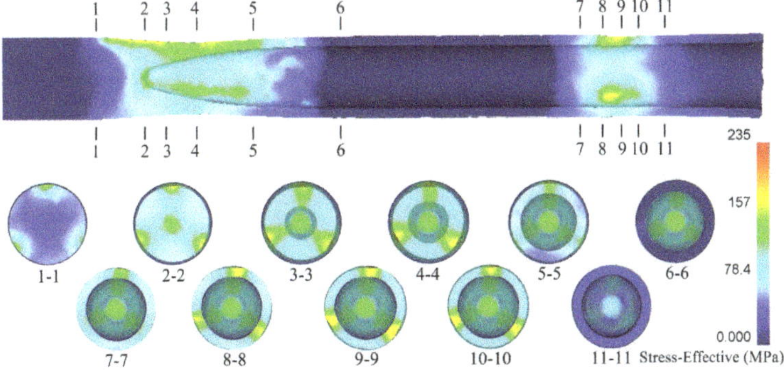

Figure 6. Stress distribution in the stable stage (Step 6000) of the TSR process.

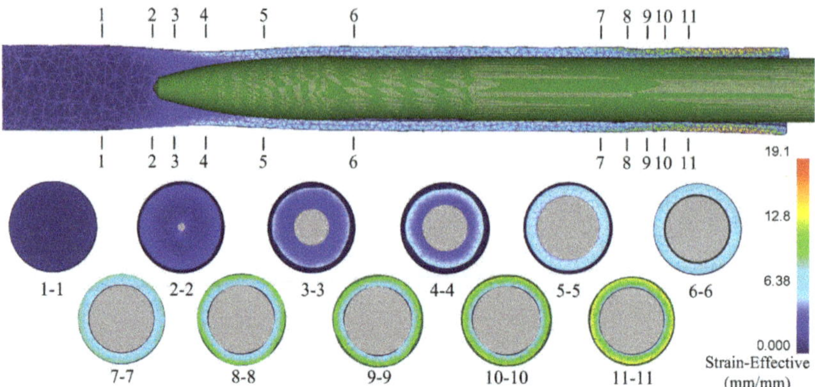

Figure 7. Strain distribution in the stable stage (Step 6000) of the TSR process.

In Figure 7, it can be observed that strains are presented in a laminar way. However, the greatest strains are present in the external layers of the formed tube. In Figure 7, the changes in the billet cross profile along the whole length of the forming area can be analyzed. Due to the influence of the rollers, the cross profile varies from a circle to a triangular circle and then to another circle in the piercing section. The cross profile continuously varies from a circle to a triangular circle and then to another circle in the rolling section. Finally, after the metal moves through the mandrel, the tube gains the required circular shape.

To clearly display the distributions of stress and strain determined for the stable stage, Figure 8 shows an isometric view and longitudinal section of distribution patterns for the seamless steel tubes during the TSR process as follows: (a) stress, (b) effective strain, (c) velocity, and (d) velocity with a vector plot, in which the distributions of the velocity field and metal flow are obtained.

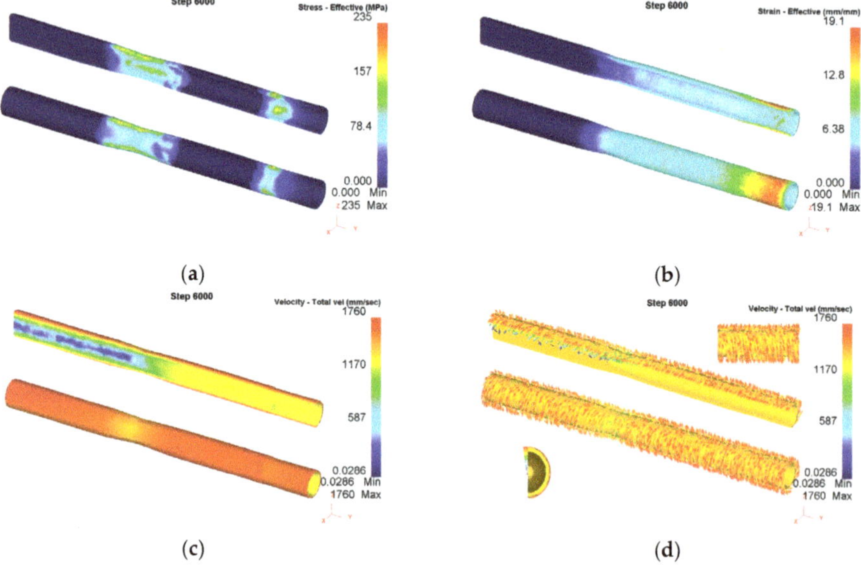

Figure 8. Distribution pattern of the producing seamless steel tube in the TSR process: (a) stress, (b) effective strain, (c) velocity, and (d) velocity with a vector plot.

In Figure 8c, distribution law is arranged in layers of the velocity fields, which show a decreasing trend from outside to inside. The highest magnitude appears at the outside surface, while the lowest magnitude appears at the central axial zone in front of the plug. When the head of the billet touches the plug, the core metal flows outward along the surface of the plug under the friction force given by the rollers. This force was enough to present the metal flow rule of the tube obtained by TSR process. Figure 8d gives the velocity vector of the tube produced in the process and shows the metal flow rule. The velocity vector is always tangential to the surface and forms an angle with the axial line, which determines the movement pattern of the billet. When touching the plug, the core metal flows outward along the surface of the plug. Note that no cavity is formed in the core metal in front of the plug.

3.2. Experimental Analysis

3.2.1. Quality of Tubes

Figure 9 shows the seamless tubes produced on the TSR testing mill. As shown in Figure 9, the tubes of carbon steel 1045 and AZ31 alloy achieved the desired cylindrical shapes. However, HSS 42CrMo could not be successfully completed with the rear-jamming of the tail of the billet present in the piercing section. The season was due to the collapsing nose of the plug, which resulted in the increase in axial resistance. Meanwhile, there were no significant defects on the outside surfaces of the tubes, except from the tail burrs of the AZ31 alloy and the trumpet-shaped heads of the tubes. In addition, the carbon steel 1045 tubes had diameters of 39.14 ± 0.35 mm, wall thicknesses of 4.67 ± 0.2 mm, and lengths of up to approximately 1360 mm, and the dimensions of the AZ31 alloy tubes were 39.85 ± 0.35 mm, 4.42 ± 0.25 mm, and 1200 mm, respectively. However, only a length of 750 mm of HSS 42CrMo tube was obtained and rolled by the TSR process, and the diameters and wall thicknesses were about 38.22 ± 0.35 mm and 4.33 ± 0.3 mm, respectively. The comparative data of these tubes are listed in Table 2.

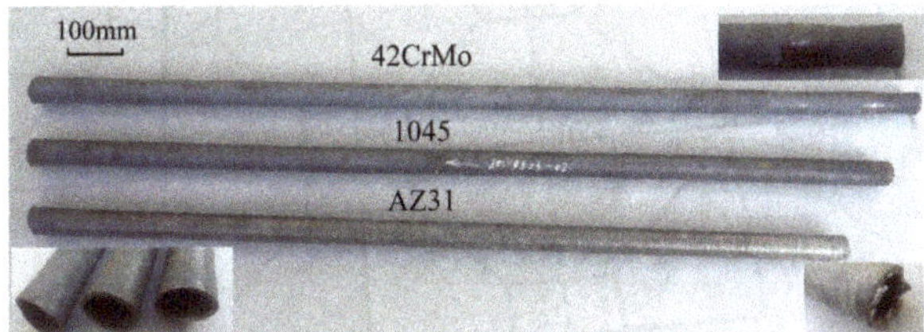

Figure 9. Seamless tubes produced in the TSR testing mill.

Table 2. The dimensional values of the tubes with different materials by experiments.

Material	Diameter (mm)	Thickness (mm)	Length (mm)	Extension Coefficient
Carbon steel 1045	39.14 ± 0.35	4.67 ± 0.20	1360	2.72
HSS 42CrMo	38.22 ± 0.4	4.33 ± 0.30	1400	2.80
AZ31 alloy	39.85 ± 0.35	4.42 ± 0.25	1200	2.40

The distribution of thicknesses of the tubes along the axial direction is shown in Figure 10. Figure 10 shows that the tubes of carbon steel 1045 and the AZ31 alloy had a more accurate thickness compared with the HSS 42CrMo tube produced in the TSR testing mill. Meanwhile, under the same process parameters in the experiment, different diameters and thicknesses were obtained with different materials and properties. The extension coefficient of the carbon steel 1045 was about 2.72, while those of HSS 42CrMo and the AZ31 alloy were about 2.8 and 2.4, respectively. The reasons for this might be linked directly to the plastic deformation properties themselves. From the condition of experiments, the color of the tandem rolled tube of the HSS 42CrMo darkened more quickly than that of the carbon steel 1045, which was sufficient to prove that HSS 42CrMo was a temperature-responsive metal. Under a proper rolling temperature, the AZ31 alloy tube was rolled more easily than others. Additionally, a uniform wall thickness of the middle section of the tubes was obtained, the reason for which might be the stable rolling in the process.

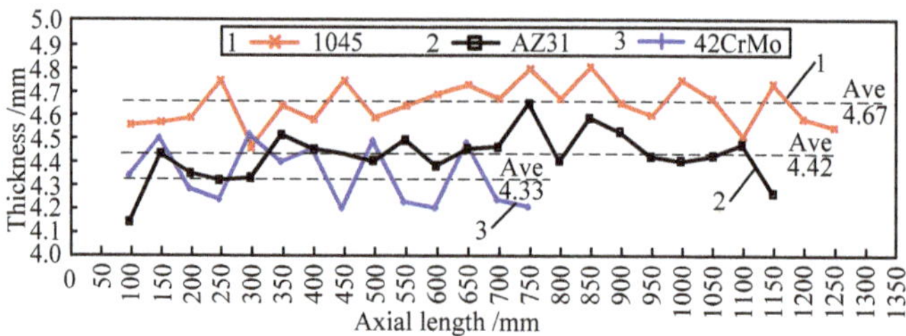

Figure 10. Distribution of the thickness in the axial direction.

3.2.2. Roll Force

Figure 11 shows the experimental data related to the roll force of the two sections and the plug axial force in the TSR process for the production of seamless tubes: (a) carbon steel 1045, (b) magnesium alloy AZ31, and (c) HSS 42CrMo, respectively. Because of the plug combined with the mandrel, the date of the axial force could be the sum of the dates on both. As shown in Figure 11a, the whole process of the rolling carbon steel 1045 can be easily divided into three stages: (i) piercing, (ii) piercing and rolling, and (iii) rolling, which is the same as the AZ31 alloy. However, there was no stage of rolling solely in Figure 11c because of the rear-jamming present during the rolling tube of HSS 42CrMo. What could be seen in Figure 11 is that the roll force values of both sections of the rolling HSS 42CrMo are greater than those of the rolling carbon steel 1045 and the AZ31 alloy, respectively. The roll force of the AZ31 alloy is the smallest. As is known, compared with carbon steel, HSS has a higher tensile strength when small amounts of alloying elements are added, while magnesium alloy has a lower one.

As shown in Figure 11a,b, with the billet pushed and contacted with the rollers of the piercing section, the roll force of the piercing section increased steeply in stage (i). In quick succession, when the head of the incoming billet made contact with and passed through the plug, the roll force of the piercing section slightly rose and then reached a relative steady state, followed by a slight decline. Meanwhile, the axial force on the plug also increased and then remained stable.

Continuously, when the pierced hollow tube moved forward in a spiral fashion and fed into the roll gap of the rolling section, it was situated in the real TSR process in stage (ii). The roll force of the rolling section increased rapidly and afterwards decreased slowly to a certain extent. Meanwhile, the force of the two sections featured a downward sloping curve on a small scale, because of the interstand tension between the two sections. The axial force also increased with the addition of force on the mandrel.

In stage (iii), as the billet exited the roll gap of the piercing section, the force of the piercing section decreased steeply, and the axial force decreased rapidly to a small value because of the lack of axial force on the plug. Meanwhile, the force of the rolling section slightly increased due to the lack of the interstand tension mentioned above. The force decreased to 0 when completed.

As shown in Figure 11c, the rolling HSS 42CrMo took longer than the carbon steel 1045, which reflects the difficulty of rolling high-deforming steel, such as 42CrMo. Based on Figure 11c, the slip between the roller and the billet in the piercing section occurred at about 12 s, accompanied by the increase in the force of the piercing section, and no axial movement occurred thereafter in the actual experiment. The force decreased to 0 after a manual stop. The reason for this is mentioned above: the collapsing nose of the plug and the increase of axial resistance must be a major factor causing the rear-jamming, which increased the force of axial resistance, even with greater friction. As is known, friction between the billet and the roller is key in the rolling process. Meanwhile, the material characteristics of HSS 42CrMo in the deformation may be an intrinsic factor, which is determined by the small amounts of alloying elements. Even so, this is enough to prove the feasibility of the TSR process for HSS seamless tubes. The rolling process thus might become smoother in two ways: the rolling velocity might increase, and the roll friction might increase. These methods, under these conditions, could therefore decrease the work time of plugs.

As shown in Figure 11b, the variation tendency of the roll force of the AZ31 alloy was similar to the one of steel 1045. Only the used time of each deformation section was shorter than steel 1045, and the average force values were lower. Based on other research on piercing experiments with the AZ31 alloy [20,21], using the TSR process to manufacture seamless tubes of AZ31 alloys was a daring attempt and has been proven to be a breakthrough.

Generally, this could be used as a reference for similar kinds of material. Conversely, there might be some distinctions for different kinds of materials, in which some factors affect the results, e.g., deformation temperature, deformation rate, and deformation degree.

Fortunately, different material productions for tubes formed by TSR have been successful, and the process has been shown to be feasible for use. Further tests where the plug is replaced should be done, with the goal of achieving a longer contact time under high temperatures. Adjusting process parameters is another effective way of increasing the rolling velocity or shortening the contact time between the plug and the billet. The emphasis should be on high-quality tubes of nonferrous and high-deforming metals, such as tubes of titanium alloy used in oil well fields, and such high quality is the ultimate goal of the TSR process.

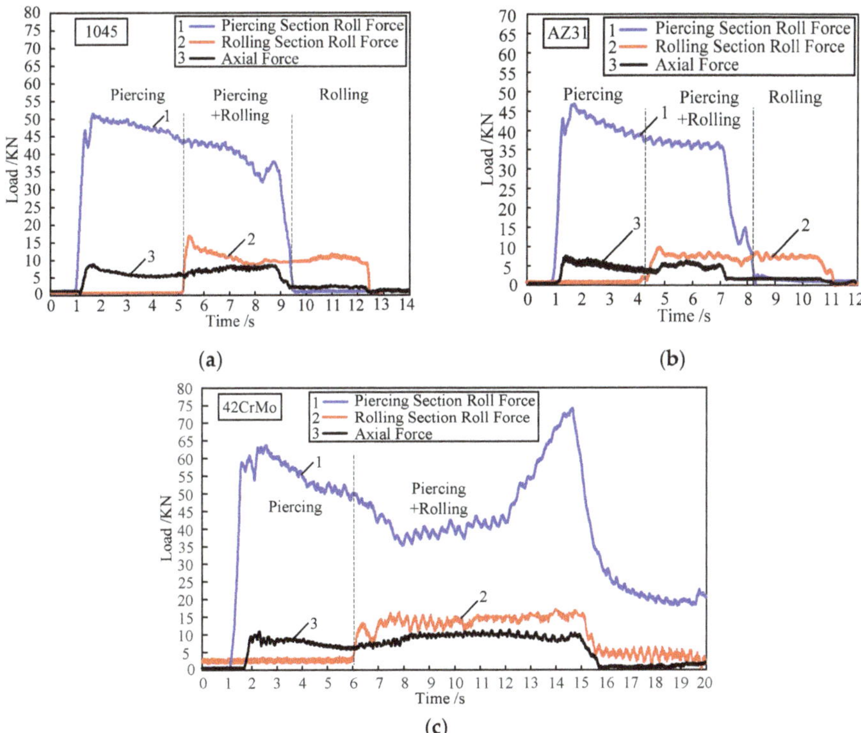

Figure 11. Experimental data relating roll force and plug axial force in the TSR process: (**a**) 1045, (**b**) AZ31, and (**c**) 42CrMo.

4. Conclusions

Based on the studies carried out, it was found that rolling on a TSR testing mill allowed TSR to be run with high efficiency in the deformation of seamless tubes pierced and rolled at the same time time, which makes continuous skew rolling more economical and practical. The following conclusions can be drawn.

(1)   The distribution of stress presented higher stress intensity magnitudes on outside surfaces contacting with the rollers.

(2)   The distribution of strain was presented in a laminar way, with higher strain intensity magnitudes occurring at the external layers of the formed tube. The cross profile varied from a circle to a triangular circle and then to another circle in the piercing section, and again, continuously, in the rolling section.

(3)   The distribution of velocity was also arranged in layers with a higher magnitude at the outside surface and the lowest one at the central axial zone in front of the plug.

(4)   Based on experiments in a TSR testing mill, the desired seamless tubes of carbon steel 1045, HSS 42CrMo, and the AZ31 alloy were obtained, with an accuracy of ± 0.2 mm in wall thickness and a diameter of ±0.35 mm.

(5)   The TSR process is suitable for manufacturing seamless tubes that are difficult to deform or that are deformed in a narrow temperature range.

Author Contributions: Y.S. and F.W. conceived and designed the experiments; F.M., J.H., and J.C. performed the experiments; F.M. and J.H. analyzed the data; F.M. performed the finite element simulations and wrote the manuscript. All authors have read and agreed to the published version of the manuscript.

Funding: This research was funded by the Scientific and Technological Innovation Programs of Higher Education Institutions in Shanxi grant number [2019L0852], the Major Project of the Ministry of Science and Technology of Shanxi Province in China grant number [20191102009], and the Start-Up Foundation for Doctors of Yuncheng University grant number [YQ–2019006]. And the APC was funded by [YQ–2019006].

Acknowledgments: The authors acknowledge the Funds for the Scientific and Technological Innovation Programs of Higher Education Institutions in Shanxi (No. 2019L0852), the Major Project of the Ministry of Science and Technology of Shanxi Province, China (No. 20191102009), and the Start-Up Foundation for Doctors of Yuncheng University (No.YQ–2019006).

Conflicts of Interest: The authors declare that there is no conflict of interest.

References

1.   Wang, F.J.; Shuang, Y.H.; Zhang, G.Q. A new type of seamless steel pipe production process-tandem skew rolling process. *Steel Pipe* **2014**, *43*, 54–58.

2.   Wang, F.-J.; Shuang, Y.-H.; Hu, J.-H.; Wang, Q.-H.; Sun, J.-C. Explorative study of tandem skew rolling process for producing seamless steel tubes. *J. Mater. Process. Technol.* **2014**, *214*, 1597–1604. [CrossRef]

3.   Wang, F.J.; Shuang, Y.H. Tandem skew rolling process for compact producing seamless steel tubes. *Iron Steel* **2016**, *21*, 44–48.

4.   Shuang, Y.H.; Wang, F.J.; Wang, Q.H. Explorative study of tandem skew rolling process and equipment for producing seamless steel tubes. *J. Mech. Eng.* **2017**, *53*, 18–24. [CrossRef]

5.   Mao, F.L.; Shuang, Y.H.; Wang, Q.H.; Wang, F.J.; Gou, Y.J.; Zhao, C.J. Theoretical and Experimental Study of the Tandem Skew Rolling Process. *Steel Res. Int.* **2018**, *89*, 1800022. [CrossRef]

6.   Niu, X.; Shuang, Y.H.; Wang, Q.H.; Wang, F.J. Discussion on velocity coordination between roll units of tandem skew rolling process. *Steel Pipe* **2016**, *45*, 54–58.

7.   Mao, F.L.; Shuang, Y.H.; Wang, Q.H.; Liu, Q.Z. Tension detecting device and its experimental study of a steel tube skew tandem rolling mill. *Hot Work. Technol.* **2017**, *46*, 150–155.

8.   Mao, F.L.; Shuang, Y.H.; Wang, Q.H.; Wang, F.J. Effect of process parameters on rolling force and dimensional precision of tubes rolled by tandem skew rolling process. *J. Plast. Eng.* **2018**, *25*, 108–114.

9.   Wang, Q.H.; Shuang, Y.H. Research on the speed setting model and control of the tandem skew rolling mill. *Process Auto. Instrum.* **2017**, *38*, 13–18.

10. Wang, Q.H.; Shuang, Y.H.; Mao, F.L. Predictive control for speed and tension system of tandem skew rolling mill. *J. Taiyuan Univ. Tech.* **2017**, *48*, 991–995, 1028.

11. Wang, F.J.; Shuang, Y.H.; Hu, J.H.; Sun, J.C.; Wang, Q.H. Numerical simulation and experimental analysis of three-roll cross piercing process. *Hot Work. Technol.* **2014**, *43*, 95–98, 105.

12. Bogatov, A.A.; Nukhov, D.S.; Toporov, V.A. Simulation of rotary piercing process. *Metallurgist* **2017**, *61*, 101–105. [CrossRef]

13. Pater, Z.; Tomczak, J.; Bulzak, T. Numerical analysis of the skew rolling process for main shafts. *Metalurgija* **2015**, *54*, 627–630.

14. Pater, Z.; Kazanecki, J.; Bartnicki, J. Three dimensional thermo-mechanical simulation of the tube forming process in Diescher's mill. *J. Mater. Process. Technol.* **2006**, *177*, 167–170. [CrossRef]

15. Pater, Z.; Kazanecki, J. thermo-mechanical analysis of piercing plug loads in the skew rolling process of thick-walled tube shell. *Met. Foundry Eng.* **2006**, *32*, 31–40. [CrossRef]

16. Stefanik, A.; Morel, A.; Mróz, S.; Szota, P. Theoretical And Experimental Analysis Of Aluminium Bars Rolling Process In Three-High Skew Rolling Mill. *Arch. Met. Mater.* **2015**, *60*, 809–813. [CrossRef]

17. Romantsev, B.A.; Skripalenko, M.M.; Huy, T.B.; Skripalenko, M.N.; Gladkov, Y.A.; Gartvig, A.A. Computer Simulation of Piercing in a Four-High Screw Rolling Mill. *Metallurgist* **2018**, *61*, 729–735. [CrossRef]

18. Toporov, V.A.; Chepurin, M.V.; Parfenov, V.A.; Stepanov, A.I. Skew rolling in the piercing of blanks. *Steel Transl.* **2014**, *44*, 452–455. [CrossRef]

19. Murillo-Marrodán, A.; García, E.; Cortés, F. Study of Friction Model Effect on A Skew Hot Rolling Numerical Analysis. In *The World Congress on Engineering*; Springer: Singapore, 2017; pp. 377–387.

20. Ding, X.F.; Shuang, Y.H.; Wang, Q.H.; Zhou, Y.; Gou, Y.J.; Wang, J.; Lin, W.L. New Rotary Piercing Technique of AZ31 Magnesium Alloy Seamless Tube. Rare Metal. *Mat. Eng.* **2018**, *47*, 357–362.

21. Ding, X.F.; Shuang, Y.H.; Liu, Q.Z.; Zhao, C.J. New rotary piercing process for an AZ31 magnesium alloy seamless tube. *Mater. Sci. Technol.* **2018**, *34*, 408–418. [CrossRef]

Article

Influence of Heat Treatment on the Workability of Modified 9Cr-2W Steel with Higher B Content

Hyeong Min Heo [1,*], Jun Hwan Kim [2], Sung Ho Kim [2], Jong Ryoul Kim [3] and Won Jin Moon [1]

[1] Gwangju center, Korea Basic Science Institute, 77 Yongbong-ro, Buk-gu, Gwangju 61186, Korea
[2] Innovative Fuel Technology Development Division, KAERI, 989-111 Daedeok-daero, Yuseong-gu, Daejeon 34057, Korea
[3] Department of Materials Engineering, Hanyang University, 55, Hanyangdaehak-ro, Sangnok-gu, Ansan-si 15588, Korea
*   Correspondence: herhm87@naver.com; Tel.: +82-62-712-4499

Received: 15 July 2019; Accepted: 16 August 2019; Published: 18 August 2019

Abstract: In this study, the effect of heat treatment on the fracture behavior of alloy B steel with boron (B) contents as high as 130 ppm was investigated. The Alloy B are derived from Gr.92 steel with outstanding creep characteristics. The amounts of minor alloying elements such as B, N, Nb, Ta, and C were optimized to achieve better mechanical properties at high temperatures. Hence, workability of the alloy B and Gr.92 were compared. An increase in the B content affected the phase transformation temperature and texture of the steel. The development of the {111}<uvw> components in γ-fibers depended on the austenite fraction of the steel after the phase transformation. An increase in the B content of the steel increased its α-to-γ phase transformation temperature, thus preventing the occurrence of sufficient transformation under the normalizing condition. Cracks occurred at the point of the elastic-to-plastic deformation transition in the normal direction during the rolling process, thereby resulting in failure. Therefore, it is necessary to avoid intermediate heat treatment conditions, in which γ-fibers do not fully develop, i.e., to avoid an imperfect normalization.

Keywords: modified 9Cr-2W steel; B content; phase transformation; texture; heat treatment

1. Introduction

Owing to its low-cost raw materials and excellent base load capacity, nuclear power generation plays a key role in meeting the energy demands of Korea. However, the large-scale commercialization of nuclear power is limited because of the large amounts of nuclear fuel spent after the operation. In order to overcome this limitation, it is imperative to develop an innovative nuclear power plant, which can contribute to sustainable development and environmental protection while producing sufficient energy to meet the increasing demands [1,2].

Sodium cooled fast reactors (SFRs) have gained immense attention in this context owing to their economic efficiency, stability, nuclear non-proliferation, and low emission of spent nuclear fuel. These reactors operate at temperatures much higher than the operating temperatures of light water reactors and exhibit high thermal efficiency [3–7]. The nuclear fuel cladding tube, which transfers the fission energy and comprises the nuclear fuel rod and fissile material, is the main component of these reactors. Hence, the development of efficient fuel cladding tubes is essential as they are crucial for the safety of nuclear reactors.

Ferritic-martensitic (FM) steels are considered to be promising candidates for fuel cladding and duct materials for SFRs owing to their high thermal conductivities, low expansion coefficients, and superior irradiation swelling compared to that of austenitic steels [8–10]. However, the creep rupture strength of FM steels abruptly decreases during the long-term creep exposure at high temperatures [11]. Advanced FM steel alloys are derived from Gr.92 steel. The amounts of minor alloying elements such as B, N,

Nb, Ta, and C were optimized to achieve better mechanical properties at high temperatures [10]. In our previous study, we prepared 36 model alloy ingots of 9Cr-2W steel by adjusting their B and N contents, and the two model alloy ingots with the best creep characteristics were used to design cladding tubes [12]. Research on the material property and the manufacturing technology of these alloys will considerably contribute to extend their application as nuclear reactor parts with excellent creep characteristics and the ability to withstand high temperatures.

The optimum heat treatment conditions and amount of cold work depend on the manufacturing process parameters such as the cold rolling rate and normalizing and tempering heat treatment temperature and time [8]. However, when an alloy B steel plate with a B content as high as 130 ppm is rolled, fracture occurs only when it is subjected to tempering after the normalization. In addition, similar phenomena occur during the cold drawing process under the same conditions. Fracture during the manufacturing process reduces the yield of alloy B steels, resulting in a low productivity.

In this study, the causes of fracture during the manufacturing process of alloy B steels with low production yield were investigated in order to develop a novel manufacturing process for alloy B steels with high formability. The specimens of the rolled alloy B and Gr.92 steels subjected to different intermediate heat-treatment conditions were observed using an optical microscope (OM), a stereoscope, nano-secondary ion mass spectrometry (Nano-SIMS), and electron backscatter diffraction (EBSD). In addition, the mechanical properties of the rolled specimens along the orientation direction were evaluated by carrying out their compression tests.

2. Materials and Methods

2.1. Materials

The chemical compositions of the two different kinds of 9Cr-2W steel, designated alloy B and Gr.92, are shown in Table 1. In order to investigate the effect the heat treatment conditions on the formability of alloy B steel, plate specimens were manufactured as follows. Ingots were prepared using a vacuum induction melting process and were fabricated into round bars, which were subjected to hot forging. The round bars had an outer diameter of 100 mm. The hot-forged specimens were normalized at 1050 °C for 6 min, followed by air cooling (AC) at 298 K. Tempering of the normalized specimens was carried out at 800 °C for 6 min, followed by AC. The reference material, Gr.92 with an outer diameter of 200 mm, was purchased and tested. Gr.92 was normalized at 1040 °C for 5 h, followed by AC. Tempering of the normalized specimens was carried out at 760 °C for 8 h, followed by AC. The alloy B and Gr.92 specimens were subjected to the same initial heat treatment and normalizing and tempering conditions. Both the steel specimens had an outer diameter of 100 mm and a thickness of 5 mm.

Table 1. Chemical compositions of the alloy B and Gr.92 steel specimens (wt. %).

Materials	C	Si	Mn	P	S	Ni	Cr	Mo	W	V	Nb	Ta	N	B	Fe
Alloy B	0.07	0.10	0.45	0.0056	0.001	0.45	9.0	0.5	2.0	0.20	0.2	0.05	0.02	0.013	Bal
Gr.92	0.1	0.45	0.44	0.02	0.009	0.49	9.0	0.46	1.60	0.20	0.07	-	0.04	0.004	Bal

Cold rolling was performed at the same reduction ratio as the cold drawing process in which fracture occurred. The specimens were rolled to a thickness of 3 mm (from 5 mm), showing a reduction ratio of 40%. The distance between rolling rolls was fixed at 3 mm, and rolling was performed until the thickness of the specimen became 3 mm.

The rolled specimens were heat-treated. The tempering treatment of the rolled specimens was carried out at 800 °C for 6 min, followed by AC. The specimens were normalized at 1050 °C for 6, 15, and 30 min, and 1150 °C for 6 min, followed by AC and were tempered at 800 °C for 6 min, followed by AC. A schematic of the experimental procedure is shown in Figure 1.

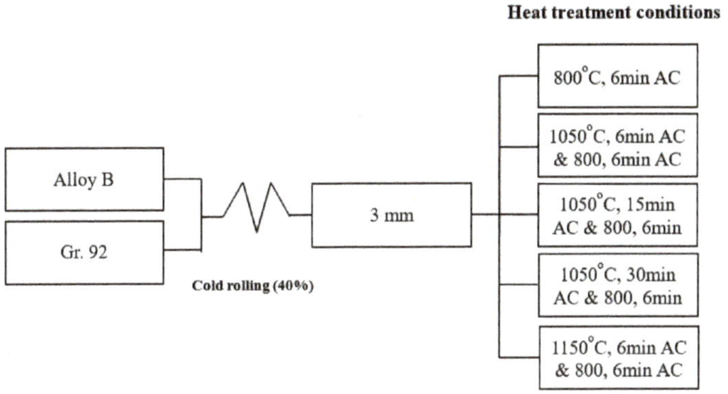

Figure 1. Schematic of the experimental procedure. AC: air cooling.

2.2. Observation of Microstructure

The microstructures of the heat-treated alloy B and Gr.92 specimens were investigated using OM and EBSD. The specimens for the microstructural observations were grinded, polished (using a diamond suspension of up to 0.25 μm, and etched (95 mL water + 3 mL nitric acid + 2 mL fluoric acid). The microstructures of the heat-treated specimens were examined by carrying out their EBSD analyses after a metallographic sample preparation step. The EBSD samples required additional one-hour polishing with a 0.04 μm colloidal silica suspension. The EBSD patterns of the specimens were obtained using a Su5000 (Hitachi High Technology, Omuta-shi, Fukuoka, Japan) instrument equipped with a Hikari detector operating with EDAX-TSL OIM DATA collection software (AMETEK Inc., Berwyn, PA, USA). The crystallographic textures of the specimens formed during the various heat treatment processes were recorded. The orientation distribution functions (ODFs) $f(g)$ of the specimens were calculated using the harmonic series expansion method based on Bunge (l = 22) from their incomplete pole figures {110}, {200}, and {211} [13,14]. The orientation g was expressed in the form of a triple Euler angle (φ_1, Φ, φ_2). All the ODF calculations were carried out assuming that the samples showed a triclinic symmetry, as given by the rolling direction (RD), transverse direction (TD), and normal direction (ND) of a plate such that $0° \leq \{\varphi_1, \Phi, \varphi_2\} \leq 90°$. Subsequently, secondary ion mass spectrometry (SIMS) imaging with a Nano-SIMS (CAMECA, Gennevilliers, Hauts-de-Seine, France) device was used to observe boron segregation at the grain boundaries. The Nano-SIMS specimens were prepared by a procedure identical to the aforementioned method and were analyzed. Nano-SIMS measurements with statistical errors of less than 0.2% were obtained by using a CS^+ gun beam with a diameter of 80 nm and impact energy corresponding to 16 keV and 0.35 pA. Prior to obtaining the distribution maps, the concentration depth profiles of $11B^+$, $12C^-$, $52Cr^+$, and $56Fe^+$ were determined by Nano-SIMS in an area of 10×10 μm^2. $^{11}B^{16}O_2{}^-$, $^{12}C^-$, $^{52}Cr^{16}O^-$, and $^{56}Fe^{16}O^-$ ions were detected and images of boron, carbon, chromium, and iron were obtained [15]. The analysis was performed at the Busan center of the Korea Basic Science Institute (KBSI).

2.3. Compression Test

The applicability of the compression test as a general method for evaluating the cold workability of steel has been investigated [16–18]. Compression tests were carried out at 298 K and the strain rate of 1 mm/min using an INSTRON-3367 (INSTRON, Norwood, MA, USA) system to evaluate the mechanical properties and workability of the alloy B and Gr.92 specimens subjected to the different heat treatment processes. The sampling and compression test procedure of the specimens are shown in Figure 2. The specimens were obtained by rolling a plate in three different directions, i.e., RD, TD, and ND. These specimens were cubes with the dimensions of 2 mm × 2 mm × 2 mm and had a smooth surface. The actual displacement and load of the specimens were recorded continuously during the compression test. In addition, the occurrence of cracks in the specimens was monitored.

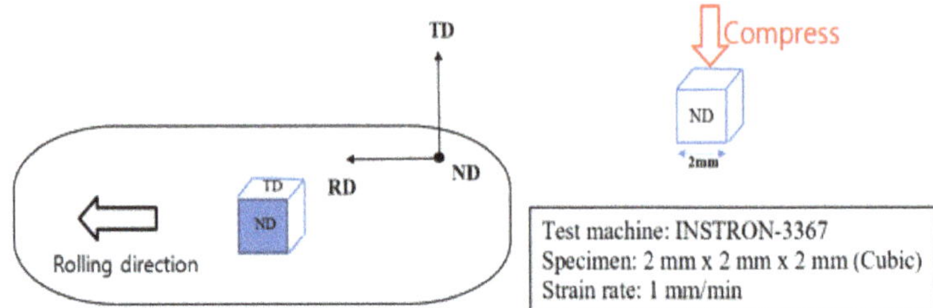

Figure 2. Sampling and compression of the specimens. RD: rolling direction; TD: transverse direction; ND: normal direction.

3. Results

3.1. Observation of the Fracture Surfaces of the Alloy B and Gr.92 Steels under Different Heat Treatment Conditions

The fracture surface of the rolled alloy B plate subjected to normalization and tempering is shown in Figure 3. The alloy B plate was characterized using scanning electron microscopy (SEM) to determine its mode of fracture, as shown in Figure 3b. The fracture surface of the plate showed transgranular cleavage facets. The fractograph clearly revealed that the mode of fracture was brittle.

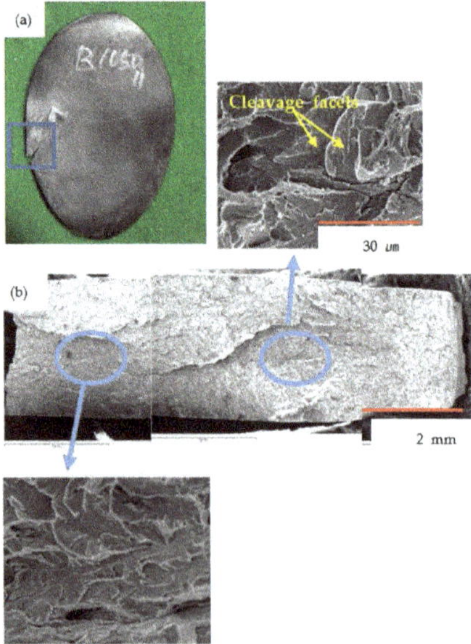

Figure 3. Alloy B steel during the rolling process under the tempering condition after the normalization: (**a**) fractograph of failed part and (**b**) scanning electron microscopy (SEM) image.

The OM images of the alloy B and Gr.92 steel specimens heat-treated under different conditions are shown in Figures 4 and 5, respectively. The rolling process resulted in the formation of an elongated grain boundary in the RD of the specimens. Elongated grain boundary was observed when the alloy B and Gr.92 specimens were only tempered (800 °C, 6 min) without normalization. However, the alloy B specimens heat-treated at 1050 °C for 6 min and 800 °C for 6 min showed elongated grain boundaries and the deformation of the structure. The remaining heat treatment conditions yielded a tempered lath martensite structure.

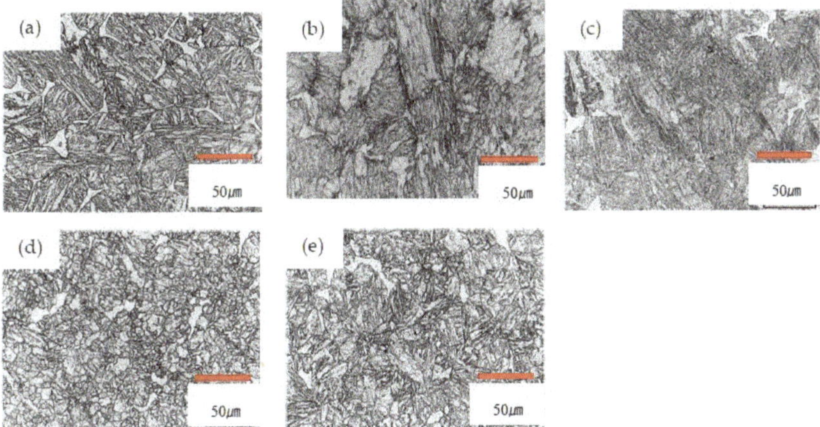

Figure 4. Optical microscope (OM) images of the alloy B steel specimens subjected to different intermediate heat treatment conditions: (**a**) as-received; (**b**) 800 °C, 6 min; (**c**) 1050 °C, 6 min and 800 °C, 6 min; (**d**) 1050 °C, 30 min, and 800 °C, 6 min; (**e**) 1150 °C, 6 min, and 800 °C, 6 min.

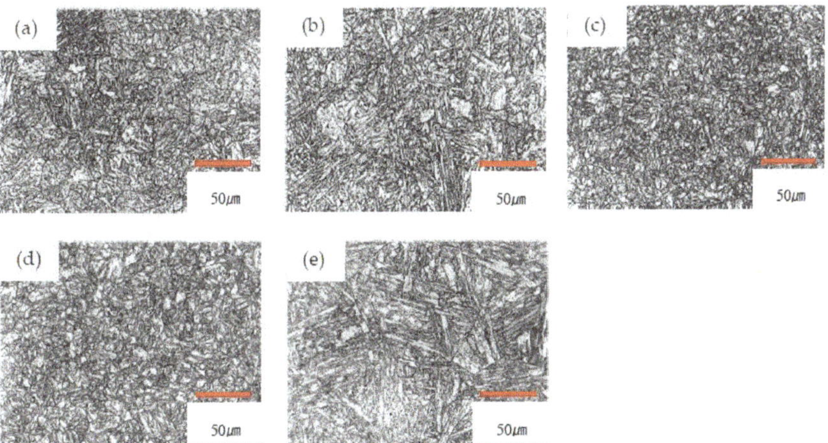

Figure 5. OM images of the Gr.92 steel specimens subjected to different intermediate heat treatment conditions: (**a**) as-received; (**b**) 800 °C, 6 min; (**c**) 1050 °C, 6 min, and 800 °C, 6 min; (**d**) 1050 °C, 30 min, and 800 °C, 6 min; (**e**) 1150 °C, 6 min, and 800 °C, 6 min.

The image quality map and grain boundary character distribution (GBCD) of the alloy B and Gr.92 specimens heat-treated under the 1050 °C, 6 min and 800 °C, 6 min are shown in Figure 6. An image quality map was derived from EBSD analysis of the ND, where the high angle grain boundary (HAGB, misorientation angle > 15, and color code: red) and coincidence site lattice (CSL, color key: Σ3—blue, Σ11—yellow, Σ17—violet, Σ19—brown, Σ25—orange) are shown in Figure 6a,b. The image quality map of the alloy B appeared as an elongated HAGB. In the contrast, the Gr.92 exhibited an equiaxed HAGB. The fraction of low angle grain boundary (LAGB) and HAGB in alloy B were 0.65 and 0.30, respectively. In the case of Gr.92, LAGB and HAGB values were 0.41 and 0.39, respectively. The mis-orientation angle distribution plot shows that the fraction of LAGB in alloy B was higher than that of Gr.92. The reason for this is that the fraction of LAGB was high due to the formation of cell and sub grain during the cold deformation when alloy B does not undergo complete phase transformation from martensite to austenite despite the normalizing heat treatment. In contrast, the fraction of HAGB in the Gr.92 was higher than that of Alloy B because the complete phase transformation formed a lot of prior austenite grain boundary (PAGB) and tempered martensite.

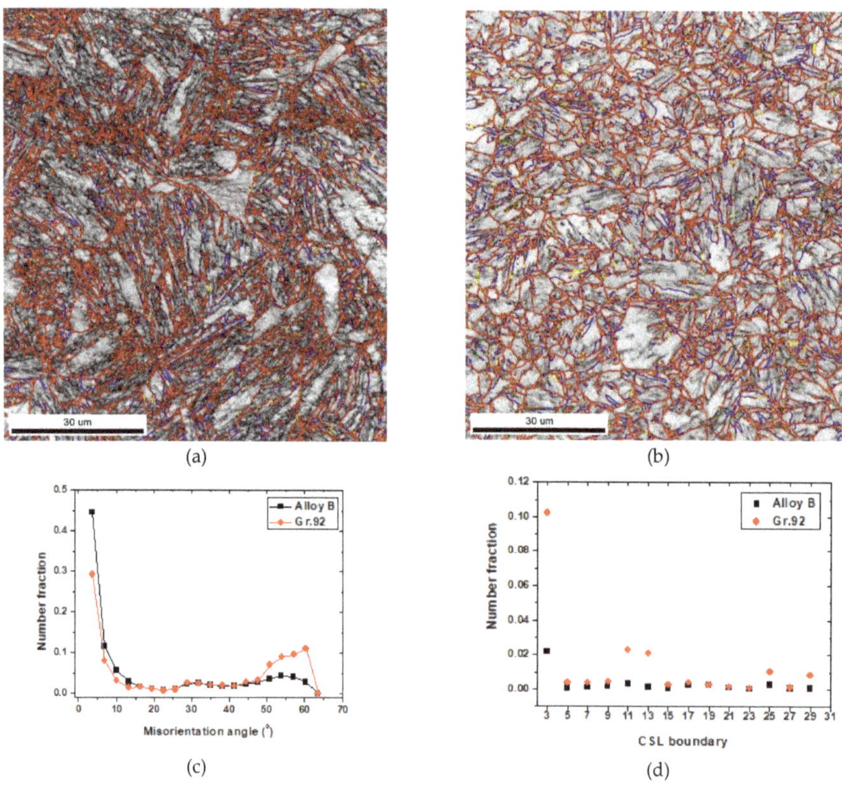

Figure 6. The image quality map and grain boundary character distribution (GBCD) of the alloy B and Gr.92 specimens heat-treated under the 1050 °C, 6 min and 800 °C, 6 min conditions: (**a**,**b**) the image quality map of alloy B and Gr.92 are denoted by the high angle grain boundary (HAGB) (mis orientation angle > 15, color code: red) and CSL boundary (color key: Σ3—blue, Σ11—yellow, Σ17—violet, Σ19—brown, Σ25—orange (**c**) misorientation angle distribution plot, and (**d**) coincidence site lattice (CSL) boundary distribution plot.

Figure 6d shows that the number fraction of the CSL boundary (Σ3–29) in alloy B and Gr.92 accounted for about 0.04 and 0.18 of the total number of boundaries, respectively. Among the CSL boundaries, the most prominent were Σ3, Σ11, Σ13, and Σ25. The CSL boundaries of Σ3 and Σ11 could be preferentially found within regions of martensite.

3.2. Texture

The typical main texture components of cold-rolled low-carbon steels in the $\varphi2 = 45°$ section of the ODF are shown in Figure 7 [19]. Body-centered cubic metals and alloys tend to form α- and γ-fibers. Typically, α-fibers exhibit the orientation and a common <110> crystal axis parallel to the RD, i.e., the {hkl}<110> orientations such as {001}<110>, {112}<110>, and {111}<110> at $(\varphi_1, \Phi, \varphi_2) = (0°, 0°, 45°)$, $(0°, 35°, 45°)$, and $(0°, 54.7°, 45°)$, respectively, in the Euler space. On the other hand, γ-fibers exhibit orientations with the {111} crystal axis parallel to the ND i.e., the {111}<uvw> orientations such as {111}<110> and {111}<112> at $(\varphi_1, \Phi, \varphi_2) = (60°, 54.7°, 45°)$ and $(90°, 54.7°, 45°)$, respectively, in the Euler space. In addition, the Goss orientation {011}<100> is typically observed at $(90°, 90°, 45°)$ in the Euler space [19,20].

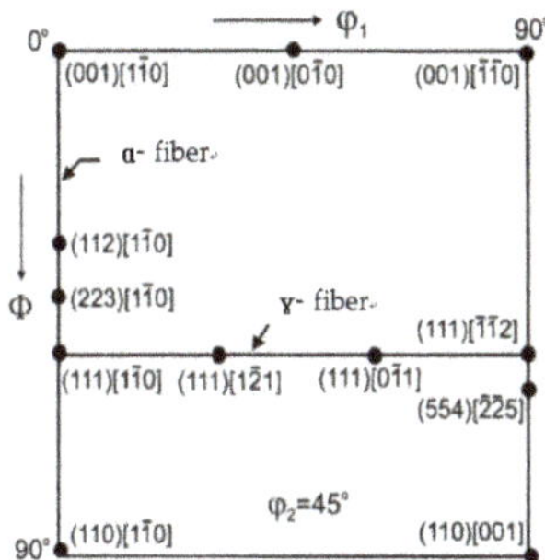

Figure 7. Main texture components of cold-rolled low carbon steels in the $\varphi_2 = 45°$ section of the orientation distribution functions (ODF) [19].

The ODFs for the $\varphi_2 = 45°$ sections of the alloy B and Gr.92 specimens heat-treated under different conditions are shown in Figures 8 and 9, respectively. The evolution of the ODF intensity of the γ-fibers in the alloy B and Gr.92 specimens under different heat treatment conditions is shown in Figure 10. Under all the heat treatment conditions, both the alloy B and Gr.92 specimens showed the development of α-fibers, which exhibited orientation and a common <110> crystal axis parallel to the RD. In the case of the alloy B steel, the as-received specimen and the specimens heat-treated at 1050 °C for 6 min and 800 °C for 6 min did not develop γ-fiber components, which is advantageous for formability. In all the other heat treatment conditions, γ-fiber components were developed. On the other hand, the Gr.92 steel showed γ-fiber components under all the conditions. This is because the γ-fiber components were developed uniformly under the initial conditions in the case of the Gr.92 steel.

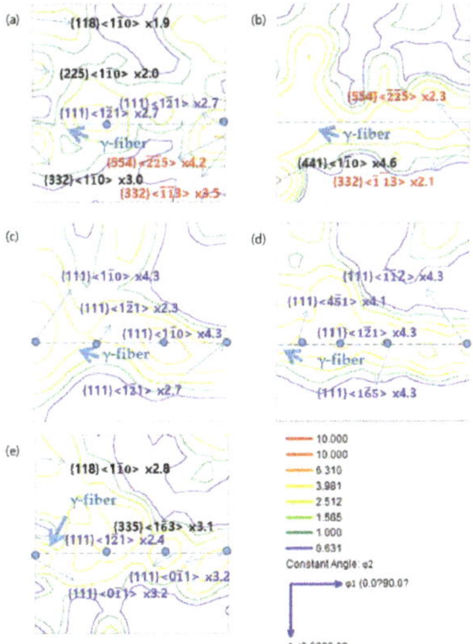

Figure 8. $\varphi_2 = 45°$ section of the ODF in the alloy B steel specimens heat-treated under different conditions: (**a**) as-received, (**b**) 800 °C, 6 min, (**c**) 1050 °C, 6 min, and 800 °C, 6 min, (**d**) 1050 °C, 30 min, and 800 °C, 6 min, and (**e**) 1150 °C, 6 min, and 800 °C, 6 min.

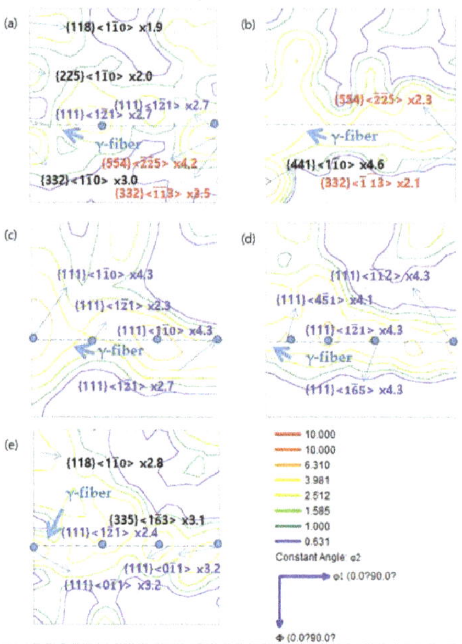

Figure 9. $\varphi_2 = 45°$ section of the ODF of the Gr.92 steel specimens heat-treated under different conditions: (**a**) as-received, (**b**) 800 °C, 6 min, (**c**) 1050 °C, 6 min, and 800 °C, 6 min, (**d**) 1050 °C, 30 min, and 800 °C, 6 min, and (**e**) 1150 °C, 6 min, and 800 °C, 6 min.

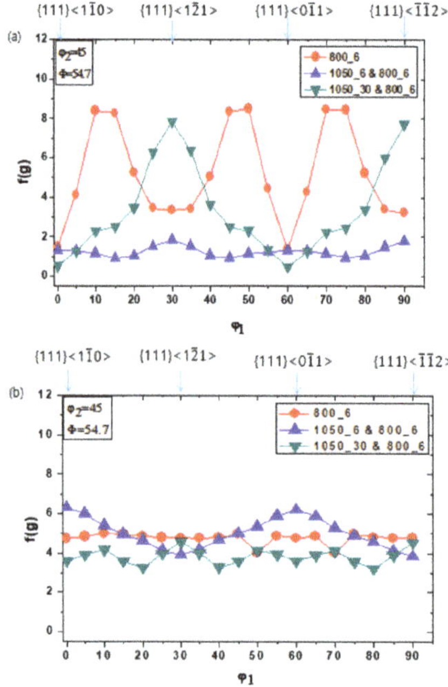

Figure 10. Evolution of the ODF intensity $f(g)$ in the γ-fiber ($\Phi = 54.7°$ and $\varphi_2 = 45°$, constant) components of the steels subjected to different intermediate heat treatment conditions: (**a**) alloy B and (**b**) Gr.92.

3.3. Effects of Heat Treatment Conditions on the Mechanical Properties of the Steels

The results of the compression tests of the alloy B and Gr.92 steels along the ND at 298 K are shown in Figure 11. Except for the 1050 °C, 6 min and 800 °C, 6 min conditions, all the conditions showed the same compression test results. The ultimate reduction in height R_c (%) is calculated as:

$$R_c(\%) = \frac{H-h}{H} \times 100 \tag{1}$$

where H is the height of the specimen before test and h is the height of the specimen after test. The compressive test was performed up to 1.7 mm due to the limitation of the test equipment. Additionally, the 0.2% yield strength and the ultimate compressive strength are obtained by recording the displacement and load. In the case of the alloy B specimens (along the ND), the hardening and strength increased rapidly in the elastic region. Therefore, fracture occurred during the elastic-to-plastic transition. The images of the alloy B and Gr.92 specimens after the compression along the ND are shown in Figure 12. In the case of the Gr.92 specimens, when the rolling was successful without fracture, cracks were absent in all the directions. In the case of the alloy B, specimens heat-treated under the 1050 °C, 6 min and 800 °C, 6 min conditions (wherein fracture was observed), cracks were observed in the ND.

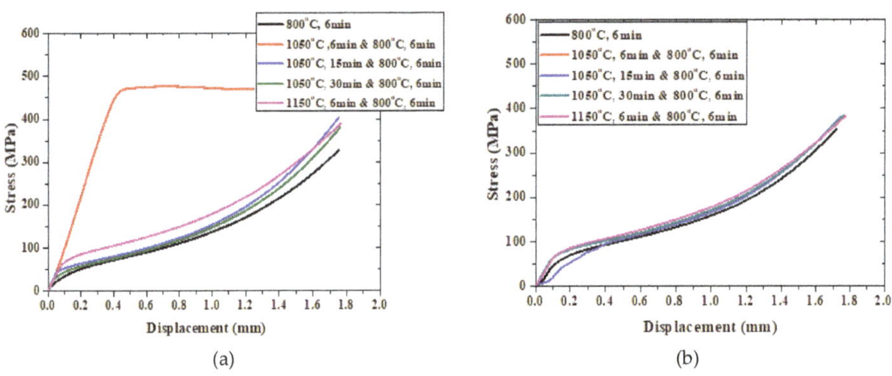

	Yield Strength (MPa)	Ultimate Compressive Strength (MPa)	Elongation (%)
800 °C, 6 min	36	306	85
1050 °C, 6 min and 800 °C, 6 min	353	475	40
1050 °C, 15 min and 800 °C, 6 min	49	404	85
1050 °C, 30 min and 800 °C, 6 min	45	380	85
1150 °C, 6 min and 800 °C, 6 min	58	388	85

(c)

	Yield Strength (MPa)	Ultimate Compressive Strength (MPa)	Elongation (%)
800 °C, 6 min	45	353	85
1050 °C, 6 min and 800 °C, 6 min	74	381	85
1050 °C, 15 min and 800 °C, 6 min	60	384	85
1050 °C, 30 min and 800 °C, 6 min	64	383	85
1150 °C, 6 min and 800 °C, 6 min	62	384	85

(d)

Figure 11. Compression test of the specimens heat-treated under different conditions along the ND: (**a**) alloy B, (**b**) Gr.92, (**c**) mechanical properties of Alloy B, and (**d**) mechanical properties of Gr.92.

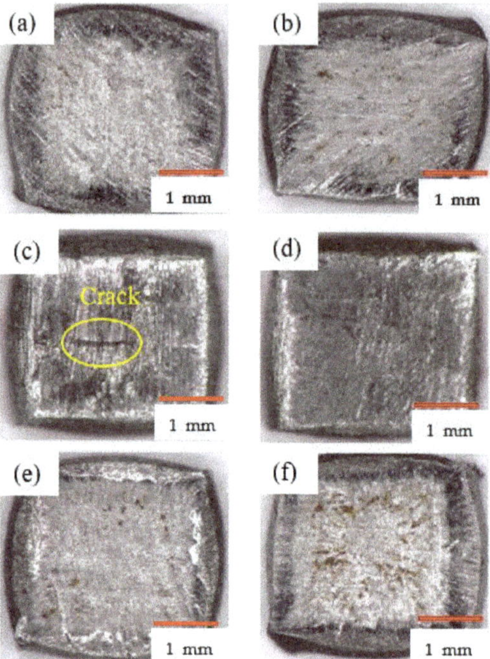

Figure 12. Images of the specimens heat-treated under different conditions after the compression test along the ND: (**a**) alloy B 800 °C, 6 min, (**b**) Gr.92 800 °C, 6 min, (**c**) alloy B 1050 °C, 6 min, and 800 °C, 6 min (**d**) Gr.92 1050 °C, 6 min, and 800 °C, 6 min, (**e**) alloy B 1150 °C, 6 min, and 800 °C, 6 min, and (**f**) Gr.92 1150 °C, 6 min, and 800 °C, 6 min.

4. Discussion

In this study, the effect of heat treatment on the fracture behavior of alloy B steel with boron (B) contents as high as 130 ppm was investigated. As shown in Figures 3 and 4, the alloy B steel showed cleavage fracture only under the 1050 °C, 6 min and 800 °C, 6 min conditions. Tempered lath martensite and deformed microstructures were obtained after the heat treatment. In the case of tempering, the austenite phase transformation does not occur, so some deformed microstructure remains [21]. On the other hand, the deformed microstructure transforms into an austenitic structure by heating at the phase transformation temperature (Ac$_3$, the temperature at which the α-to-γ transformation commences) or higher temperatures under the normalized condition [22]. The deformed microstructure disappears during the austenitization and a lath martensite structure is formed with a high dislocation density in the PAGB upon cooling [23]. PAGBs are divided into packet boundaries parallel to a group of laths with the same habit plane, and each packet boundary is further divided into block boundaries with the same orientation as that of the lath group. The lath and packet are parallel to the {111} plane of the austenite phase and exhibit a K-S orientation relationship [24]. In a packet, there are six variants with different direction parallel relationships on the same conjugate parallel close-packed plane (e.g., (111)γ // (011)α) [24,25].

However, the alloy B steel specimens heat-treated under the 1050 °C, 6 min and 800 °C, 6 min conditions showed deformed microstructure. This can be attributed to the segregation of B at the grain boundaries. It is well known that grain boundary segregation of trace element (such as B, As, Sn, Cu etc.) affect the chemical composition of interfaces, causing increased material embrittlement due to a reduction of cohesion at the interfaces, or it can stabilize the nanocrystalline structure through reduced mobility of the grain boundaries [26–30].

In Figures 13 and 14, the alloy B specimens that were heat-treated to 1050 °C for 6 min and 800 °C for 6 min were used to conduct a Nano-SIMS analysis of the precipitate and the distribution of boron at grain boundaries.

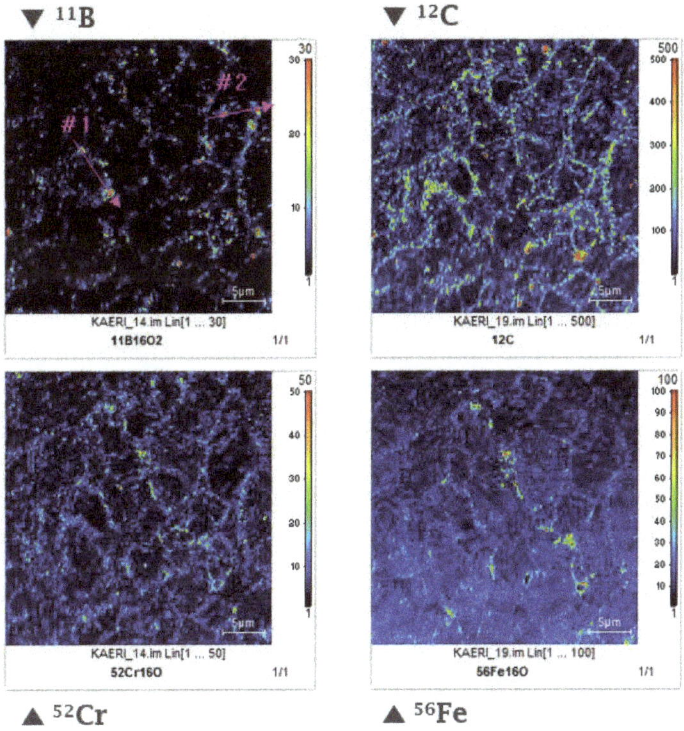

Figure 13. Nano-secondary ion mass spectrometry (Nano-SIMS) Images of B, C, Cr, and Fe of the alloy B specimens heat-treated under the 1050 °C, 6 min and 800 °C, 6 min conditions.

The segregation behavior of Alloy B specimens heat-treated to 1050 °C for 6 min and 800 °C for 6 min is explained by precipitation. Cr-based carbide $(Cr, Fe)_{23}(B, C)_6$ was the main precipitate of 9–12% Cr FM steel. When the content of boron is 0.004 wt % or more, a brittle Fe_2B precipitate is formed [31,32]. Fe_2B precipitate was not observed frequently. Equilibrium phase diagrams predicted by thermodynamic calculations demonstrate that the Fe_2B phase constitutes a fraction of 0.001 at temperatures exceeding 800 °C [32]. These are also similar to the Nano-SIMS results. The result of Nano-SIMS analysis demonstrated that the B and C distributions produced a behavior almost identical to the Cr and Fe distribution. If the observed Cr and Fe elements are assumed to have the distribution of $(Cr, Fe)_{23}(B, C)_6$, this indicates that the amount of boron incorporated into the $(Cr, Fe)_{23}(B, C)_6$ precipitate was highly dependent on the location of the precipitate associated with the PAGB [3,4]. Also, trace element Nb, Ta, and V affect to form of $(Cr, Fe)_{23}(B, C)_6$. The combination of boron and these elements suppress the α-to-γ transformation by stabilizing austenite due to formation of MX carbides with high thermal stability and reducing the rate of carbon diffusion and the inhibition of $(Cr, Fe)_{23}(B, C)_6$ [31]. The Ac3 of the alloy B steel was measured by thermomechanical analysis and was compared with that of the Gr.92 steel [32]. The alloy B specimens showed higher phase transformation temperatures than the Gr.92 specimens, as can be observed from Table 2. We believe that the precipitate is not the cause of the failure that occurred during processing. This is because there was no precipitate around the fracture surface.

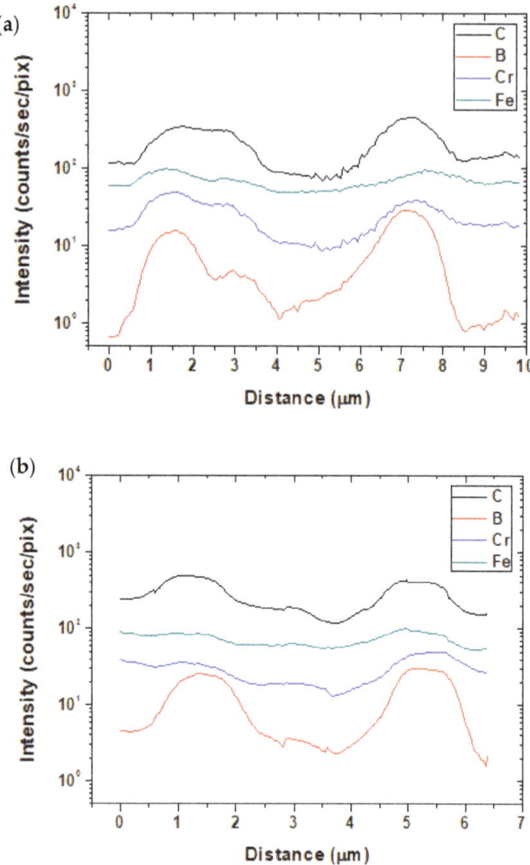

Figure 14. Nano-SIMS distribution map of sputtered carbon, boron, chromium, and iron ions in the areas denoted by the arrows of the alloy B specimens heat-treated under the 1050 °C, 6 min and 800 °C, 6 min conditions: (**a**) #1 (**b**) #2.

Table 2. Phase transformation temperatures for alloy B and Gr.92 steels [32].

Materials	Ac$_1$ (°C)	Ac$_3$ (°C)
Gr.92	844	879
Alloy B	852	909

The different textures of the two steels can be attributed to their different B contents. The texture of both the steels depended on the fraction of the austenite phase present in them during the intercritical annealing. During the $\alpha \rightarrow \gamma \rightarrow \alpha$ transformation, a texture with a strong {111} component develops by intercritical annealing with the transformation of a certain amount of the γ phase [33–35]. Therefore, in the case of the alloy B steel specimens, γ-fibers, which become parallel to the ND during the $\alpha \rightarrow \gamma \rightarrow \alpha$ transformation because of the incomplete normalization, were not formed.

Figure 11 shows the compression test results of the alloy B steel along the ND. Fracture commenced at the point where hardening occurred rapidly in the elastic region and the strength increased suddenly, resulting in the transformation of the elastic region into the fracture region. This is because the γ-fiber component did not develop on the ND plane and the plastic deformation did not occur in the ND plane.

Crack initiation occurs along the slip bands in a grain or at the grain boundary on the surface. In the crystallographic mode of ferrite steels, cleavage involves the separation of atomic bonds along the low index {001} crystal plane [36]. The separation process along the low index plane prefers to lower the surface energy. The low index plane lies on the {001} plane and other low-index planes such as {110}, {112}, and {123}. It has been reported that cleavage fracture is caused by the lack of slip bands [37]. This phenomenon was observed in the 1050 °C, 6 min and 800 °C, 6 min alloy B specimens because the ND // {111} plane (γ-fiber), which ensured that formability does not develop. Thus, the slip system did not function properly during the rolling and plastic deformation does not occur on the ND plane and fracture occurs on it.

5. Conclusions

In this study, a modified 9Cr-2W steel (Alloy B) was developed for application as the fuel cladding with good creep performance for SFRs. It is expected that the investigation of the properties and manufacturing processes of these alloys will significantly contribute to expanding the applicability of components with high-temperature strength and creep characteristics for nuclear reactors in the future.

The analysis of the fracture of the alloy B steel sample during the manufacturing process showed that an increase in the B content resulted in an increase in the $\alpha \rightarrow \gamma$ phase transformation temperature such that sufficient transformation did not occur under the normalizing condition. The effect of heat treatment on the alloy samples depended on their B contents. Under the incomplete heat treatment conditions, the γ-fiber components, which are advantageous for formability, did not develop in the samples. It was assumed that cracks occurred when the elastic-to-plastic deformation transition occurred in the ND during the rolling process, thereby resulting in failure. Therefore, it is necessary to avoid intermediate heat treatment conditions in which γ-fibers do not fully develop, i.e., an imperfect normalization should be avoided.

Author Contributions: Writing—original draft preparation, review and editing, H.M.H.; conceptualization; J.H.K.; project administration; S.H.K.; supervision; J.R.K. and W.J.M.

Funding: This research was funded by the Ministry of Science and ICT of South Korea (No. 2012M2A8A2025646).

Acknowledgments: This work was supported by the National Nuclear R & D program through the Sodium-cooled Fast Reactor Development Agency funded by the Ministry of Science and ICT of South Korea (2012M2A8A2025646). We would like to thank Editage (www.editage.co.kr) for editing and reviewing this manuscript for English language.

Conflicts of Interest: The authors declare no conflict of interest.

References

1. Kim, J.H.; Baek, J.H.; Kim, S.H.; Lee, C.B.; Na, K.S.; Kim, S.J. Effect of hot rolling process on the mechanical and microstructural property of the 9Cr-1Mo steel. *Ann. Nucl. Energy* **2011**, *38*, 2397–2403. [CrossRef]
2. Heo, H.M. Effect of Boron on the Formability of Modified 9Cr-2W steel. Ph.D. Thesis, Hanyang University, Ansan, Korea, February, 2019.
3. Jeong, E.H.; Park, S.G.; Kim, S.H.; Kim, Y. Do Evaluation of the effect of B and N on the microstructure of 9Cr–2W steel during an aging treatment for SFR fuel cladding tubes. *J. Nucl. Mater.* **2015**, *467*, 527–533. [CrossRef]
4. Heo, H.M.; Jeong, E.H.; Kim, S.H.; Kim, J.R. Comparison between effect of B and N on the microstructure of modi fi ed 9Cr-2W steel during aging and creep. *Mater. Sci. Eng. A* **2016**, *670*, 106–111. [CrossRef]
5. Kim, J.H.; Heo, H.M.; Kim, S.H. Effect of an intermediate process on the microstructure and mechanical properties of HT9 fuel gadding. *J. Korean Inst. Met. Mater.* **2013**, *51*, 893–900. [CrossRef]
6. György, H.; Czifrus, S. Investigation on the potential use of thorium as fuel for the Sodium-cooled Fast Reactor. *Ann. Nucl. Energy* **2017**, *103*, 238–250. [CrossRef]
7. Kim, J.T.; Jae, M.S. A study on reliability assessment of a decay heat removal system for a sodium-cooled fast reactor. *Ann. Nucl. Energy* **2018**, *120*, 534–539. [CrossRef]
8. Kim, T.K.; Kim, S.H.; Lee, C.B. Effects of an intermediate heat treatment during a cold rolling on the tensile strength of a 9Cr-2W steel. *Ann. Nucl. Energy* **2009**, *36*, 1103–1107. [CrossRef]

9. Garner, F.A.; Toloczko, M.B.; Sencer, B.H. Comparison of swelling and irradiation creep behavior of fcc-austenitic and bcc-ferritic/martensitic alloys at high neutron exposure. *J. Nucl. Mater.* **2000**, *276*, 123–142. [CrossRef]

10. Lee, C.B.; Cheon, J.S.; Kim, S.H.; Park, J.Y.; Joo, H.K. Metal Fuel Development and Verification for Prototype Generation IV Sodium-Cooled Fast Reactor. *Nucl. Eng. Technol.* **2016**, *48*, 1096–1108. [CrossRef]

11. Kim, S.H.; Song, B.J.; Ryu, W.S.; Hong, J.H. Creep rupture properties of nitrogen added 10Cr ferritic/martensitic steels. *J. Nucl. Mater.* **2004**, *329–333*, 299–303. [CrossRef]

12. Jeong, E.H.; Kim, J.H.; Kim, S.H.; Kim, Y. Do Influence of B and N on the microstructural characteristics and high-temperature strength of 9Cr-2W steel during an aging treatment. *Mater. Sci. Eng. A* **2017**, *700*, 701–706. [CrossRef]

13. Bunge, H.J. *Texture Analysis in Materials Science*; Butterworths: London, UK, 1982; ISBN 9780408106429.

14. Choi, S.H.; Jin, Y.S. Evaluation of stored energy in cold-rolled steels from EBSD data. *Mater. Sci. Eng. A* **2004**, *371*, 149–159. [CrossRef]

15. Seol, J.B.; Lee, B.H.; Choi, P.; Lee, S.G.; Park, C.G. Combined nano-SIMS/AFM/EBSD analysis and atom probe tomography, of carbon distribution in austenite/ε-martensite high-Mn steels. *Ultramicroscopy* **2013**, *132*, 248–257. [CrossRef]

16. Abe, H.; Furugen, M. Evaluation Method of Workability in Cold Pilgering of Zirconium-based Alloy Tube. *Mater. Trans.* **2010**, *51*, 1200–1205. [CrossRef]

17. Abe, H.; Furugen, M. Method of evaluating workability in cold pilgering. *J. Mater. Process. Technol.* **2012**, *212*, 1687–1693. [CrossRef]

18. Heo, H.-M.; Kim, J.-H.; Kim, S.-H.; Kim, J.-R. Evaluation of workability on the microstructure and mechanical property of modified 9Cr-2W steel for fuel cladding by cold drawing process and intermediate heat treatment condition. *Metals (Basel)* **2018**, *8*, 193. [CrossRef]

19. Humphreys, F.J.; Hatherly, M. *Recrystallization and Related Annealing Phenomena*; Sleeman, D., Ed.; Elsevier: Kidlington, UK, 2012; ISBN 9788578110796.

20. Huh, M.Y.; Engler, O. Effect of intermediate annealing on texture, formability and ridging of 17%Cr ferritic stainless steel sheet. *Mater. Sci. Eng. A* **2001**, *308*, 74–87. [CrossRef]

21. Birosca, S.; Ding, R.; Ooi, S.; Buckingham, R.; Coleman, C.; Dicks, K. Nanostructure characterisation of flow-formed Cr–Mo–V steel using transmission Kikuchi diffraction technique. *Ultramicroscopy* **2015**, *153*, 1–8. [CrossRef]

22. Klueh, R.L.; Harries, D.R. *High-Chromium Ferritic and Martensitic Steels for Nuclear Applications*; ASTM International: West Conshohocken, PA, USA, 2001; ISBN 978-0-8031-2090-7.

23. Saroja, S.; Vijayalakshmi, M.; Raghunathan, V.S. Influence of cooling rates on the transformation behaviour of 9Cr-1 Mo-O.07C steel. *J. Mater. Sci.* **1992**, *27*, 2389–2396. [CrossRef]

24. Morito, S.; Huang, X.; Furuhara, T.; Maki, T.; Hansen, N. The morphology and crystallography of lath martensite in Fe-C alloy steels. *Acta Mater.* **2006**, *54*, 5323–5331. [CrossRef]

25. Maki, T. Morphology and substructure of martensite in steels. *Phase Transform. Steels* **2012**, *2*, 34–58.

26. Lejček, P.; Hofmann, S.; Paidar, V. The significance of entropy in grain boundary segregation. *Materials (Basel)* **2019**, *12*, 492. [CrossRef]

27. Watanabe, T. An approach to grain-boundary design for strong and ductile polycrystals. *Res Mech.* **1984**, *11*, 47–84.

28. Kirchheim, R. Grain coarsening inhibited by solute segregation. *Acta Mater.* **2002**, *50*, 413–419. [CrossRef]

29. Briant, C.L. *Impurities in Engineering Materials: Impact, Reliability and Control*; Dekker, M., Ed.; Routledge: New York and Basel, 1999.

30. El-Shennawy, M.; Farahat, A.I.; Masoud, M.I.; Abdel-Aziz, A.I. Effect of Boron Content on Metallurgical and Mechanical. *Int. J. Mech. Eng.* **2016**, *5*, 1–14.

31. Hara, T.; Asahi, H.; Uemori, R.; Tamehiro, H. Role of Combined Addition of Niobium and Boron and of Molybdenum and Boron on Hardnenability in Low Carbon Steels. *ISIJ Int.* **2008**, *44*, 1431–1440. [CrossRef]

32. Wang, S.S.; Peng, D.L.; Chang, L.; Hui, X.D. Enhanced mechanical properties induced by refined heat treatment for 9Cr-0.5Mo-1.8W martensitic heat resistant steel. *Mater. Des.* **2013**, *50*, 174–180. [CrossRef]

33. Vivas, J.; Capdevila, C.; Jimenez, J.; Benito-Alfonso, M.; San-Martin, D. Effect of Ausforming Temperature on the Microstructure of G91 Steel. *Metals (Basel)* **2017**, *7*, 236. [CrossRef]

34. HASHIMOTO, O.; SATOH, S.; TANAKA, T. Development of {111} Texture in Intercritical Annealing of Low Carbon Steels. *Trans. Iron Steel Inst. Jpn.* **1987**, *27*, 746–754. [CrossRef]

35. Obasi, G.C.; Birosca, S.; Quinta Da Fonseca, J.; Preuss, M. Effect of β grain growth on variant selection and texture memory effect during $\alpha \rightarrow \beta \rightarrow \alpha$ phase transformation in Ti-6Al-4V. *Acta Mater.* **2012**, *60*, 1048–1058. [CrossRef]

36. Tu, M.Y.; Hsu, C.A.; Wang, W.H.; Hsu, Y.F. Comparison of microstructure and mechanical behavior of lower bainite and tempered martensite in JIS SK5 steel. *Mater. Chem. Phys.* **2008**, *107*, 418–425. [CrossRef]

37. Gross, D.; Seelig, T. *Fracture Mechanics With an Introduction to Micromechanics*; Ling, F.F., Ed.; Spring: Heidelberg, Germany, 2011; ISBN 9781118147757.

Article

Springback Calibration of a U-Shaped Electromagnetic Impulse Forming Process

Xiaohui Cui [1,2,3,*], Zhiwu Zhang [1], Hailiang Yu [1,2,3], Xiaoting Xiao [4] and Yongqi Cheng [4]

[1] Light Alloy Research Institute, Central South University, Changsha 410083, China; 163812020@csu.edu.cn (Z.Z.); yuhailiang@csu.edu.cn (H.Y.)
[2] State Key Laboratory of High Performance Complex Manufacturing, Central South University, Changsha 410083, China
[3] College of Mechanical and Electrical Engineering, Central South University, Changsha 410083, China
[4] School of Material and Energy, Guangdong University of Technology, Guangzhou 510000, China; xiaoxt@gdut.edu.cn (X.X.); chengyq@gdut.edu.cn (Y.C.)
* Correspondence: cuixh622@csu.edu.cn

Received: 31 March 2019; Accepted: 17 May 2019; Published: 24 May 2019

Abstract: A three-dimensional (3D) finite-element model (FEM), including quasi-static stamping, sequential coupling for electromagnetic forming (EMF) and springback, was established to analyze the springback calibration by electromagnetic force. Results show that the tangential stress at the sheet bending region is reduced, and even the direction of tangential stress at the bending region is changed after EMF. The springback can be significantly reduced with a higher discharge voltage. The simulation results are in good agreement with the experiment results, and the simulation method has a high accuracy in predicting the springback of quasi-static stamping and electromagnetic forming.

Keywords: electromagnetically assisted forming; springback control; numerical simulation

1. Introduction

In recent years, lightweight products are more commonly used because of the requirement for energy conservation. For this reason, aluminum and magnesium alloys as well as high-strength steel applications have been produced in increasing numbers. However, the common springback behavior during sheet-metal bending greatly affects both size and shape precision of produced parts.

During bending, the sheet deformation under an external load consists of both plastic and elastic deformation. When the external load is removed, the elastic strain is recovered. This results in a geometrical deviation from the ideal target product [1,2]. More specifically, the springback is the deviation between the shape before and after unloading, which is caused by the elastic recovery of the bending region. Since all metal materials have elastic modulus, springback is an unavoidable problem in sheet forming processes. In comparison with traditional low-carbon steel, the springback tendency of aluminum alloy is very pronounced. Therefore, the effective prediction and control of springback is important to improving the precision of bent aluminum-alloy parts.

Electromagnetic forming (EMF) is a high-speed forming method that can deform a workpiece through magnetic force using the electromagnetic induction theorem. The deformation speed of electromagnetic forming is extremely fast, often reaching 300 m/s. Compared with traditional stamping, this can be seen as a high-speed deformation process. EMF can significantly enhance plastic strain due to the inertia effect [3,4]. Therefore, the EMF process can increase the formability of metal sheets and reduce both springback and wrinkling [5,6]. It was reported that EMF technology could be used for aluminum sheet/tube forming, bending, joining, welding, cutting, and calibration of parts. For example, Guo et al. [7] used electromagnetic incremental forming (EMIF) to obtain an integral curvature panel with the grid ribs stiffened. Xiong et al. [8] designed a V-bending electromagnetic forming (EMF) method though experimentation and simulation to solve the shape difference between the formed part and the desired part. However, only a few simple-shaped parts can be manufactured using these methods.

Thus, electromagnetically assisted sheet-metal stamping (EMAS), which combines the advantages of high-speed forming into the traditional stamping process, has been used widely. For EMAS, the sheet is bent by quasi-static stamping firstly, then the working coils are setup at the sheet bending area to generate a high magnetic force. Finally, an increased formability was found in EMAS compared with the traditional stamping process [9]. Many researchers have studied the effect of EMAS on springback in various experiments. For example, Shang et al. [10] established the V-bending method and evaluated the effect of pulsed magnetic force on springback using 2024-T3 aluminum sheets. Liu et al. [11] and Sun et al. [12] established the U-shaped parts by electromagnetically assisted bending using 5052-O aluminum sheets, respectively. Iriondo et al. [13,14] established the forming of L-shaped parts and analyzed the effect of pulsed electromagnetic force on the springback of 5754 aluminum alloy and DP600 high-strength steel. Woodward et al. [15] designed disposable actuators to control and minimize springback. Therefore, it was recognized in the field of electromagnetic forming EMF can reduce springback. However, the question of why the angle can be controlled after EMF process was not clearly answered in those papers.

In this paper, the three-dimensional (3D) finite-element model (FEM) utilizing ABAQUS and ANSYS software was used to describe the mechanism of electromagnetic impulse calibration.

2. FEM of Electromagnetic Impulse Calibration

2.1. Simulation Strategy

There are two main simulation methods for electromagnetic forming: (1) the loose coupling method, and (2) the sequential coupling method. In [16], we analyzed the differences between the two methods. If the magnetic forces calculated are not based on the updated EM model, the simulation approach can be deemed the loose coupled method. For the EMF, the sequential coupling method has higher calculation accuracy because it considers the effect of workpiece deformation on magnetic field calculation. In this paper, quasi-static stamping, sequential coupling for electromagnetic forming, and springback have been all considered, and the simulated flowchart can be seen in Figure 1. The effects of local electromagnetic momentum are neglected and the hypothesis of a symmetric Cauchy stress tensor is used. The ANSYS/emag software, which uses Maxwell electromagnetic equations to calculate the electromagnetic force, was used for electromagnetic field simulation. The ABAQUS/explicit software was used to calculate the quasi-static stamping and dynamic deformation process. When the deformation is terminated, the springback of the sheet is calculated by ABAQUS/implicit software.

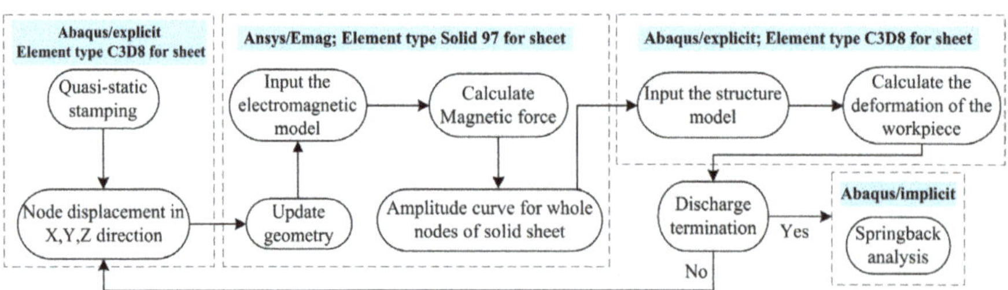

Figure 1. Simulation strategy for springback control by EMF.

2.2. Structure Field Model

The simulation data in this paper were taken from published studies by Liu et al. [11] and Sun et al. [12]. The main parameters of the EMF equipment were: rated voltage (5 kV), and capacitance (774 µF). The detailed experimental conditions were as follows: (1) The elasticity modulus of a

AA5052-O aluminum-alloy sheet was set as 68 GPa, the Poisson's ratio was set as 0.33, the density was set as 2.7×10^3 kg/m^3, and the yield strength was set as 90 MPa. The length, width, and thickness of the sheets were $120 \times 40 \times 1$ mm. (2) The length and the corner radius of the punch were 60 and 10 mm, respectively, as shown in Figure 2a. The gap between the punch and die was set to 1.05 mm. The quasi-static constitutive behavior of the AA5052-O sheet is described by Equation (1).

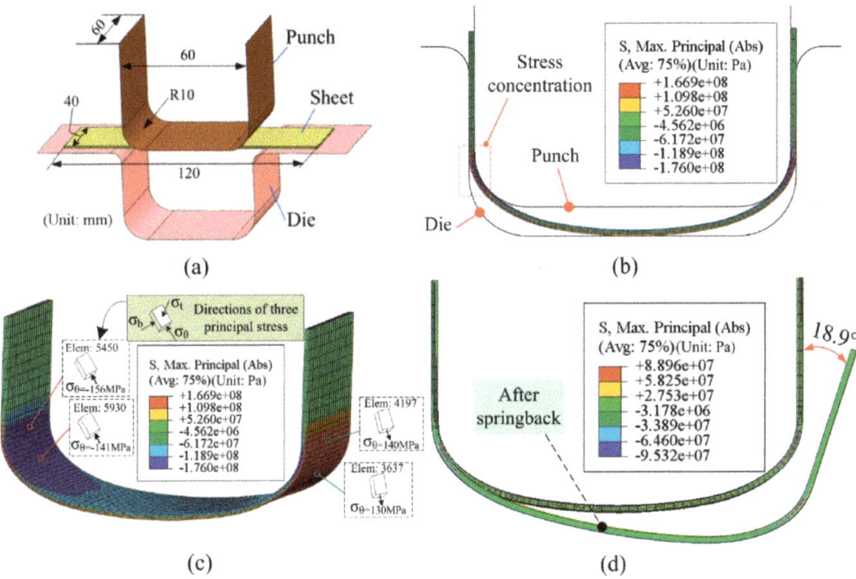

Figure 2. Structure-field model. (**a**) Geometric dimension; (**b**) 2D view of stamping; (**c**) 3D view of stamping; (**d**) angle in 2D view after springback.

For electromagnetic forming, inaccurate simulation results could be obtained if the strain rate effect is not considered. To consider the effect of a high strain rate on the forming process, the behavior of viscoplastic material was modeled with a rate-dependence law (Cowper–Symonds power law) in ABAQUS finite-element analysis software. Cui et al. [17] produced large-scale ellipsoid parts using electromagnetic incremental forming combined with stretch forming. The Cowper–Symonds power law was used to accurately predict the high-speed forming process with multi-steps of coil discharge and stretching. Thus, the Cowper–Symonds constitutive model was used in this paper as shown in Equation (2).

$$\sigma_s = 364.16\varepsilon^{0.25} \tag{1}$$

$$\sigma = \sigma_s \left(1 + \left(\frac{\dot{\varepsilon}}{P}\right)^m\right) \tag{2}$$

where σ is the dynamic flow stress, σ_s is the quasi-static constitutive behavior of the sheet, $\dot{\varepsilon}$ is the strain rate, P = 6500 s^{-1}, and m = 0.25 is the specific parameter of aluminum alloy.

The ABAQUS/explicit software was used to create a three-dimensional (3D) U-bending model, as shown in Figure 2a. To enhance the calculation speed, the punch and die were treated as rigid bodies. The element type of the punch and die was R3D4. The sheet was meshed using C3D8, an eight-node hexahedral element. Contact conditions between the punch and sheet, as well as the sheet and die, have been considered in this paper. The friction coefficient was set to 0.1. The sheet-center bulged during the quasi-static stamping. According to the experimental condition in [11,12], we set a certain displacement for the punch moves to ensure that the sheet center would just make contact with the die. The deformed sheet can be seen in a two-dimensional (2D) view in Figure 2b. Based on the stamping results, it can be found that the maximum principal stress (Abs) occurred predominantly at the sheet corner, as shown in Figure 2c. According to the principle of plastic forming, the three principal stresses during sheet bending were in the tangential direction, the width direction, and the thickness direction, respectively. In order to better understand the changes of stress during forming, the three principal stress components were defined as following: σ_θ, tangential component; σ_b, width component; and σ_t, thickness component. As the principal stresses along the tangential direction have a great effect on springback, the distribution of tangential stress (σ_θ) on the sheet at different times should be analyzed. However, the distribution of tangential stress (σ_θ) cannot be directly selected in the Abaqus software. Thus, the tangential stress was judged as special elements based on element stress vector. As shown in Figure 2c, the tangential compression and tensile stress occurred at the inner and outer layer of the sheet corner region after stamping, respectively. Figure 2d shows the 2D view before and after the springback. The springback angle was about 18.9°, and the radius of the bend region changed to about 12 mm.

2.3. Electromagnetic Field Model

Based on the geometric dimensions of the coil and the sheet, a 3D electromagnetic field model was established, as shown in Figure 3a. The electromagnetic field model consisted of the far-air region, air region, coil, and sheet. The element types for the far-air region, air region, coil, and sheet were infin111, solid97, solid97, and solid97, respectively. Based on the deformed sheet in Figure 2c, the electromagnetic field model was updated. The updated models contained sheet and coil, as shown in Figure 3b. Figure 3c shows the induced current distribution on the sheet metal after coil discharge. The current on both sides of the width of the sheet metal and the current at the corner region of the sheet metal constitute a current loop. The direction of induced current on the sheet metal was opposite to the current loaded into the coil.

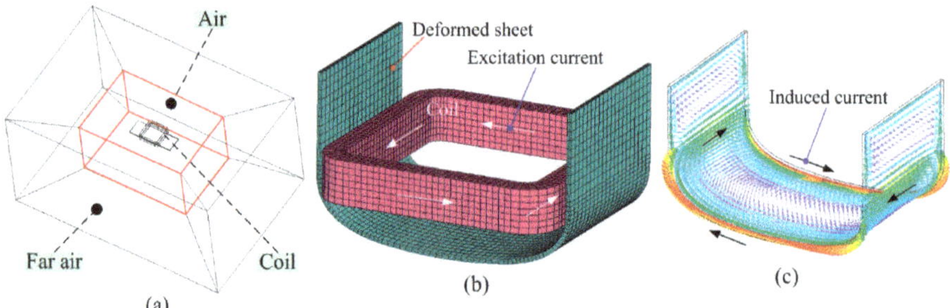

Figure 3. Field analysis of the (**a**) electromagnetic field model; (**b**) updated model of the sheet and the coil; and (**c**) induced current on the sheet metal.

Sun et al. [12] reported the current density flowing through the coil for a 1×10 mm section at 3 kV. According to our calculation, we can obtain the current curve during the coil discharge at 3 kV. At 60 µs, the current reaches its maximum value of 42 kA. In EMF, the discharge current, I (t), flowing in the coil is approximately described by Equation (3). For an EMF device, the capacitance (C), inductance (L), and resistance (R) were always kept consistent. Figure 4 shows the current flowing through the coil at different voltages.

$$I(t) = U\sqrt{\frac{C}{L}}e^{-\frac{R}{2L}t}\sin\frac{1}{\sqrt{LC}}t \tag{3}$$

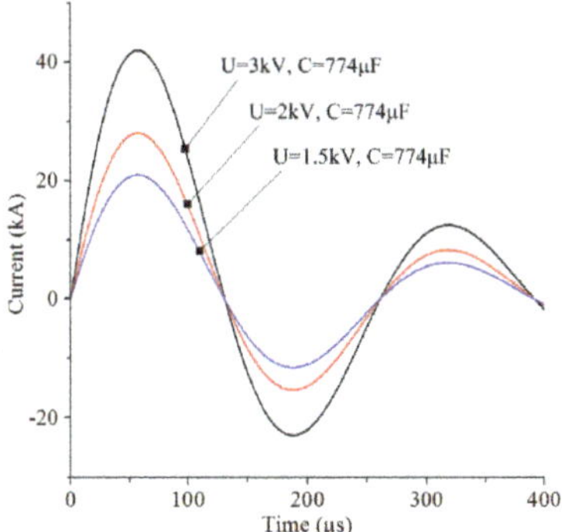

Figure 4. Current through the coil as a function of time.

3. Results

3.1. Discharge Voltage U = 1.5 kV

Figure 5 shows the shape deformation of sheet metal at different times for discharge voltage U = 1.5 kV. In order to facilitate the observation of different sheet regions deformed, the displacement of the sheet metal at the initial discharge time is defined as 0. When the coil is discharging, the material at the sheet corner region moves away from the punch. Thus, the contact area between the sheet and the die increases and the middle part of the sheet becomes flat during EMF. After the discharge time exceeds 1500 μs, the bottom of the sheet metal is separated from the die. At the time of 3000 μs, it can be considered that the sheet metal deformation is terminated.

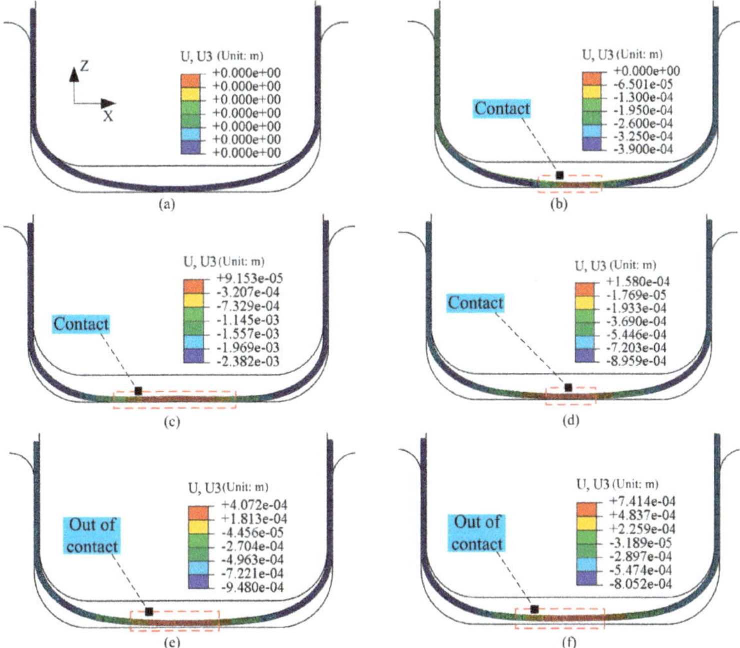

Figure 5. Distribution of displacement in the sheet for 1.5 kV at different times: (**a**) 0 μs, (**b**) 75 μs, (**c**) 300 μs, (**d**) 600 μs, (**e**) 1500 μs, and (**f**) 3000 μs.

Figure 6 shows the distribution of maximal principal stress in the sheet at different times for the discharge voltage U = 1.5 kV. In Figure 6a–d, the stress concentration area on sheet metal gradually moves towards the sheet metal center. From the time of 0–75 µs, the value of tangential stress at elements 5450 and 4197 decreases, while the tangential stress at elements 5930 and 3637 increases. From the time of 600–3000 µs, there are small tangential stresses at the corner region. Because the tangential compressed stress exists in the inner corner region and the tangential tensile stress exists in the outer corner region at 3000 µs, springback should occur.

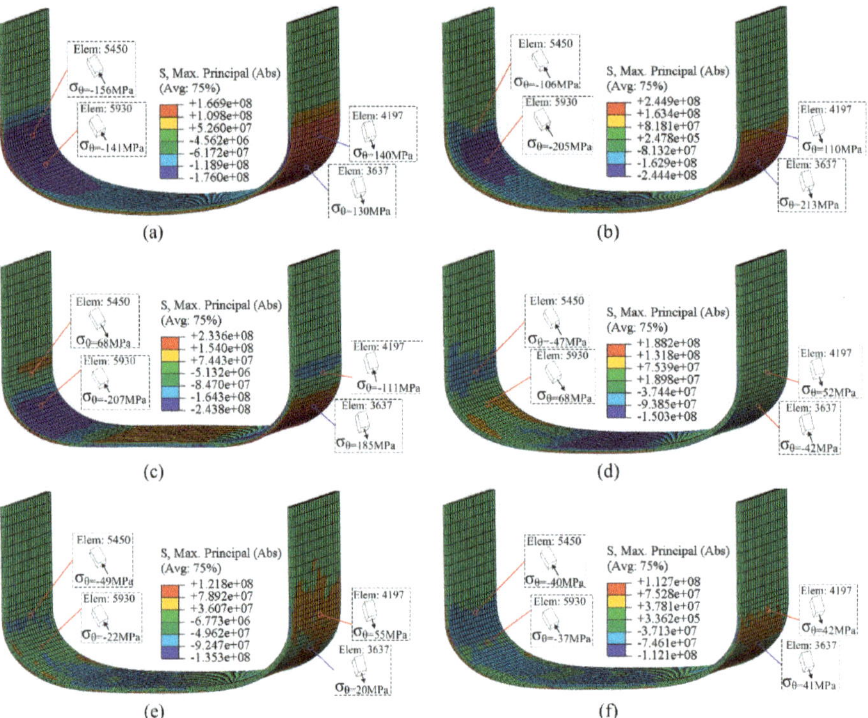

Figure 6. Distribution of tangential stresses in the sheet for 1.5 kV at different times: (**a**) 0 µs, (**b**) 75 µs, (**c**) 300 µs, (**d**) 600 µs, (**e**) 1500 µs, and (**f**) 3000 µs.

Comparing the results of Figure 2 with Figure 6, the tangential stress on the sheet corner after EMF is smaller than after quasi-static stamping. The springback angle after EMF with 1.5 kV should be smaller than the one after stamping. Figure 7 shows the springback results after a coil discharge at 1.5 kV. The springback angle after coil discharge at 1.5 kV is about 4.9°.

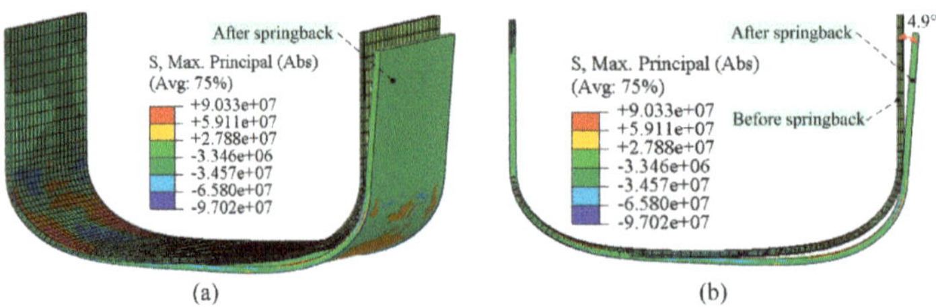

Figure 7. Springback after coil discharge at 1.5 kV: (**a**) 3D view; (**b**) 2D view profile.

Figure 8 shows the changes of three principal stresses over time at nodes 3694 and 7893. The position of the special nodes on elements is shown in Figure 8. The position of the special elements on the sheet is shown in Figure 6. At the discharge time of 0 μs, the values of three principal stresses are obtained by quasi-static stamping. When the discharge voltage is 1.5 kV, the three principal stresses at special nodes undergo disordered high frequency oscillation. At the end of the EMF process of 3000 μs, the tangential stresses at the two special nodes are obviously less than the one at the moment of 0 μs.

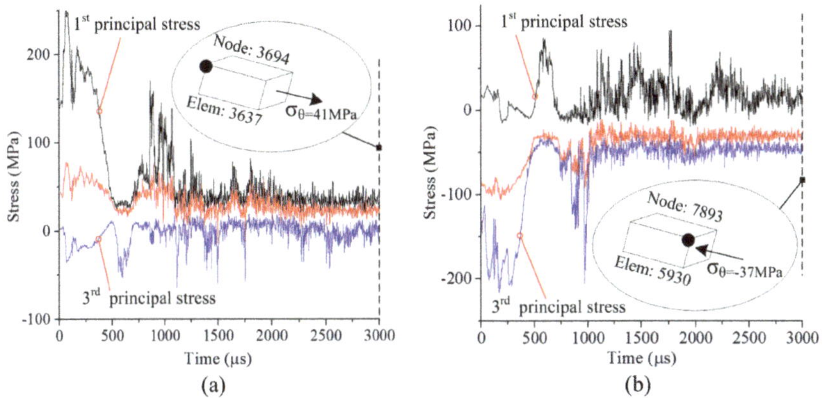

Figure 8. Stress changes at special nodes: (**a**) node 3694; (**b**) node 7893.

3.2. Discharge voltage U = 2 kV

Figure 9 shows the distribution of maximal principal stress on the sheet at different times when the discharge voltage U = 2 kV. In comparison with the results in Figure 6, the value of the tangential stress at elements 5450 and 4197 decreases more sharply, while the tangential stress at elements 5930 and 3637 increase more sharply with a higher discharge voltage from the time of 0 to 75 μs. At the time of 600 μs, tangential tensile stresses occurs at elements 5450 and 5930, positioned at the inner layer of the sheet corner. The direction of tangential stress for elements 5450 and 5930 is different from the results in Figures 2c and 9d. At 3000 μs, both tangential compression stress and tensile stress exist on inner and outer layer of sheet corner. Thus, the springback angle could further decrease with 2 kV discharging, compared with the results of 1.5 kV discharging.

Figure 10 shows the springback after coil discharge at 2 kV; the springback angle after coil discharge at 2 kV is about 0.9°. In comparison with the results in Figure 2, Figure 7, and Figure 10, the springback angle after coil discharge decreases more sharply with a higher discharge voltage.

3.3. Discharge Voltage U = 3 kV

Figure 11 shows the distribution of maximal principal stress on the sheet at different times for the discharge voltage U = 3 kV. It is found that the two sides of the sheet metal width deform faster than the sheet center. This is because, a large induced current appears at the side of the sheet metal after coil discharge, and a larger induced current corresponds to a larger magnetic field force. Therefore, a pit appears at the sheet corner region after deformation, which reduces the deformation uniformity compared with 1.5 and 2 kV.

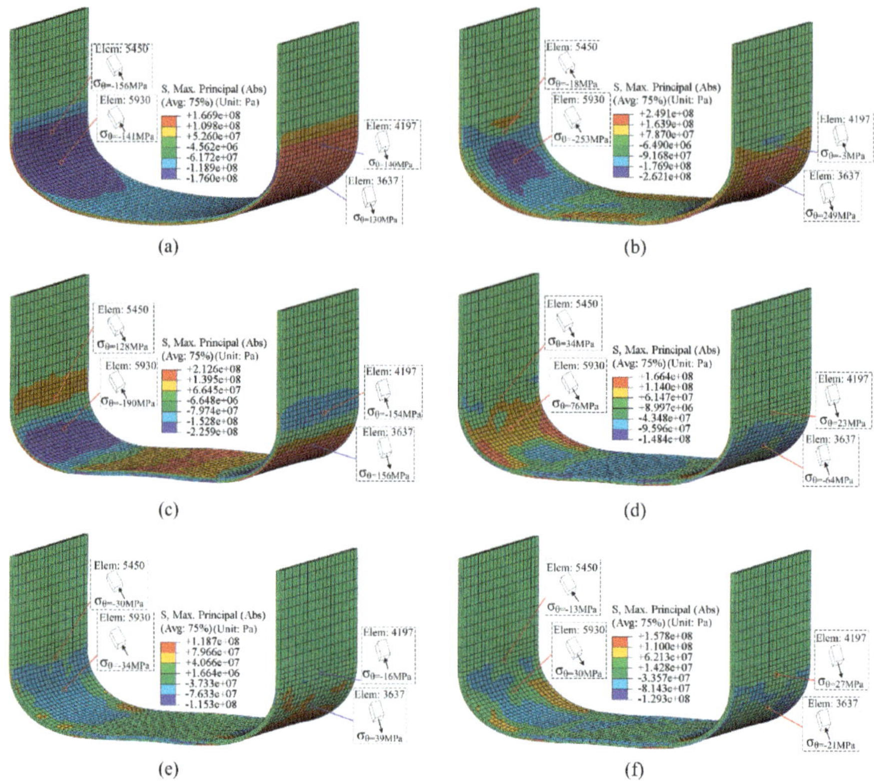

Figure 9. Distribution of tangential stresses in the sheet for 2 kV at different times: (**a**) 0 µs, (**b**) 75 µs, (**c**) 300 µs, (**d**) 600 µs, (**e**) 1500 µs, and (**f**) 3000 µs.

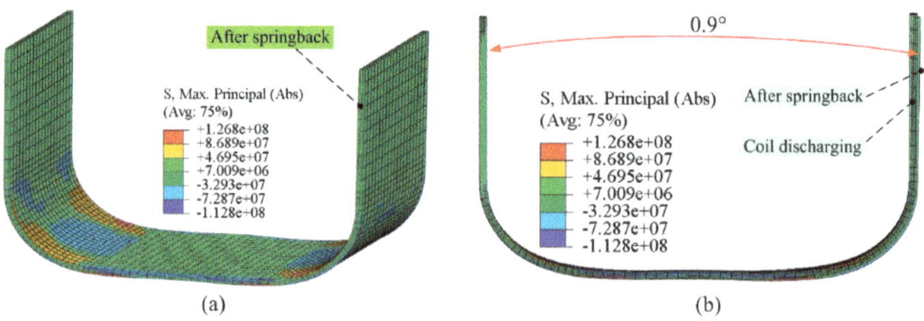

Figure 10. Springback after the coil discharge at 2 kV: (**a**) 3D view; (**b**) 2D view.

Figure 12 shows the springback results after a coil discharge at 3 kV. The springback angle after coil discharge at 3 kV is about −1.4°. However, a pit appears at the sheet corner region after springback, which causes the deformation uniformity to be reduced. The corner radius is not the same along the entire width of the sheet.

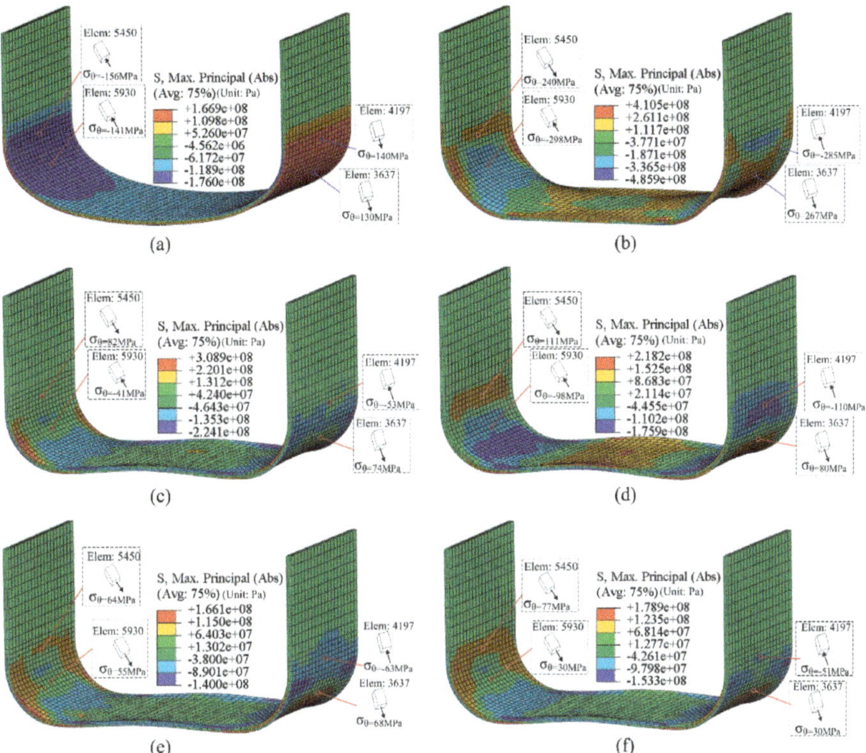

Figure 11. Distribution of tangential stresses in the sheet for 3 kV at different times: (**a**) 75 μs, (**b**) 150 μs, (**c**) 300 μs, (**d**) 600 μs, (**e**) 1500 μs, and (**f**) 3000 μs.

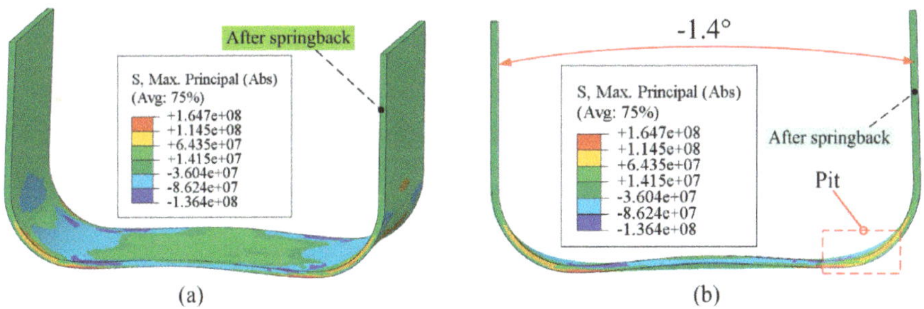

Figure 12. Springback after coil discharge at 3 kV: (**a**) 3D view; (**b**) 2D view.

3.4. Plastic Strain and Strain Rate

Figure 13 shows the distribution of equivalent plastic strain on the sheet. Under the condition of quasi-static stamping, the plastic strain mainly occurs at the corner region of the sheet, and the plastic strain in the middle region of the sheet is close to zero. After the coil discharge, the maximum plastic strain on the sheet increases due to the sheet becoming re-deformed under the action of electromagnetic force. Plastic deformation occurs at the middle part of the sheet. In order to facilitate subsequent analysis, special nodes 7284 and 7158 were chosen. It can be found that the plastic strain of nodes 7284 and 7158 gradually increases as the discharge voltage increases.

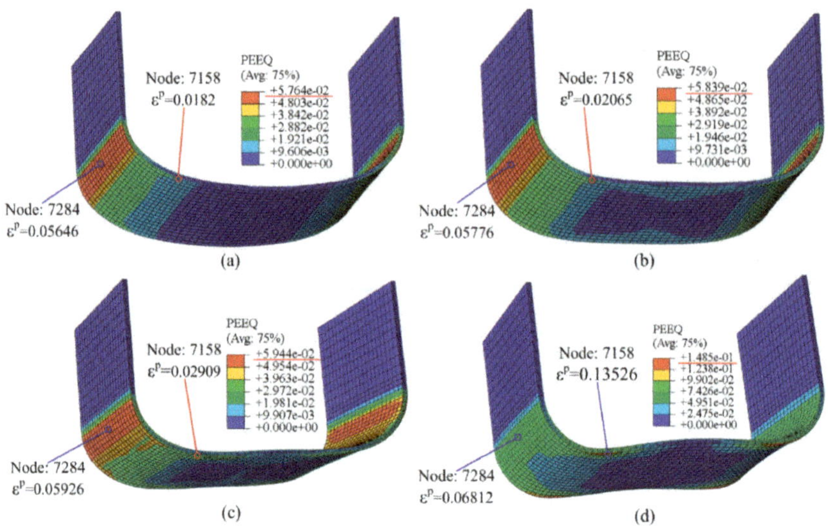

Figure 13. Equivalent plastic strain. (a) Quasi-static stamping; (b) 1.5 kV; (c) 2 kV; (d) 3 kV.

Figure 14 shows the strain rate versus time for nodes 7158 and 7284. In the quasi-static stamping, the punch is assumed to bend the sheet at a speed of 2 mm/s. Thus, the maximum strain rates of nodes 7284 and 7158 are 0.06 and 0.01 s^{-1}, respectively. If the coil is discharged at 1.5, 2 or 3 kV, the plastic strain at node 7284 is only slightly increased, while the strain rate at node 7284 is greatly increased. The maximum strain rate at node 7284 at 3 kV is 341 s^{-1}. As shown in Figure 13, the plastic strain at node 7158 with 3 kV discharge is significantly increased compared to the quasi-static condition. This results in a significant increase in the strain rate at node 7158 after discharge, to a maximum of 2511 s^{-1}.

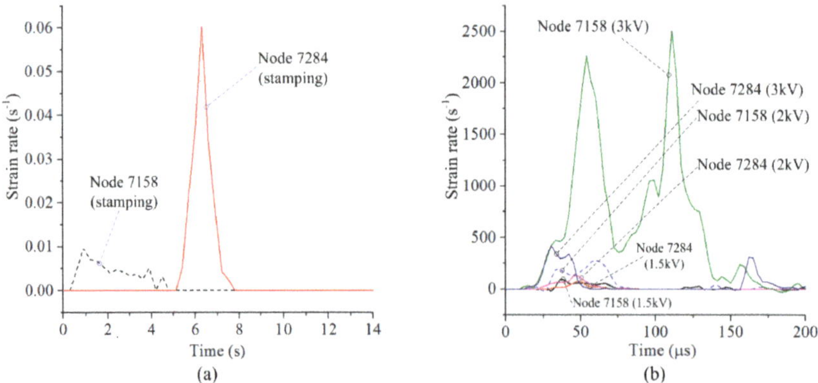

Figure 14. Strain rate versus time for (a) quasi-static stamping, and (b) coil discharge with different voltages.

3.5. Simulation Verification

The changes in the angle after EMF are shown in Table 1. Both experimental data [11,12] and the simulation suggest that the method is accurate at predicting electromagnetically assisted bending. At 3 kV there exists a pit at the sheet corner region after springback, which coincides with the simulation results shown in Figure 12.

Table 1. Simulation and experiment.

Discharge Voltage	0	1.5 kV	2 kV	3 kV
Experiment	20.8°	4.3°	1°	−0.2°
Simulation	18.9°	4.9°	0.9°	−1.4°

4. Conclusions

In this study, a 3D finite-element model (FEM) with U-shaped bending was developed using the ABAQUS and ANSYS software package. It can be concluded that:

(1) The simulation results are in good agreement with the experimental results. It is proven that the simulation method is accurate at predicting the springback for quasi-static stamping and electromagnetically assisted bending.

(2) After the coil discharges with 1.5 and 2 kV, the tangential stress on the sheet corner will be greatly reduced, and the springback will also be significantly reduced. If discharge voltage is 3 kV, the tangential stress distribution on the sheet corner region is more uneven and a higher plastic strain on the sheet. Thus, the springback can be further reduced at 3 kV. However, the deformation uniformity is reduced.

Author Contributions: The research was conceived by X.C. and Z.Z.; X.C. and Z.Z. planned and performed all the simulations; Z.Z. collected all data; Theoretical and experimental analysis were performed by H.Y., Q.C., and X.X.; The manuscript was reviewed by X.C. and H.Y., The manuscript was written by X.C.; with support from all co-authors.

Funding: This work was supported by the National Natural Science Foundation of China (Grant Nos. 51775563 and 51405173), Innovation Driven Program of Central South University (Grant number: 2019CX006), State Key Laboratory of Materials Processing and Die & Mould Technology, Huazhong University of Science and Technology (No. P2017-013) and the Project of State Key Laboratory of High Performance Complex Manufacturing, Central South University (ZZYJKT2017-03), Guangzhou Science and Technology Plan Project (201707010472) and The project was supported by Open Research Fund of State Key Laboratory of High Performance Complex Manufacturing, Central South University (No. Kfkt2018-02).

Conflicts of Interest: The authors declare no conflict of interest.

References

1. Yan, Y.; Wang, H.B.; Wan, M. Prediction of stiffener buckling in press bend forming of integral panels. *Trans. Nonfer. Met. Soc. China.* **2011**, *21*, 2459–2465. [CrossRef]

2. Wang, T.; Platts, M.J.; Wu, J. The optimisation of shot peen forming processes. *J. Mater. Process. Technol.* **2008**, *206*, 78–82. [CrossRef]

3. Psyk, V.; Risch, D.; Kinsey, B.L.; Tekkaya, A.E.; Kleiner, M. Electromagneticforming—A review. *J. Mater. Process. Technol.* **2011**, *211*, 787–829. [CrossRef]

4. Oliveira, D.A.; Worswick, M.J.; Finn, M. Electromagnetic forming of aluminum alloy sheet: Free-form and cavity fill experiments and model. *J. Mater. Process. Technol.* **2005**, *170*, 350–362. [CrossRef]

5. Li, Z.G.; Li, N.; Jiang, H.W.; Xiong, Y.Y.; Liu, L. Deformation texture evolution of pure aluminum sheet under electromagnetic bulging. *J. Alloy. Compd.* **2014**, *589*, 164–173. [CrossRef]

6. Xu, J.R.; Yu, H.P.; Cui, J.J.; Li, C.F. Formability of AZ31 magnesium alloy sheets during magnetic pulse bulging. *Mater. Sci. Eng. A* **2013**, *569*, 150–158.

7. Guo, K.; Lei, X.P.; Zhan, M.; Tan, J.Q. Electromagnetic incremental forming of integral panel under different discharge conditions. *J. Manuf. Process.* **2017**, *28*, 373–382. [CrossRef]

8. Xiong, W.R.; Wang, W.P.; Wan, M.; Li, X.J. Geometric issues in V-bending electromagnetic forming process of 2024-T3 aluminum alloy. *J. Manuf. Process.* **2015**, *19*, 171–182. [CrossRef]

9. Kiliclar, Y.; Vladimirov, I.N.; Wulfinghoff, S.; Reese, S.; Demir, O.K.; Weddeling, C.; Tekkaya, A.E.; Engelhardt, M.; Klose, C.; Maier, H.J.; et al. Finite element analysis of combined forming process by means of rate dependent ductile damage modeling. *Int. J. Mater. Form.* **2017**, *10*, 73–84. [CrossRef]

10. Shang, J.H. Electromagnetically Assisted Sheet Metal Stamping. Ph.D. Thesis, The Ohio State University, Columbus, OH, USA, 2006.

11. Liu, D.H.; Zhou, W.H.; Li, C.F. Springback control and deformation analysis for electromagnetically assisted bending of U-shaped parts. *Chin. J. Nonfer. Met.* **2013**, *23*, 3075–3082.

12. Sun, C. Deformation Analysis of U Shape Parts Formed by Electromagnetically Assisted Sheet Metal Bending. Master's Thesis, Harbin Institute of Technology, Harbin, China, June 2009.

13. Iriondo, E.; Gutiérrez, M.A.; González, B.; Alcaraz, J.L. Daehn GS. Electromagnetic impulse calibration of high strength sheet metal structures. *J. Mater. Process. Technol.* **2011**, *211*, 909–915. [CrossRef]

14. Iriondo, E.; Alcaraz, J.L.; Daehn, G.S.; Gutiérrez, M.A.; Jimbert, P. Shape calibration of high strength metal sheets by electromagnetic forming. *J. Manuf. Process.* **2013**, *15*, 183–193. [CrossRef]

15. Woodward, S.; Weddeling, C.; Daehn, G.; Psyk, V.; Carson, B.; Tekkaya, A.E. Agile production of sheet metal aviation components using disposable electromagnetic actuators. In Proceedings of the 4th International Conference on High Speed Forming, Columbus, OH, USA, 9 March 2010; pp. 35–46.

16. Cui, X.H.; Mo, J.H.; Li, J.J.; Huang, L.; Zhu, Y.; Li, Z.W.; Zhong, K. Effect of second current pulse and different algorithms on simulation accuracy for electromagnetic sheet forming. *Int. J. Adv. Manuf. Technol.* **2013**, *69*, 1137–1146. [CrossRef]

17. Cui, X.H.; Mo, J.H.; Li, J.J.; Xiao, X.T.; Zhou, B.; Fang, J.X. Large-scale sheet deformation process by electromagnetic incremental forming combined with stretch forming. *J. Mater. Process. Technol.* **2016**, *237*, 139–154. [CrossRef]

Review

Recent Developments and Trends in the Friction Testing for Conventional Sheet Metal Forming and Incremental Sheet Forming

Tomasz Trzepiecinski [1],* and Hirpa G. Lemu [2],*

[1] Department of Materials Forming and Processing, Rzeszow University of Technology, al. Powst. Warszawy 8, 35-959 Rzeszów, Poland
[2] Department of Mechanical and Structural Engineering, University of Stavanger, N-4036 Stavanger, Norway
* Correspondence: tomtrz@prz.edu.pl (T.T.); hirpa.g.lemu@uis.no (H.G.L.); Tel.: +48-17-743-2527 (T.T.); +47-518-32173 (H.G.L.)

Received: 18 November 2019; Accepted: 20 December 2019; Published: 25 December 2019

Abstract: Friction is the main phenomenon that has a huge influence on the flow behavior of deformed material in sheet metal forming operations. Sheet metal forming methods are one of the most popular processes of obtaining finished products, especially in aerospace, automobile, and defense industries. Methods of sheet forming are carried out at different temperatures. So, it requires tribological tests that suitably represent the contact phenomena related to the temperature. The knowledge of the friction properties of the sheet is required for the proper design of the conditions of manufacturing processes and tools. This paper summarizes the methods used to describe friction conditions in conventional sheet metal forming and incremental sheet forming that have been developed over a period of time. The following databases have been searched: WebofKowledge, Scopus, Baztool, Bielefield Academic Search Engine, DOAJ Directory of Open Access Journals, eLibrary.ru, FreeFullPdf, GoogleScholar, INGENTA, Polish Scientific Journals Database, ScienceDirect, Springer, WorldCat, WorldWideScience. The English language is selected as the main source of review. However, in a limited scope, databases in Polish and Russian languages are also used. Many methods of friction testing for tribological studies are selected and presented. Some of the methods are observed to have a huge potential in characterizing frictional resistance. The application of these methods and main results have also been provided. Parameters affecting the frictional phenomena and the role of friction have also been explained. The main disadvantages and limitations of the methods of modeling the friction phenomena in specific areas of material to be formed have been discussed. The main findings are as follows—The tribological tests can be classified into direct and indirect measurement tests of the coefficient of friction (COF). In indirect methods of determination, the COF is determined based on measuring other physical quantities. The disadvantage of this type of methods is that they allow the determination of the average COF values, but they do not allow measuring and determining the real friction resistance. In metal forming operations, there exist high local pressures that intensify the effects of adhesion and plowing in the friction resistance. In such conditions, due to the plastic deformation of the material tested, the usage of the formula for the determination of the COF based on the Coulomb friction model is limited. The applicability of the Coulomb friction model to determine the COF is also very limited in the description of contact phenomena in hot SMF due to the high shear of adhesion in total contact resistance.

Keywords: coefficient of friction; bending under tension; draw bead; friction; friction testing; material properties; mechanical engineering; sheet metal forming; strip drawing; surface properties; tribology

---

1. Introduction

Conventional sheet metal forming (SMF) methods are widely used for making shaped components across the many industries. In an SMF process, a thin piece of metal sheet, commonly referred to as the workpiece, is stretched by stamping tool into the desired shape without excessive thinning or wrinkling. Finished products fabricated by SMF processes have good quality, are geometrically accurate, and commonly do not require finishing procedures. The main problem in SMF techniques is the appropriate compensation of the springback phenomenon [1,2]. However, sheet forming processes are complex and require expensive tooling that is economically feasible only for mass production [3]. Goals of experimental investigations in sheet metal forming operations are classified into two [4,5]:

(1)　understanding the contact conditions during sheet metal forming and
(2)　assessing the influence of specific variables in the forming operations.

Incremental sheet forming (ISF) is a relatively recent technique that allows solving many problems of the conventional sheet forming process. ISF is a dieless forming process suitable for small batches and has demonstrated its great potential to form complex three-dimensional parts with using a relatively simple and low-cost tools. Potential application areas of ISF include the aerospace industry, biomedical applications, rapid prototype production, and metal forming in the automotive industry. ISF processes are characterized by much less forming force compared to conventional stamping, no need to manufacture the dies, a higher value of the sheet deformation in relation to die-forming, and the ability to shape elements on a conventional computer numerical control (CNC) milling machine [6].

In general, ISF includes several sheet flexible forming approaches, such as Single Pint Incremental Forming (SPIF), Two Point Incremental Forming (TPIF), also known as Double-Sided Incremental Forming (DSIF), Incremental Forming with Counter Tool (IFWCT), or Water Jet Incremental Sheet Metal Forming (WJISMF) [6–8].

In SMF, the shape, dimensional accuracy, and surface roughness of the final part mainly depend on the character of sheet deformation due to material properties, tool design, forming conditions (temperature, forming speed, etc.), initial surface topography, friction conditions (dry or lubricated contact). Former studies [9,10] showed that the surface properties of the tool have a significant effect on friction as well as wear behavior of sheet metal forming under lubricated conditions. The existence of friction forces at the deformed material and tool interface, in particular, determines the non-uniformity of the deformation of the sheet metal.

Therefore, it is necessary to have sufficient knowledge of the frictional resistance in order to properly design the die tool and to forecast the flow of the material under forming, especially in the case of complex shapes. Moreover, finite element-based numerical simulations that allow for an accurate analysis of deformation of the sheet require a proper description of the contact phenomena. However, for proper operation, one requires knowledge of the frictional resistance and its evolution during the forming process. This justifies the work on the proper planning of the friction tests using proper simulators that are developed to model the friction phenomena simulation tests. In this paper, the methods used to represent the friction conditions in conventional SMF and ISF have been summarized. Furthermore, the main disadvantages and limitations of the modeling methods of the friction phenomena in specific areas of the material to be formed have been discussed.

2. Methods

To fulfil the aim of the article, the main scientific databases have been explored. The English language is selected as the main source of review. However, in a limited scope, Polish and Russian languages are also used in exploring the Polish language and Russian language-based scientific databases. To identify the potential relevant documents, the following bibliographic databases were searched: WebofKowledge, Scopus, Baztool, Bielefield Academic Search Engine, DOAJ Directory of Open Access Journals, eLibrary.ru, FreeFullPdf, GoogleScholar, INGENTA, Polish Scientific Journals Database, ScienceDirect, Springer, WorldCat, WorldWideScience. No restriction has been made on the publication years. The search strategy was limited to scientific articles and scientific theses distributed in open access and distributed under the access available at the authors' Universities. The duplicated articles found by various databases were not considered. Papers were also excluded if they did not fit into the conceptual framework of the study. References available in found articles were also considered. Many years of experience of the authors allow drawing the hypothesis that there is a very limited number of friction tests in sheet metal forming, which causes that no special numerical screening algorithm was used. The articles were reviewed "manually". It should be pointed out that the results found by different databases were consistent. The keywords used to search in the databases include, but not limited to:

I. English language: "friction coefficient sheet forming", "friction coefficient metal forming", "friction sheet forming, "friction deep drawing", "tribology sheet metal forming", "friction bending under tension", "friction blank holder", "coefficient of friction", "tribometer coefficient of friction", "friction simulative tests", "strip drawing test", "hot forming friction testing", "friction in draw bead", "draw bead friction test", "advances in friction testing", "advances in friction measuring", "friction apparatus for sheet metal forming";

II. Polish language: "tarcie podczas kształtowania blach", "współczynnik tarcia", "tarcie przeróbka plastyczna metali", "tribotester", "pomiar tarcia kształtowanie blach", "opory tarcia", "tarcie pod dociskaczem", "metody wyznaczania oporów tarcia", "metody określania współczynnika tarcia", "postępy w pomiarze współczynnika tarcia", "pomiar sił tarcia w procesie tłoczenia", "test przeciągania paska blachy", "test zginania blachy z rozciąganiem", "test przeciągania w warunkach ściskania blachy", "test zginania w warunkach ściskania blachy", "test zginania blachy z przeciąganiem", "test zginania blachy z ciągnieniem";

III. Russian language: "триботестер", „коэффициент трения формование листового металла", „трение при формовании листового металла", „методыопределения сопротивления трения", „сопротивление трения", „прогресс в измерении коэффициента трения", „фрикционно-пластическое формование металлов", „измерение трения", „методыопределения коэффициента трения".

IV. The next section shortly describes the fundamentals of tribology in sheet metal forming. The results of the review are presented in Sections 4 and 5.

3. Frictional Resistance

Frictional resistance depends on diverse parameters such as lubrication, normal pressure, the surface roughness of the sheet metal and the tool, type of materials of the contact pair, sliding speed, and temperature [11]. In forming at elevated temperatures, the coefficient of friction (COF) is usually higher than in cold forming. This is because higher temperature forming increases adhesion at the contact interface. The interaction between the two surfaces and the oxide and debris layers formed is constantly changing and strongly affected by the forming conditions [12]. It is well known [13,14] that friction changes as a result of plastic deformation of the material and the increased surface roughness. This leads to the transition from a hydrodynamic to a mixed friction regime, and the metal to metal contact increases.

The unfavorable effects of frictional resistance include [15,16]: non-uniform workpiece deformation, worsening of workpiece surface roughness, and the increased tool wear. The existence of frictional resistance between workpiece and punch in the stamping process has an advantage that the maximum allowable drawing force can be increased. As it was concluded by Menezes et al. [17], the variations of the COF between the tool and workpiece surface directly affect both the stress distribution and the shape of the workpiece. This also has direct impacts on the material microstructure.

Friction in sheet metal forming is a complex function of the physico-mechanical properties of the tool and workpiece, forming parameters, the topography of both metallic sheets and tools, and contact conditions. In order to fully characterize the role of friction in sheet metal forming, a tribological system may be introduced, which consists of four main elements (Figure 1)—friction pair, lubricant, micro- and macroenvironment, and processing parameters.

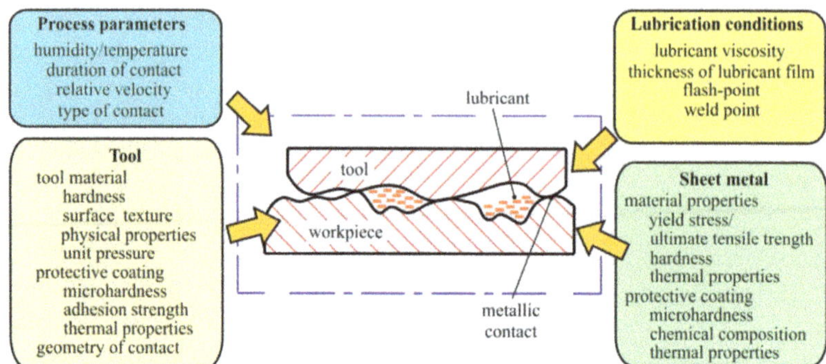

Figure 1. Scheme of a tribological system in sheet metal forming.

Tribology plays a key role in the formability of material and quality of products. The surface roughness of sheet and tooling surfaces, surface coating, and lubrication type are the fundamental parameters of the tribological performance of elements in the automotive industry [18]. The most economical and effective way to reduce the harmful effects of frictional resistances in stamping operations is lubrication. The most important properties of the lubricant are viscosity and extreme pressure load [19]. The pressure between the die and the workpiece is a crucial factor determining the effectiveness of lubrication. At a relatively low contact pressure, the frictional interaction takes place through mixed lubricated asperity contacts that are dominated by boundary effects [20]. In such conditions, the total applied load is partially carried by the hydrodynamic action of the lubricant film. The contact normal stresses in deep drawing and hot sheet metal forming may reach a value of 750–800 MPa. So, the proper selection of lubricant type is crucial to realize the tool-workpiece separation and friction reduction. The application of good boundary lubricants containing hard coatings and extreme-pressure additives could increase the frictional resistance of the die-workpiece contact surfaces [21]. Results of Kim et al. [22] also showed that the combined uses of good boundary lubrication and hard coatings reduce heat generation at the die-workpiece contact.

In SMF, the friction coefficient is controlled by two different friction components [23]: (1) an adhesive force acting in the real contact areas and (2) a deformation force acting when the harder tool surface asperities plow into the surface of the softer sheet metal. Surface plowing refers to the deformation of the surface of a softer material due to a harder material sinking into it. In other words, the friction effect is observed due to the adhesion effects between asperities, the plowing at contacting asperities, and the hydrodynamic friction stresses that appear when the lubrication regime is applied. The impact of adhesion and plowing on the frictional resistance is a function of the real area of contact. Therefore, frictional resistance depends on the real area of contact, which can be influenced by two mechanisms [24–26]: (1) roughening and (2) flattening. These mechanisms occur due to sliding, normal load and stretching, where flattened surfaces exhibit a higher COF. On the other hand, roughening of asperities decreases the real area of contact resulting in lower friction.

In addition, in hot forming processes, the temperature may dynamically change in the forming process, so a constant COF cannot reflect friction and galling characteristic of the interface [27]. Galling, in particular, accelerates with rising temperatures in both lubricated and dry friction conditions [28], thus affecting the quality of parts formed.

Many studies were devoted to investigating the effectiveness of friction reduction by different types of lubricants. A lubricant with high viscosity can form a thicker film, enlarging the distance between the interacting asperities and hence reduces abrasion. In contrast, lubricants with an extremely high value of pressure load may maintain an oil film under higher contact pressure between asperities and minimize the asperity adhesion [19,20].

4. Friction Testing in Conventional Sheet Metal Forming

4.1. Background

The first quantitative investigations of friction, lubrication, and wear were made by inventor Leonardo da Vinci (1452–1519) almost 450 years ago. Guillaume Amontons (1663–1705) in 1699 enunciated fundamental laws of friction, with those names not associated. One of the two main statements is that the friction force is independent of the apparent area of contact between the two surfaces. Secondly, the friction force acting between two sliding surfaces is proportional to the applied load. The concept of experimental apparatuses and development of da Vinci's understanding of the friction laws have been comprehensively examined by Hutchins [29]. Many of the concepts of friction measurement presented in da Vinci's manuscripts are still up-to-date. Furthermore, they formed the basis for creating new concepts of friction tests.

Due to the presence of different conditions with regard to the pressure forces, state of stresses, and speed of displacement in the individual areas of the draw piece, a series of tribological tests were developed to model the friction conditions for the needs of the plastic sheet forming processes [16,30]. Over the years, two main groups of tribological tests were developed. The tests simulating tribological conditions model the geometrical interactions of the friction pair, often without maintaining the process kinematics. The second group, the process simulation tests, are designed to manipulate plastic-forming operations while maintaining process kinematics. Both groups of tests can be divided into tests with direct or indirect measurement of the values of the COF. A review of the effects of coating, lubrication, and temperature on the tribological characteristics in hot forming and the tribometers for different metal forming processes at elevated temperatures has been presented by Dohda et al. [31]. Moreover, the tribological behaviors of oxides in hot forming were discussed. An overview of the simulation tests representing the friction conditions in the specific areas of the draw piece is also shown in Figure 2.

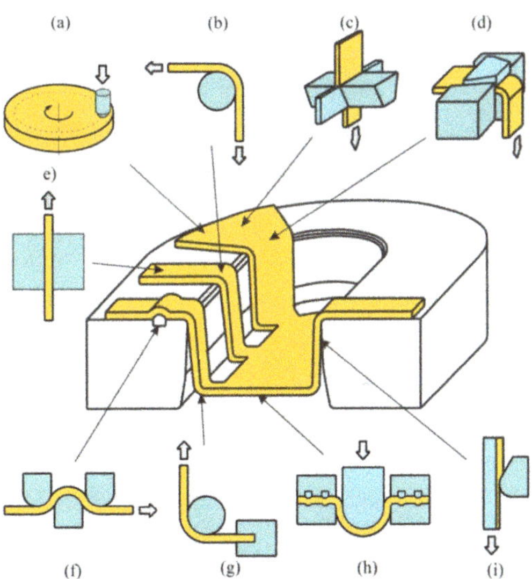

Figure 2. Tribological tests representing the friction conditions in the specific areas of the draw piece: (**a**) pin-on-disc tribometer, (**b**) bending under tension, (**c**) drawing with tangential compression, (**d**) bending with tangential compression, (**e**) strip-drawing test, (**f**) draw-bead test, (**g**) strip-tension test, (**h**) hemispherical stretching, (**i**) strip-reduction testing; prepared based on [16].

Most tests allow indirect determination of the value of the COF based on the measurement of the changes in other physical parameters like strains or forces. Besides, in order to study the interactions in the sheet-tool interface, the industry is nowadays using different numerical simulation-based tests [32–34].

Bay et al. [16] divided the friction tests with regard to the normal pressure value, surface expansion, and sliding length into three groups:

(1) bending under tension tests with mild tribological conditions with normal medium pressures, low-sliding speeds, and no surface expansion,
(2) draw-bead tests with medium-to-high normal pressures, medium sliding lengths, and no surface expansion and
(3) strip-reduction tests with normal high pressures, low-sliding lengths, and surface expansion.

Even until the present, no universal method has been developed for determining the COF. This is attributed to the variety of the geometry of tool contact with the deformed material as well as the existence of different stress and strain states in specific areas of the draw piece in the sheet-forming processes (Figure 3).

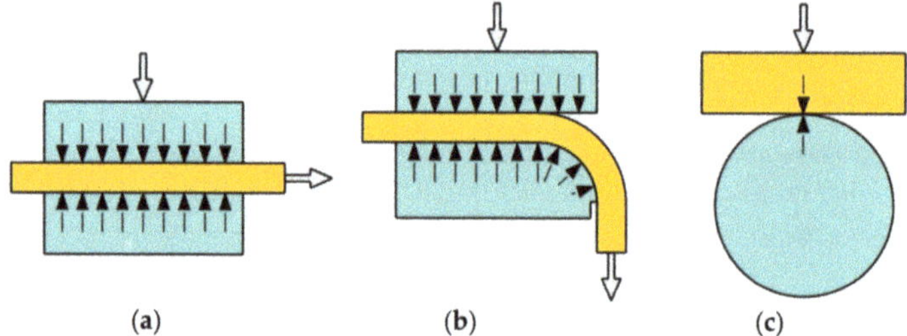

Figure 3. Types of sheet-tool contact: (**a**) flat contact, (**b**) sliding over a curved part of the tool, and (**c**) linear sliding contact.

4.2. Friction Between Flat Dies

Strip drawing test (SDT) is used to model friction condition in metal forming, i.e., between the punch and the die wall. In this test, a plane specimen of the tool material is pressed against a strip of sheet metal, which is drawn simultaneously [15]. The strip is drawn between non-rotating counter-samples, which are mostly flat [14] or have cylindrical shape [13,15] (Figure 4). The existence of frictional forces between the contact surfaces assists in achieving better accuracy of measurement of the COF. Several parameters such as the pressure force between counter-samples, lubrication conditions, drawing speed, the surface roughness of the counter-samples and temperature do affect the change in frictional resistance.

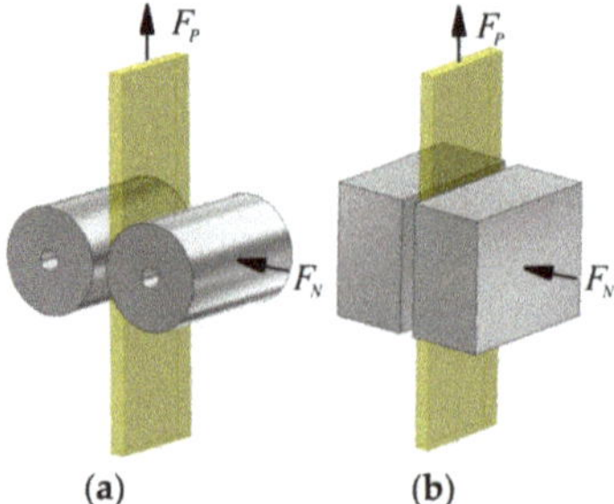

Figure 4. Schematic diagram of the strip-drawing test: (**a**) cylindrical dies and (**b**) flat dies.

The value of the COF is determined according to the formula:

$$\mu = \frac{F_P}{2F_N} \tag{1}$$

where F_P is the pulling (friction) force, F_N is the normal force.

The SDT has been studied by many authors to determine the COF in a wide range of process parameters. The experimental results determined by Trzepieciński et al. [15,35] have determined several relationships that show the effect of lubricant conditions, surface roughness and the orientation of the sheet metal on the value of the COF in the forming processes. Vollertsen and Hu [36] used the SDT to identify the tribological size effects and develop a size-dependent friction function integration in finite element-based numerical simulations. Masters et al. [14] studied the friction behavior of metallic sheets at different levels of specimen strain. They found that the COF increases with the level of strain. The problem of the influence of the drawing speed and the directional surface topography on the value of friction resistance and the change of surface topography in mixed lubrication conditions in the SDT was discussed by Roizard et al. [37]. During the testing of sheets whose roughness ridges were oriented perpendicular to the direction of drawing, the friction force was 30% lower when drawing inversely oriented specimens. The analysis of surface roughness changes in aluminum sheets during a sheet-drawing test of cylindrical and flat counter-samples [38] showed a similar character of surface topography change, and comparable value of the COF determined in both tests.

The mean contact pressure between pins and sheet metal strip may be estimated by using the Hertz relation for a line contact under the elastic deformation with smooth surfaces [39]:

$$p_n = \frac{\pi}{4} \sqrt{\frac{\frac{F_n}{b} E^*}{2\pi R^*}} \tag{2}$$

where b is the specimen width, F_n is the applied load, E^* and R^* are the combined elasticity modulus and the combined radius, respectively, determined according to:

$$E^* = \frac{2E_1 E_2}{E_2\left(1 - v_1^2\right) + E_2\left(1 - v_2^2\right)} \tag{3}$$

$$R^* = \frac{R_1 R_2}{R_1 + R_2} \tag{4}$$

where R_1 and R_2, E_1, and E_2, v_1 and v_2 are radiuses, Young's moduli and Poisson's ratios of bodies in contact, respectively.

Parameters that can influence friction and lubrication are also extensively studied in the last decade. Payen et al. [40] investigated mainly the role of contact pressure and of a chemical coating on the tribological behavior and the roughness changes of sheets surfaces in SDT, which they called the plane friction test. Applying an anti-adhesive coating reduced the local friction shear stress at the boundary contacts. As a consequence, the character of asperities deformation is changed in such a way that large levels of normal crushing of asperities can be reached with reduced debris generation [40]. Coelho et al. [41] tested EBT (Electron-Beam Texturing) zinc-coated trip 700 steel sheet under the strip drawing test with flat dies considering a real contact area in the die-sheet-die system instead of the apparent one. The results for flat contact friction demonstrate that a suitable combination of velocity and pressure is required to establish the micro-hydrodynamic effect under typical boundary lubrication conditions. The micro-hydrodynamic conditions of the boundary lubrication are caused by the plastic deformation of coating, which produces high hydrostatic pressure on the entrapped lubricant.

Venema et al. [42] used the strip-drawing test with flat dies to characterize tribological interactions between the tool and the sheet metal during hot stamping using hot friction draw tester with roller hearth furnace. The essence of the friction test is based on the SDT with flat dies. Their results show that the tool materials worn down by friction could be embedded in relatively soft sheet coating which subsequently causes plowing of the tool surface. Xue et al. [43] designed the strip drawing test, which consist in a combination of flat and convex dies (Figure 5). The authors did not describe the advantage of such a configuration over a flat and rounded counter-sample (Figure 4). The COF value may be determined using the commonly used Equation (1).

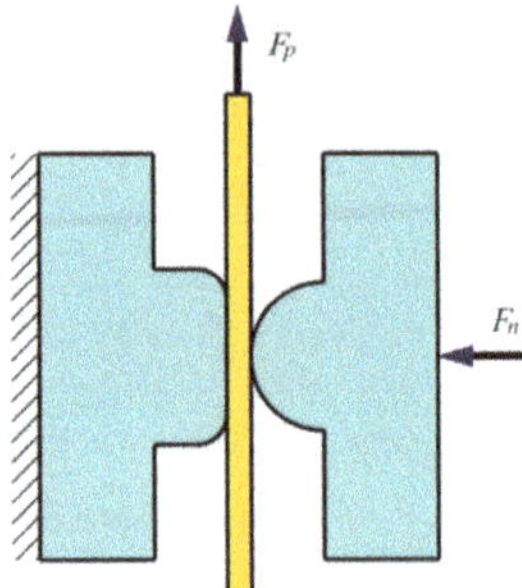

Figure 5. Strip-drawing test setup with combined shapes of counter-samples.

The motivation of the experimental investigations of Recklin et al. [44] was observed if different test stands of friction investigation show a significant quantitative difference in the friction value. The main reason for the deviation of the tribological behavior in different stands is the different stiffness. Moreover, the contact area of each test stand may be different. Therefore, Recklin et al. [44] compared two commonly used strip drawing tests and detected the causes of the deviation. It was found that the quantitative value of the mean COF over the contact pressure depends on the area of the contact surface. The second important conclusion is that the variation of COF over the contact pressure is dependent on the stiffness of the stand.

Kondratiuk and Kuhn [45] tested the COF of coatings in hot strip drawing experiments (Figure 6). The basic setup consists of a drawing device, a continuous electric furnace, a feeder section to transport the sheet metal strip. After placing a steel strip manually onto the chain drive, the strips are automatically transported into the furnace using a motor-powered chain drive. The strip is then transported into a hydraulic holding device when the pre-specified dwell time has been reached. Next, the tool closes, and the strip is compressed by applying a load acting between the upper and lower tool and is drawn through the drawing device.

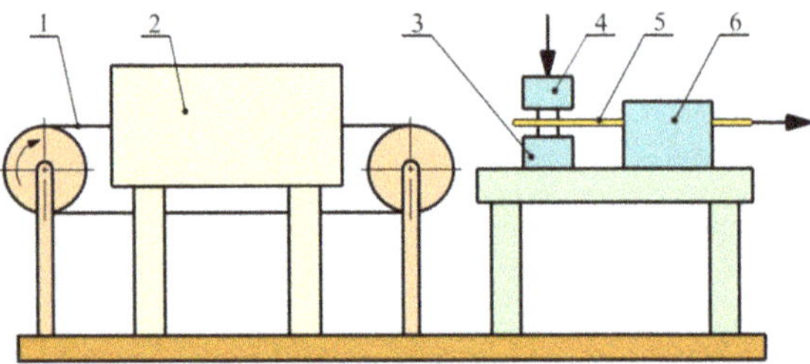

Figure 6. Schematic representation of a hot strip drawing simulator: 1—chain, 2—electric furnace, 3—tool holder, 4—movable tool holder, 5—heated sheet metal strip, 6—sliding bed with a hydraulic holding down device; prepared on the basis of [45].

The COF can be calculated using Equation (1) as a function of drawing distance. The results of investigations allowed the individual tribological characterization of the hot-dip-aluminum–silicium and electro-plated zinc alloy coatings for hot forming applications. The effectiveness of the use of lubricants to decrease the forming load in hot stamping of Al-coated 22MnB5 steel was studied by Uda et al. [46] in hot strip drawing test. Al-coated 22MnB5 steel is widely used as the sheet material in hot stamping. Technically, the used device is very similar to that used by Kondratiuk and Kuhn [45]. Based on the results of the experimental tests, a new high-performance lubricant for hot stamping was developed. Tokita et al. [47] studied the effects of the temperature deviation due to contact with the tools and the COF at high temperatures on stretch formability at a high temperature using a high temperature sliding test. Experimental testing of HR steel sheets conducted under warm forming conditions shown that, in forming in the low-temperature range (200–400 °C), the influence of the COF on sheet formability is dominant in spherical stretch forming. The COFs of both the hot rolled and galvannealed coated steel sheets at a high temperature were higher than those at room temperature.

The above-mentioned test (Figure 6) was performed outside the furnace, and hence it is not exactly isothermal. Shi et al. [48] designed a friction rig intended to make isothermal tests at elevated temperature. Using this test rig, the frictional properties of the sheets in SDT was studied. The rig is universal and can be fitted to any universal testing machine equipped with the furnace chamber. The vertical force of pulling the sheet between counter-samples is measured by testing machine, and the normal pressure is applied to test-piece by the gravitational force of the weights via a lever mechanism.

The correlations between the tool surfaces and friction in SDT with flat dies were determined by Merklein et al. [49]. In this study, diverse finishing strategies were employed on the surface of the tool. Tribological performance of tool surfaces fabricated using different finishing strategies was investigated by conducting strip drawing tests under dry conditions. From the results with lubricated strips, it was found that the preferential direction orientation has a lower influence on the friction compared to the roughness. Moreover, the orientation of grinding marks in the tool surface has a notable influence on the friction under dry conditions.

Tests with the rotary movement of the tool consist of placing a specimen in the form of a ring or a disc between the anvil of a testing machine and the counter-sample performing a rotary movement along the axis of the specimen. During the test, the tangential force and normal force are measured in the contact area. As a counter-sample, flat (pin-on-disc) [50] or spherical (ball-on-disc, Figure 7a) [51–53] pins, as well as ring-shaped tools (Figure 7b) known as twist compression tests (TCTs) are used. The frictional resistance can be determined for different slip speeds, lubrication conditions, and pressure forces.

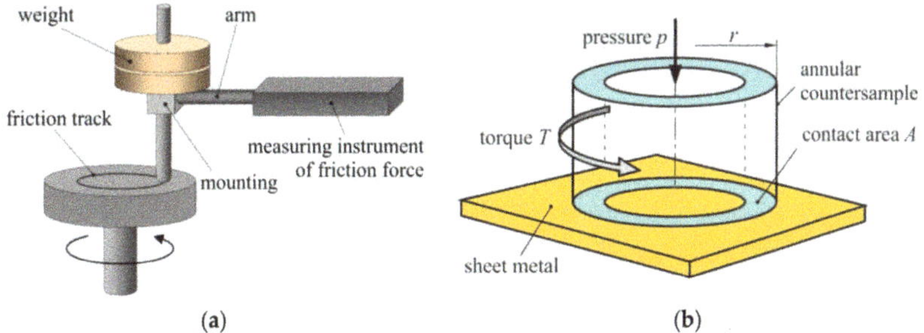

Figure 7. Illustration of rotary counter-sample tests (**a**) pin-on-disc, and (**b**) twist compression (prepared on the basis of [54]).

During the pin-on-disk tests, the friction force and the slip distance for the first disk revolution are measured using a computer program that registers and controls the values of the friction force as a function of time. The COF value was determined from the values of both normal F_N and friction F_T forces using the formula in Equation (5):

$$\mu = \frac{F_T}{F_N} \tag{5}$$

The apparent pressure p_c in the pin-on-disk test with flat pin may be evaluated from the friction band width, b_f, which can be assumed to be equal to the diameter of a circular contact section. Thus, the apparent pressure for each test was calculated from [41]:

$$p_c = \frac{4F_N}{\pi b_f^2} \tag{6}$$

where F_N is the normal force, and b_f is the width of friction band.

The results of investigations using pin-on-disc tribometers where the length of the friction track is arbitrary, cannot be compared with measurements carried out on the work stations with the to-and-fro motion of the counter-sample. Due to the cyclic contact of the counter-sample with the surface of the sheet being tested and the concentrated nature of the contact, the COF value representative for a given surface can only be determined for the initial stage of the friction process. In this manner, one can determine the anisotropy of the value of the COF according to the rolling direction of the sheet [51,53].

The use of tribotesters to determine the COF of plastically deformed sheets is limited due to the nature of the concentrated contact, whose presence in the real deep-drawing processes is limited. Moreover, the low ratio of the yield strength of the workpiece material and the counter-sample material allows the rapid initiation of wear processes. The superficial contact of the counter-sample with the sheet surface being tested is assured by a method using an annular counter-sample with a radius of external cylindrical surface r loaded with pressure p (Figure 7b) [10]. The value of the COF in the TCT is determined based on the following formula:

$$\mu = \frac{T}{A \times p \times r} \tag{7}$$

where T is the torque, A is the surface area of contact, p is the normal pressure, and r is the radius of the external cylindrical surface.

Kim et al. [22] used TCT to establish guidelines on how to select the optimum combinations of die materials, coatings, and lubricants for stamping of galvanized AHSS (DP600, TRIP780, and DP980) for automotive structural parts. Ball-on-disc tests were used to a large extent in describing the tribological behavior of materials in the contact region and to determine the COF [13,14]. It is important to note that these tests are quite similar to scratch tests and are different from ring compression tests and pin-on-disc tests. They generate apparent friction that may contain two integrated components, i.e., shear friction and plowing friction [52].

Pin-on-disc tester was used in the study reported by Dou and Xia [55], which described the effects of the normal loads and the sliding velocity on the friction characteristics between the stamping die and the sheet metal under the boundary lubrication. From this study, it was observed that the friction mechanism between sheet metal and the die is mainly abrasive wear and plowing wear, with slight adhesive wear.

Reports from some researches [56,57] indicate that the measured COF is usually higher when the sliding direction is parallel to the grinding lay. Ajayi et al. [53] used a pin-on-disc tribometer and investigated the role of surface texture, especially the effect of grinding lay on frictional behavior of the boundary lubrication regime. The results indicated frictional anisotropy of the directionally ground AISI 8620 specimens in the entire speed range of the test, although it is more apparent at low speed, and that the magnitude of the friction variation spikes is significantly reduced. At such high speeds, where the spikes are lower and closer together as a function of time, the frictional anisotropy may be mistaken for noise in the friction measurement system.

Interface friction of the ball-on-disc test under a high temperature is studied by Wu et al. [58]. A shematic illustration of the test stand is shown in Figure 8. A full contact model is developed to assess the lubrication of this testing mechanism and to decouple the contributions of solid-solid contact asperity to the interface friction stress. It was found that the temperature increase can significantly influence the lubricant performance. This can be due to the effect of lubricant additives. Furthermore, in the boundary lubrication regime, friction is sliding-speed dependent because the interaction of the solid layer may be formed by lubricant additives when the temperature of the lubricant increases.

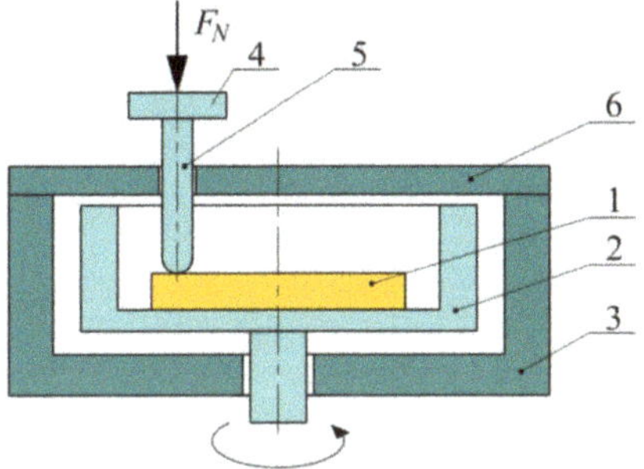

Figure 8. Schematic configuration of the tribological test: 1—specimen, 2—liquid container, 3—heat chamber, 4—load cell, 5—pin, 6—cover; prepared on the basis of [58].

Experimental testing of high steel sheets under hot stamping conditions using a high-temperature pin-on-disc test carried out by Ghiotti et al. [59] reveals that the blank temperature and contact pressure have a larger influence on the COF that typically can reproduce the condition in industrial processes. Moreover, the cooling rate does not affect the surface characteristics, in the range variations typical of the hot stamping. It was found that with increases in stamping temperature, the COF decreases. This can be explained by a reduction of shear strength of coating with the temperature.

Ball-on-plate Sliding Test (BST) is designed by UMT-TriboLab. The apparatus allows high-temperature reciprocating wear and is equipped with a resistance furnace. Figure 9 shows an illustration of the test set-up. A special clamping fixture was designed to ensure that there exists no slipping or movement of specimens while testing. The electro-mechanical drive allows us to carry out the ball-on-plate tests and assuring the existence of constant pressure at the contact interface [27]. The COF under different conditions can be calculated using the classical Coulomb's law (Equation (5)).

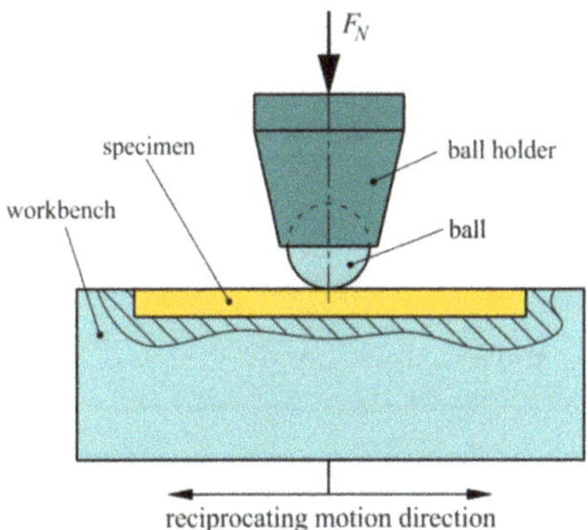

Figure 9. Schematic of UMT (Universal Mechanical Tester)-Tribolab tribometer.

Lu et al. [27] conducted experimental investigations on the unlubricated reciprocating sliding friction and galling behavior of 7075 aluminum alloy sheets at different temperatures. The results allow connecting the process temperature with the friction mechanism. Conventional scratch tests may be modified in order to obtain a well-controlled slipping contact as in BST. For example, Carlsson et al. [60] used a steel ball of diameter 7.5 mm instead of the Rockwell C diamond stylus with a radius of

200 µm. Based on the experimental test of friction of Dogal® and Aluzink® sheets, they concluded that compared with the BUT (Bending Under Tension) test, the modified scratch test, and the pin-on-disc test are easy and rapid to perform.

Kirkorn et al. [9] conducted investigations on the frictional behavior of a number of conventional tool materials that are used in the sheet forming industry employing a novel method of COF measuring based on flat-die strip drawing. According to the authors, the uniqueness of this device (Figure 10), compared with existing tribotesters used in sheet forming applications, is the ability to fully control the applied normal force and drawing velocity during experimental studies. A more comprehensive description of the functionality of the tribotester used can be found in [61]. The developed device enabled comparison of different tool steels, surface treatments, coatings, and lubricants in terms of galling and friction under real SMF process conditions. The normal and shear forces were measured, and the COF is calculated using Equation (31).

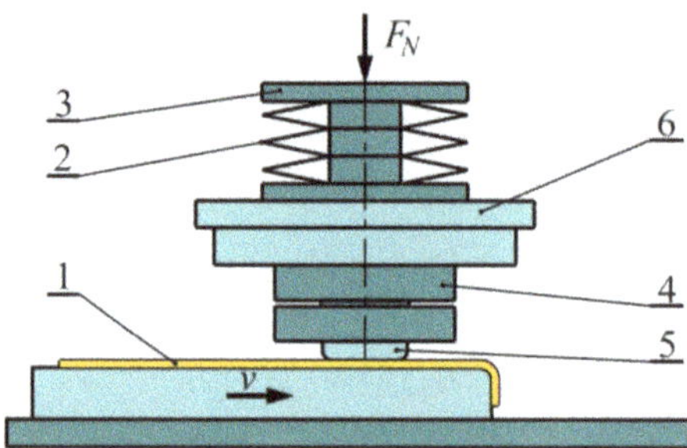

Figure 10. Schematic illustration of flat-die strip drawing: 1—specimen, 2—spring, 3—spring buffer, 4—tool holder, 5—tool, 6—force sensor.

4.3. Friction in Ironing in a Conical Die

Strip Reduction Tests (SRTs) represent severe tribological conditions with low slipping length, high normal pressure, and surface expansion. Some types of SRTs are tests carried out with a reduction of the sheet thickness between flat [62] or cylindrical [38,63] counter-samples. One of the first experimental approaches for the simulation of the ironing process was developed by Fukui et al. [64], who made direct measurements of friction force in metal-strip drawing where the strip was made to reduce between two stationary dies. The COF is defined as the ratio of normal force and tangential force occurring during sheet drawing between cylindrical surfaces (Figure 7a), and this relation is expressed as [65]:

$$\mu = \frac{F_T - 2F_N \tan\left(\frac{\alpha}{2}\right)}{2F_N + F_T \tan\left(\frac{\alpha}{2}\right)} \tag{8}$$

where F_T is the pulling (friction) force, F_N is the normal force and α is an angle of the arc contact area (Figure 11a), which may be determined according to the formula:

$$\tan\frac{\alpha}{2} = \sqrt{\frac{g_0 - g_1}{4R - (g_0 - g_1)}} \tag{9}$$

where R is the radius of the rounded edge of the counter-sample, g_0 is an initial sheet thickness, and g_1 is the final sheet thickness.

From the conditions of the equilibrium of the elementary section of the sheet between flat dies (Figure 11b), the relationship between the drawing force F_T of the sheet and the pressure force F_N is as follows [66]:

$$\frac{F_T}{F_N} = \frac{2(\mu + \tan\alpha)}{1 - \mu\tan\alpha} \tag{10}$$

Assuming that the value of the COF does not change along the contact surface between the dies and the plate, the value of the COF is given by:

$$\mu = \frac{F_T - 2F_N \tan \alpha}{2F_N + F_T \tan \alpha} \tag{11}$$

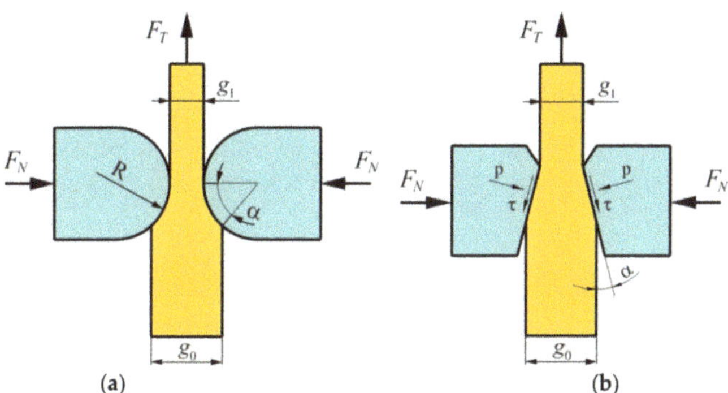

Figure 11. Illustrative diagram of the strip-reduction test carried out between (**a**) cylindrical dies (**b**) flat dies.

The SRT test has also been conducted by several researchers to simulate the ironing process [33,67]. This test is capable of simulating varying conditions such as drawing speed, reduction, sliding length, and the tool temperature [68]. The apparatus developed by Andreasen et al. [68] uses a round, non-rotating tool-pin instead of a wedge-shaped die (Figure 12). The test was conducted in such a way that the thickness of the sheet metal strip is reduced between a non-rotating tool pin (representing the conical die) and a supporting plate (representing the cylindrical punch).

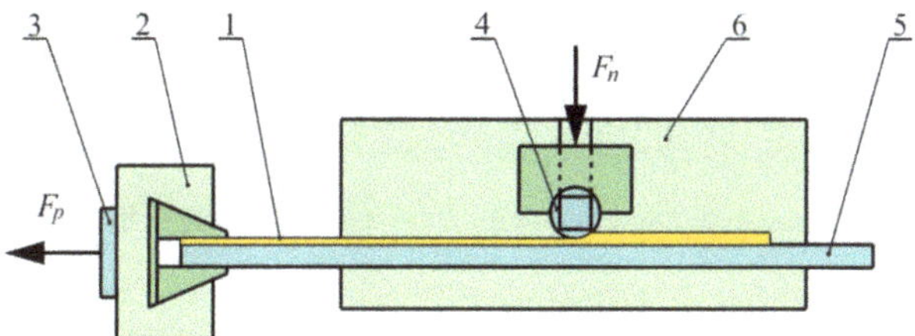

Figure 12. Schematic representation of the SRT: 1—specimen, 2—jaws, 3—piezoelectric transducer, 4—tool pin, 5—supporting plate, 6—vertical guide; prepared on the basis of [68].

The strip and the supporting plate are clamped together at the front end (corresponding to the punch nose in ironing). The two ends of the tool pins have square cross-sections that are designed to be mounted on two vertical guides that allow the change of the gap between the tool pin and the support the plate [68].

Recently, Moghadam et al. [63] developed the method of detection of the onset of galling in strip reduction testing using acoustic emission. They investigated the same apparatus as Andreasen et al. [68]. It was concluded that acoustic emission could be applied as a non-destructive measuring technique for assessment of galling in the strip reduction test emulates the ironing process. A direct correlation between the contact conditions and the generated acoustic emission events was found. Dohda and Kawai [69] proposed an alternative to Fukui et al.'s [64] strip-ironing type tribometer consisting of a bottom plate that is drawn with the strip through a wedge-shaped ironing die. Measurements carried out by Le and Sutcliffe [66] show that, in SDT, the change in friction is associated with a change in contact ratio between the tool and strip. Analysis of the surface drawn with rougher dies suggests that the lubrication has broken down, leading to the plowing of material picked up onto the tool surface.

Üstünyagiz et al. [33] proposed a new strip reduction test (SRT) for an industrial ironing process in order to replicate the severe tribological conditions involving sliding length, tool temperature, lubricant film and drawing speed. Sulaiman et al. [62] studied lubrication performance in SRTs with tools provided with longitudinal and shallow pockets that are oriented perpendicular to the sliding direction (Figure 13). They found that the distance between the pockets should be larger than the width of the pocket. By doing so, a topography similar to the flat table mount is created to avoid mechanical interlocking in the valleys. If the distance between the pockets is 2–4 times larger than the pocket width, the textured tool surface reduces friction and improves the lubrication effect.

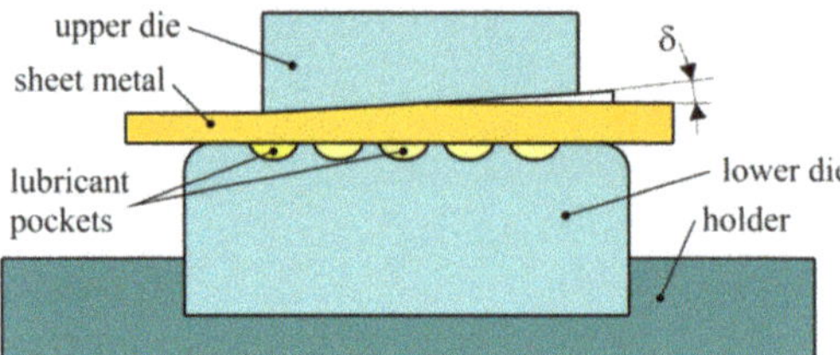

Figure 13. Schematic diagram of an SRT with a textured tool surface, prepared based on [62].

In addition, Üstünyagiz et al. [33] proposed a new Continuous Strip Reduction (CSR) test that can replicate the extreme tribological conditions under forward stroke and backward retraction of the punch (Figure 14). To replicate the forward stroke, two stationary cylindrical tool pins are used to draw the strip from right to left while its thickness is reduced in section I (Figure 15).

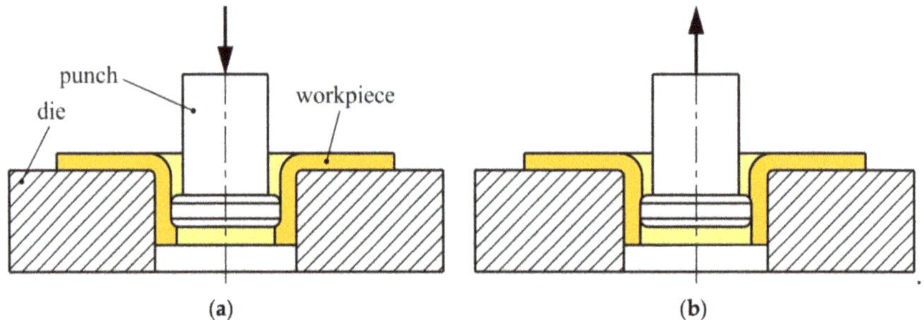

Figure 14. Schematic of the ironing process during (**a**) forward and (**b**) backward strokes.

The lower cylindrical tool pin is mounted on a heater block, which was used to adjust the tool temperature by an electric cartridge. The backward stroke of the ironing process is replicated in section II (Figure 15). This section provides a further reduction in the strip thickness, and the gap between the two stationary pins can, thus, be adjusted to a value slightly smaller than that utilized in section I. The test can simulate process parameters such as drawing speed, sliding length, strip thickness reduction and tool temperature. It is also possible to quantify the onset of the breakdown of the lubricant film and the resulting galling effect after several strokes during the forward and backward strokes.

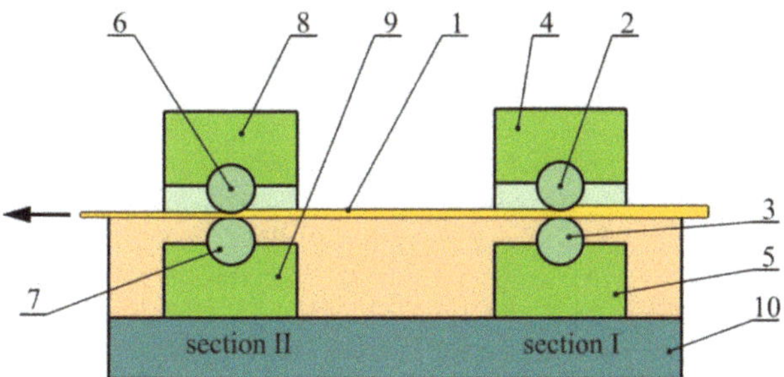

Figure 15. Schematic diagram of an SRT with a textured tool surface: 1—specimen; 2, 3, 6, 7—stationary cylindrical tool pins; 4, 8—housing; 5, 9—heater blocks; 10—base; prepared based on [33].

4.4. Friction Conditions in Die Curvature

The Bending Under Tension (BUT) test developed by Littlewood and Wallace [28] is attributed to the friction modeling on the edge of the die. The test consists of drawing the strip of metal around the cylindrical counter-sample (Figure 16). The traditional way of performing these BUT tests is by differential measurements using two sequential tests, one by drawing over a fixed circular cylindrical tool-pin, the other over a freely rotating pin, implying that no sliding takes place. Then an estimate of the friction is obtained from the difference in the front tension measured in the two tests.

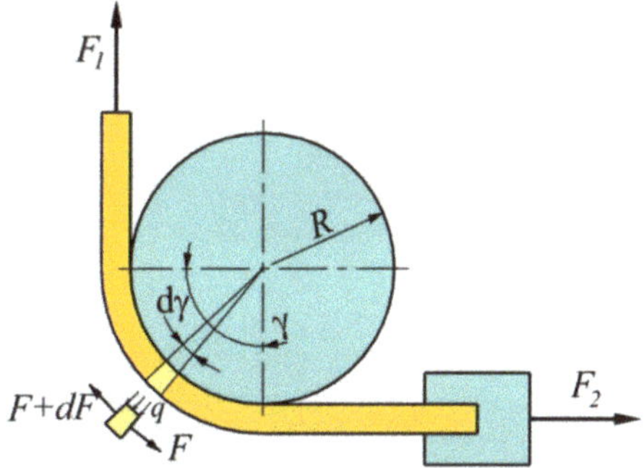

Figure 16. Forces acting on an elementary section of the strip.

Assuming that there is a constant COF μ in the contact region and the wrap angle γ (Figure 16) is constant during the test according to the equilibrium of all forces acting on an elemental cut of the strip $d\gamma$, it can be shown that:

$$F + q\mu wRd\gamma - (F + dF) = 0 \tag{12}$$

$$qwRd\gamma - Fsin\frac{d\gamma}{2} - (F + dF)sin\frac{d\gamma}{2} = 0 \tag{13}$$

For a very small $d\gamma$, one can assume that $sin\frac{d\gamma}{2} \approx \frac{d\gamma}{2}$ and dF << F. Thus, combining Equations (12) and (13) gives:

$$\mu d\gamma = \frac{dF}{F} \tag{14}$$

Integrating Equation (14), and taking into account $\gamma = \pi/2$ the COF is determined to be:

$$\mu = \frac{2}{\pi}ln\left(\frac{F_1}{F_2}\right) \tag{15}$$

The average contact stress (unit contact pressure, q) in this case is determined from the following equation:

$$q = \frac{F_1 + F_2}{2wR} \tag{16}$$

where w is the width of the strip, and R is the radius of the pin.

Trzepieciński and Lemu [32] developed a tribological simulator based on the concept of the BUT test. The test device is schematically shown in Figure 17. As illustrated, a grip supported by a load cell was used to hold the test strip at one end of the device. The specimen was wrapped around a fixed cylindrical roll and loaded in a tensile testing machine, ensuring contact over an angle of approximately 90°. Applying the fixed pin allows setting up the rolls in four positions, which enables the utilization of the full circumference of the roll. The tensile forces F_1 and F_2 can be measured simultaneously during the test. A major benefit of this test setup is that strain it is not necessary to measure the strain in order to determine COF. For some tests, the effect of strain on COF can be of interest, while the use of an extensometer may not be reasonable or warranted in some other cases.

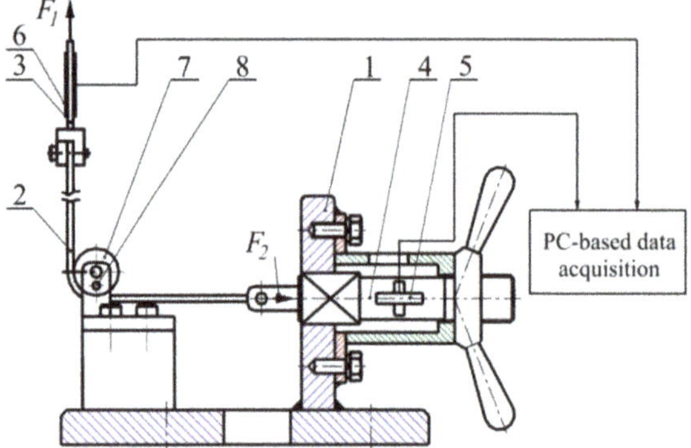

Figure 17. Schematic view of the testing device: 1—device frame, 2—specimen, 3 and 4—tension members, 5 and 6—extensometers, 7—working roll, 8—fixing pin.

Furthermore, the BUT test allows determination of the change of COF during the process of stretching the specimen over the roller. This change may be related to the change in sheet topography and increased normal pressure due to sheet deformation [15,70]. Furthermore, it may be related to a change in the contact conditions as a result of the strain hardening phenomenon. As a result of the friction resistance existing between the roller and the sheet, one gets $F_1 > F_2$ (Figure 17). Weinmann and Kernosky [71] developed a transducer that can be used to measure tension in metal strip while letting it move over the die shoulder. The transducer portion that is simulating the die shoulder is cylindrical and mounted on the body on a pair of bearings. Similar to the BUT test, the friction was determined based on both the front and back-tensions. So, the friction is not measured directly. Bay et al. [16] developed a test design that is variant of that of Weinmann and Kernosky [71], which measures not only back and front tensions but also the torque on the counter-sample (pin) directly. The friction apparatus is designed in such a way that the sheet strip clamped in two claws is bent around a non-rotating pin. Drawing the strip with the front claw while breaking the back claw ensures sliding of the strip around the pin under the controlled back-tension. To simulate the influence of varying tool temperature, authors designed electric heaters which can be inserted in the tool-pin holder. Moreover, in order to control the temperature, water-cooling channels were drilled in elements of friction apparatus.

Dilmec and Arap [72] proposed a test apparatus to measure the COFs for both the flange and the radius regions using only a single experiment. The operating procedure of this apparatus is based on the principles of strip drawing test and BUT methodology. Since the press moves vertically, the flat die simulator and the radial strip drawing test could not be individually designed. The section of the system that represents friction conditions at the die radius region, is shown in Figure 18. The COF for the radius region can be calculated using the front tension force registered by the press control system and the back-tension force using the Euler equation (Equation (15)). Then the analysis of variance method was used to investigate the effect of the blank holder force, die radius, drawing speed, the surface roughness of the tools, and the type of lubrication on dynamic COF between the sheet metal flange and the radius regions of the tools.

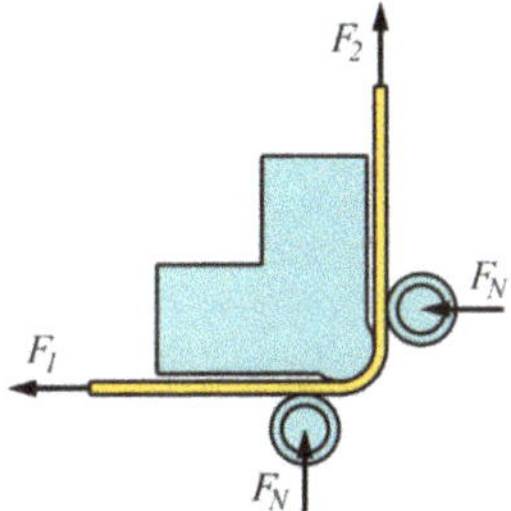

Figure 18. Schematic view of BUT device.

Many experimental and numerical works have contributed to the knowledge about tribological phenomena on the edge of a rounded die [70,73,74]. The BUT tests consist of making differential measurements by carrying out two tests, one after the other, one by drawing over a fixed circular roller, the other over a freely rotating roller [74]. A major drawback in all studies of BUT test is that friction measurement requires repeated measurements of the front and the back forces.

In another variant of the BUT test, the bending force can be determined by conducting the test with a free rotating roller or in conditions of minimized friction between the roller and the plate. The difference in the values of tensile forces for fixed and rotating rollers is attributed to the bending force of the sample around the roll. The test is carried out in two stages (Figure 19) [75]—(1) with a blocked roller and (2) with a mobile roller.

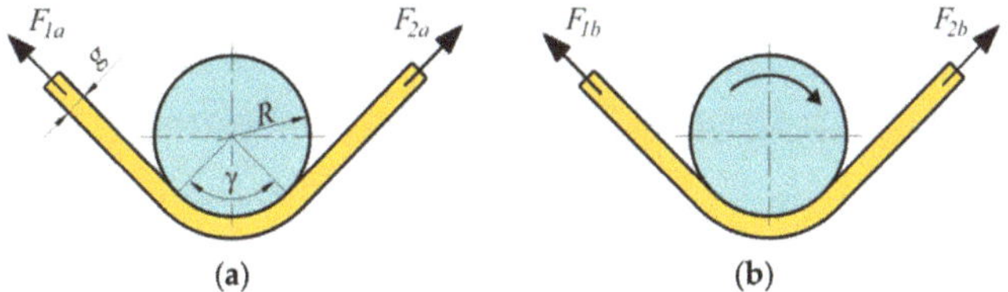

Figure 19. Schematic diagram of the bending under tension test realized by (**a**) fixed and (**b**) rotatable pin.

Taking into account the influence of sheet thickness and the radius of the counter-samples, the dependence relationship determining the COF based on the measurement of tensile forces in both stages is as follows [75]:

$$\mu = \frac{1}{\gamma}\frac{2R+g}{2R}\ln\frac{F_{1a}-(F_{1b}-F_{2b})}{F_{2a}} \tag{17}$$

where R is the radius of the counter-sample, γ is the contact angle, g is the sheet thickness, F_{1a} is a pulling force on a fixed counter-sample, F_{1b} is a pulling force on a rotatable roller, F_{2a} is a back-tension force on a fixed counter-sample, F_{2b} is a back-tension force on a rotatable roller, $F_b = F_{1b} - F_{2b}$ is the bending force of the specimen over the roller.

Sniekers and Smits [76] used a torque sensor on the pin in the BUT test to avoid the second step of the test when the pin is free to rotate. In this method, the COF is determined by measuring the torque from:

$$\mu = \frac{\frac{F_0 D}{R}}{\sqrt{F_1^2 + F_2^2 - \left(\frac{F_0 D}{R}\right)^2}} \tag{18}$$

where $F_0 D$ represents the torque on the pin (Figure 20), F_1 and F_2 are the pull force and the back-tension force, respectively.

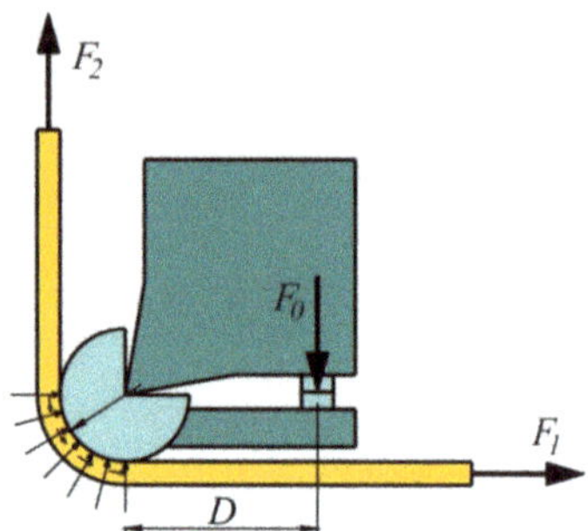

Figure 20. Equilibrium of the torque in the BUT, prepared based on [76].

The ratio of the friction stress τ [77] and the contact pressure p given by Equation (16) giving the COF are given by Equations (19) and (20), respectively.

$$\tau = \frac{2T}{\pi w R^2} \tag{19}$$

$$\mu = \frac{4T}{\pi R (F_1 + F_2)} \tag{20}$$

where w is the strip width and R is the pin radius.

Sube [78] proposed a variation of the Equation (20) as:

$$\mu = \frac{(F_1 + F_2) - F_b}{2wR} \tag{21}$$

where F_b is the bending force obtained as a difference of the pull force from the back force (Equation (22)), represented by F_1^* and F_2^* in Equation (22), respectively. These forces are obtained when the test is conducted with the pin free.

$$F_b = F_1^* - F_2^* \tag{22}$$

Direct measurement of the central force induced in the specimen is not possible in the BUT test [79]. This can be a serious problem when the specimen is made of a material that is strain-rate sensitive because it may lead to force estimation errors and, as a result, substantial errors in COF calculations. One of the limitations of the BUT test is the difficulty of isolating the bending effect based on theoretical calculation from the approximated equation of Swift's bending force. To overcome this limitation, Hassan et al. [79] developed a BUT punch friction test in which an improved analytical friction model that accounts for the bending was proposed. This proposed method enables direct measurement of tensile forces from F_1 and F_2 (Figure 20). Punch displacement is then measured using displacement capacitive transducer. The pin holder of the rig is supported by a specially designed head that can replace the chuck.

In general, the friction coefficient value can be determined from [79]:

$$\mu = \frac{1}{\theta} ln\left[\frac{F_1}{F_1 + F_b}\right] \tag{23}$$

where F_b is the bending force.

In order to isolate the effect of friction from that of bending and hence compensate the bending effect, two test methods are commonly employed: (1) using a fixed pin and (2) using a rolling pin. The actual friction coefficient is then calculated using an equation proposed by Wagoner et al. [80]:

$$\mu = \left[\frac{1}{\theta} ln\frac{F_1}{F_2}\right]_{pin} - \left[\frac{1}{\theta} ln\frac{F_1}{F_2}\right]_{roller} \tag{24}$$

where θ is the angle of contact between specimen and roller/pin.

Gali et al. [81] developed the hot forming friction measurement setup (Figure 21). This tribometer is capable of measuring the COF between the pin and a metallic strip contact stretching at different strain rates and temperatures. The apparatus consisted of a friction measurement assembly and a loading system, with two synchronized linear actuators that apply a constant strain rate to the sample. A strip was placed around the rotating heater and pulled on both ends by a tensile force. The strip was also compressed by the steel pin, which has an embedded load cell to measure the normal load. The tribometer was developed at the University of Windsor, and the principles of its operation, including the details of the strain measurements, are described in an earlier paper of Das et al. [82]. They additionally installed a camera system over the top face of the strip to measure changes in the initial diameter of circular grids inscribed on the strip's surface and then used it for in-situ evaluation of the major and minor strains in the hot zone. The results of testing AA5083 aluminum alloy sheets at different temperatures showed that COF increased, and this can be attributed to the effects of softening tribolayer at high temperature and abrasion at low temperature [81].

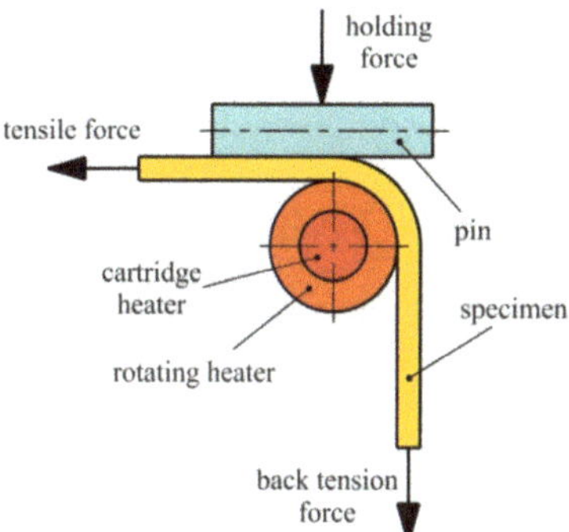

Figure 21. Schematic of the hot bending under the tension test.

While forming the draw pieces with complex shapes, such as car bodies, it is important to utilize dies having varying outline curvatures. When cylindrical counter-samples are used to model friction conditions on dies with varying outline curvatures and varying radii of edge fillets, it is difficult to get results on the edge of the die that fully reflect the friction conditions. To overcome this limitation, Trzepieciński and Lemu [32] proposed an experimental and numerical method to describe the friction in sheet metal forming. Due to the rounded shape of the proposed counter-samples (Figure 22), it was found necessary to employ a different method of determining the forces can eliminate friction through the steps listed below [32]:

- determining the pulling and back-tension forces experimentally, i.e., by friction test, using a counter-sample having a non-cylindrical profile,
- using numerical methods to determine the pulling and back-tension forces under frictionless conditions ($\mu = 0$) employing a counter-sample with a non-cylindrical profile.

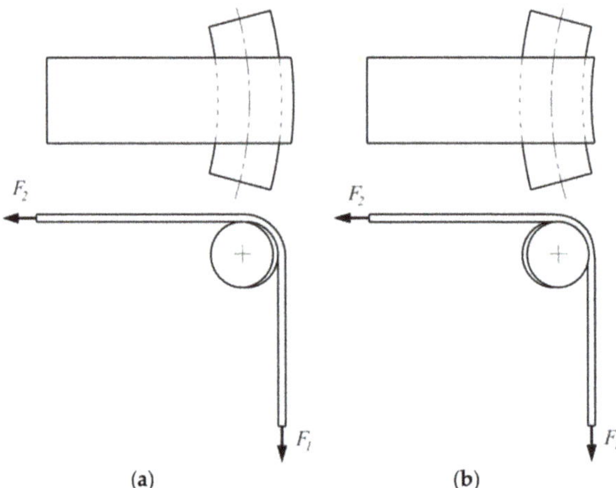

Figure 22. Proposed methods for the determination of the COF on a die edge with a (**a**) convex and (**b**) concave profile.

Considering that the sample is bent at an angle of 90° while conducting the friction test, the COF is determined by [32]:

$$\mu = \frac{1}{\gamma}\frac{2R+g}{2R}\ln\frac{F_{1exp} - (F_{1num} - F_{2num})}{F_{2exp}} \tag{25}$$

where R is the counter-sample radius, g id the thickness of the sheet, F_{1exp} is the pulling force (experimentally determined), F_{1num} is the pulling force (numerically determined), F_{2exp} is the back-tension force (experimentally determined), and F_{2num} is the back-tension force (numerically determined).

The knowledge of intensity and distribution of the contact and friction stresses over the die surface in SMF is important both from the tool wear analysis and the die design. Jurkovic et al. [83] proposed a method of defining contact stresses and determination of COF. A more detailed figure can be found in the aforementioned reference [83]. For the determining method of direct contact stresses, the authors elaborated and designed sensors for the measurement of normal stresses and tangential contact stresses. The contact stresses are measured by means of pins with transducers.

4.5. Friction Testing in the Draw Bead Region

The draw bead test (DBT) is a concept developed by Nine [84] for friction modeling at the draw bead in sheet metal forming. Draw beads compensate for the material flow resistance around the perimeter of the draw piece or to change the stress state in specific regions of the draw piece. The curvature of the metallic sheet passing the draw bead model (Figure 23) changes frequently. Thus, the aim of the method is to enable separation of the deformation resistance in the sheet from the frictional resistance. In DBT, the values of the pulling and the clamping forces are measured when pulling the strip over fixed (Figure 23a) and rotating (Figure 23b) rollers. Drawing the sheet over the set of rotating rolls minimizes the frictional resistance.

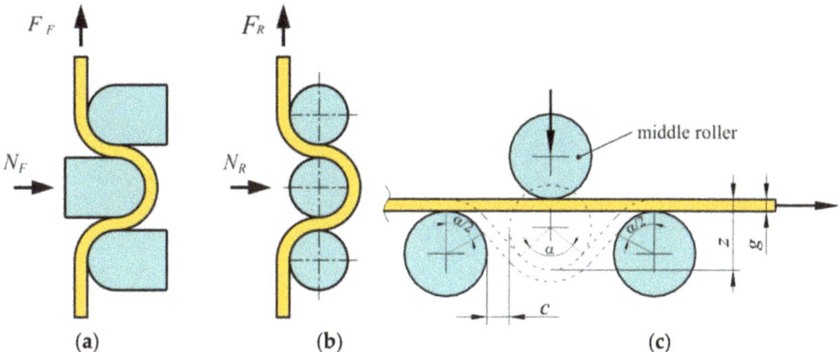

Figure 23. The concept of determining frictional resistance: (**a**) fixed rollers, (**b**) rotatable rollers, (**c**) geometrical parameters of draw bead.

When the wrap angle of the middle roller is 180°, the difference in the drawing force for the rotating and fixed rollers may be attributed to the friction process and used to calculate the value of the COF according to the relationship [84]:

$$\mu = \frac{F_F - F_R}{\pi \cdot N_F} \tag{26}$$

where F_F is the pulling force, F_R is the pulling force, and N_F is the normal force or clamping force. These forces are obtained from the fixed rolls, the freely rotating rolls, and the fixed beads, respectively.

The change of friction conditions may be obtained by changing the lubrication conditions, shape, and dimensions of the draw bead model, wrap angle of the counter-sample, and the drawing speed of the sheet [85]. The influence of the manufacturing tolerance of the sheet metal on the disturbance of the recorded forces is often omitted in DBTs. The influence of the values of surface roughness parameters on the value of the COF was discussed in detail in Skarpelos and Morris [86], which found that metallic sheets with a higher *Ra* parameter showed a lower value of COF.

The value of the COF for a wrap angle α different from 180° is determined from [87]:

$$\mu = \frac{sin\alpha}{2a} \frac{P_F - P_R}{N_F} \tag{27}$$

where α represents a quarter of contact angle of the actual engagement of the strip over the middle roll.

The difference between forces P_F and P_R in the numerator of Equation (27) is the force that expresses the frictional resistance. This force is a component of frictional resistance occurring in the direction of the sample drawing. This resistance is associated with the occurrence of friction on the entire length of the contact arc on each of the three rollers equal to $2\alpha = \pi$ rad (Figure 23c). The value of the friction force $F_T = P_F - P_R$ is burdened with the error resulting from the assumption that the use of rotating rollers eliminates friction. The normal force N_F is a component of the resultant pressure on the direction normal to the direction of drawing the sheet, and thus, the value of the COF determined according to Equation (27) cannot be identified with the classical Amontons–Coulomb friction coefficient.

When the middle roller's maximum displacement becomes equal to its diameter, its wrap angle becomes different from 180°, because it is necessary to ensure adequate clearance *c* (Figure 23c) between the counter-samples. This prevents the blocking of the specimen. The value of the wrap angle on all rollers is related nonlinearly to the value of clearance *c* and the displacement of the middle roller, as shown in Figure 24. Excessive drawing resistance may occur, and as a result, test samples of mild steel sheets may break when the wrap angles get larger. In contrast, the use of excessive clearance *c* between the rollers can cause an unfavorable change in the way the sheet deforms [88]. According to the recommendations in Nanayakkara et al. [87], the side clearance c should be equal to approximately twice the sheet thickness:

$$c = g + s \qquad (28)$$

where s is an additional margin equal to 0.77 mm for $g = 0.8$ mm sheet thickness.

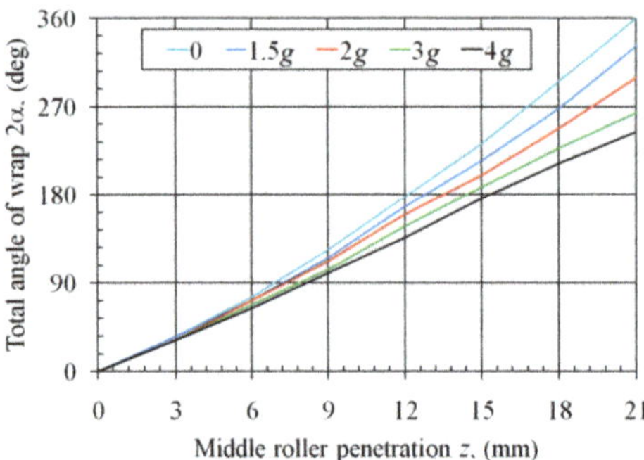

Figure 24. The relation between the total angle of wrap 2α and the middle roller penetration z.

Figueiredo et al. [4] used the draw bead simulation (DBS) test to investigate the tribological properties of DP600 dual-phase steel sheets. Results of experimental tests allowed to conclude that the die material surface roughness has a significant effect on COF. A major drawback in all studies on friction measurement in the draw bead region is that the determination of friction demand repeated measurements of the drawing force using fixed and rotating counter-samples. Such a way of the determination of the value of COF leads to large uncertainties due to scattering [89]. In order to avoid this problem, Olsson et al. [90] proposed a new draw bead test that can measure the friction force acting directly on the tool radius using a build-in piezoelectric torque transducer. The method enables recording the breakdown of lubricant film as a function of the drawing distance. For coated sheet steels, the friction coefficient is assumed constant with contact pressure. One of the first comprehensive tests of frictional resistances of zinc-coated sheets using the BUT and DBS tests have been conducted by Vallance and Matlock [91].

The conventional methodology of the DBS test is only valid to measure an average COF over the pressure range. However, the new methodology proposed by Kim et al. [92] can be employed to determine friction coefficients from draw-bend friction tests involving pressure non-uniformity. In this methodology, contact pressure maps that are obtained from numerical simulations, contrary to the uniform pressure assumption, are used to build the COF model expressed as a 2nd order polynomial of pressure.

4.6. Friction Modeling in the Punch Curvature

In the tensile strip test (Figure 25), developed by Duncan et al. [93], a strip specimen of sheet metal is pulled over two rollers. This is done to simulate the frictional phenomena between the punch and sheet metal in sheet forming. While the pulling force and the strain are measured on one side of the roller, the back-tension force on the second side of the roller is calculated from the measured strains, i.e., using the stress-strain relationship of the test material [94]. Assuming that the COF is constant over the rollers, its value may be determined using the formula [95]:

$$\mu = \frac{2}{\pi} ln \frac{F_p}{F_b} \qquad (29)$$

where F_p is a pulling force and F_b is a back-tension force (Figure 25).

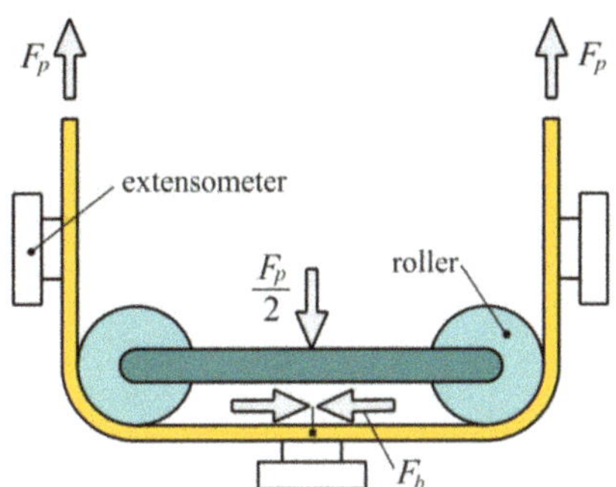

Figure 25. Schematic diagram of the tensile strip test.

Similar to that introduced by Duncan et al. [93], the apparatus developed by Lovell and Deng [19] is a "plane-strain bend/unbend tension" device that stretches and bends the test specimens of the metal strip. COF between the die and the workpiece, and the surface roughness of the deformed sheets, could be measured under different lubrication conditions, strip widths, testing speeds and pin radii.

Hao et al. [96] studied the friction at the tool-workpiece interface using the "U" shaped friction test. As depicted in Figure 26, the strip specimen is pulled over two pins and around a rounded die, which is mounted on a load cell. The pulling force, F_1, and the strip tension on the other side of the pins, F_2, are measured during the test. In the" U" shaped strip friction test, the COF is estimated from measurements of strip tensions on each side of the die. The force and moment equilibrium on an element of the strip give the following formulas that can be used in determining the COF μ, the average contact pressure p_s, and the average friction stress f [22] as given in Equations (30), (31), and (32) respectively:

$$\mu = \frac{1}{\theta} ln \frac{F_1}{F_2} \tag{30}$$

$$p_s = \frac{F_1 + F_2}{2wR} \tag{31}$$

$$f = \frac{F_1 - F_2}{wR\theta} \tag{32}$$

where F_1 and F_2 are the strip tensions (Figure 26), R is the radius of the die, w is the strip width, and θ is the wrap angle of the strip on the die.

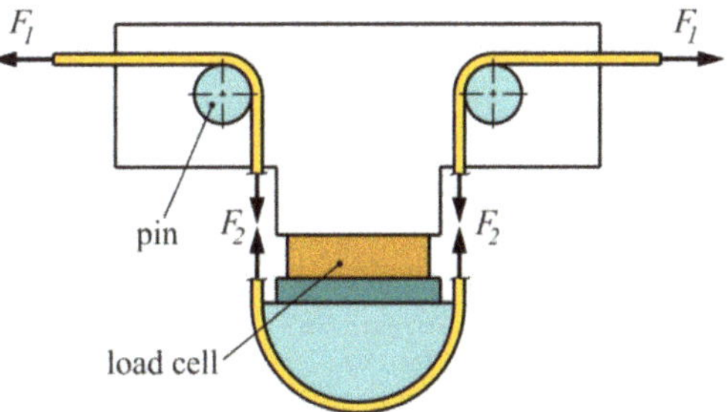

Figure 26. Schematic of the "U" shape strip test, prepared based on [96].

This test is a modification of the tensile strip test, which was developed at Ohio State University (OSU) [80]. It was designed for the experimental determination of the COF in conditions of variable wrap angle of the sheet over a cylindrical counter-sample (Figure 27). As in the case of the bending under tension test, the resistance associated with sample bending and drawing can be separated by running the OSU test on fixed and rotating rollers. The rotating rollers allow the elimination of the effects of friction and allow the measurement of forces required to bend the specimen. The undeniable advantage of the OSU test is to ensure the realism of the deformation process of the sheet and contact conditions:

- simultaneous occurrence of the bending and stretching of the sheet,
- the wrap angle of the counter-samples changes gradually with the increase in their displacement.

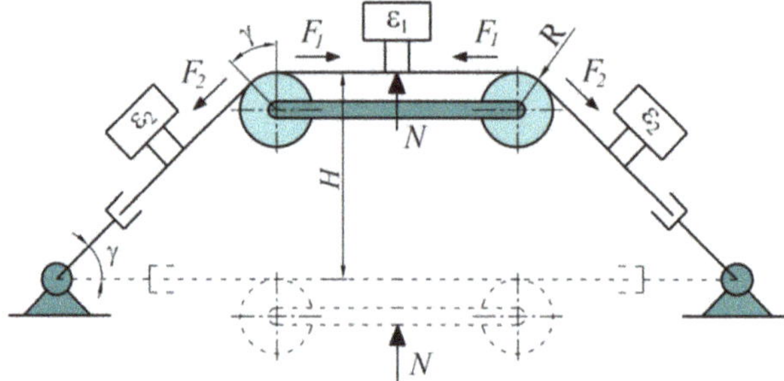

Figure 27. Schematic diagram of the OSU friction test: F_1—the force acting on the center of the specimen, F_2—is the force acting in side walls, H—punch height, N—pressure force, R—radius of cylindrical counter-sample, γ is the contact angle, ε_1 and ε_2—strains of the specimen in the center and side walls, respectively.

Apart from the effect of bending and straightening the sheet, the relationship between the tensile forces F_1 and F_2 is equal:

$$\frac{F_1}{F_2} = e^{\gamma\mu} \tag{33}$$

where γ is the wrap angle of the counter-sample, μ is the COF.

After rearranging Equation (33), the COF takes the form:

$$\mu = \frac{1}{\gamma} ln \frac{F_1}{F_2} \tag{34}$$

The plastic deformation of the sheet caused by bending around the counter-sample causes changes in the topography of the surface, and consequently, the local pressures relative to the actual contact surface is smaller than the nominal pressures taking into account the nominal contact surface [70].

4.7. Friction Conditions Under Sheet Stretching

Figure 28 illustrates a schematic view of an apparatus developed by Azushima and Zhu [97] for the Tension-Bending Test (TBT) and represents stretching over the punch radius of curvature and under the punch nose. The sheet metal strip is clamped with the end of the actuator ram head, and the other end is subjected to a tension load. The drawing load is then measured by means of a strain gauge load cell inserted between the ram load and the chuck. The test may be conducted under different drawing speeds, the geometry of the die, surface topography of die, and lubrication conditions.

The COF is determined using following equation [97]:

$$\mu = 2\frac{F_D - F_T - F_B}{\theta(F_D - F_T)} \tag{35}$$

where F_D is drawing load, F_T is back-tension force, F_B is the Swift bending tension [98], and θ is the die angle.

The average contact pressure is defined as [97]:

$$p_{av} = \frac{F_T + F_D}{2wR} \tag{36}$$

where R is the die radius, and w is the specimen width.

TBT has been used by Azushima and Zhu [97] to study the lubrication characteristics at the interface between the A1100 aluminum sheet and tool in the die corner of deep drawing. They found that the surface roughening becomes predominant when the average contact pressure gets lower, and thus, the COF becomes constant. In addition, increasing average contact pressure leads to the flattening of surface asperities for higher average contact pressure, and hence, the COF decreases. Azushima et al. [98] measured the COF using TBT, and they observed that the asperity flattening occurred for the specimen with a rough surface, which dominated the boundary lubrication regime, and the COF remained constant. It is well known that, due to the geometry of contact in TBT, the asperity flattening and the surface roughening took place due to the plastic strain.

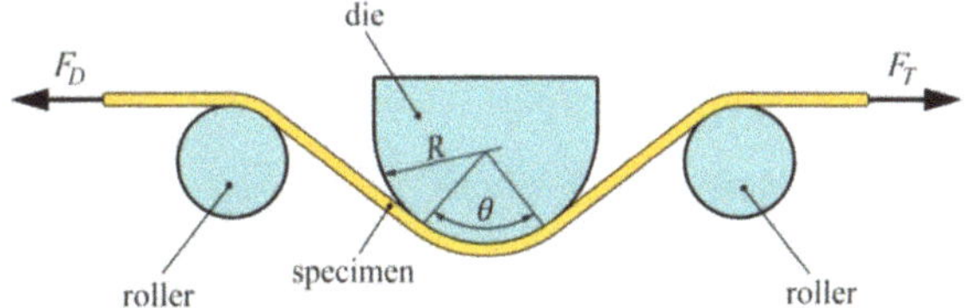

Figure 28. Schematic illustration of the flat-die strip drawing.

5. Friction Testing in Incremental Sheet Forming

In the ISF process, a rigid tool with a rounded tip is widely used do deform the sheet metal. To avoid possible scratch on the sheet surface, an Oblique Roller-Ball (ORB) tool in combination with the path generation algorithm, was developed by Lu et al. [99]. To better understand the frictional effect, the authors also developed an analytical model of friction based on the stress analysis. In the design (Figure 29), a ball cap is clamped by the tool arm to a certain angle instead of clamped vertically in the conventional design. The rotation of the spindle will not change the position of the rollerball. The measured forming loads, i.e., horizontal and vertical forces, show the cyclical change in tool forces tends to be a "U" shape. The force difference after passes suggests the varied friction conditions.

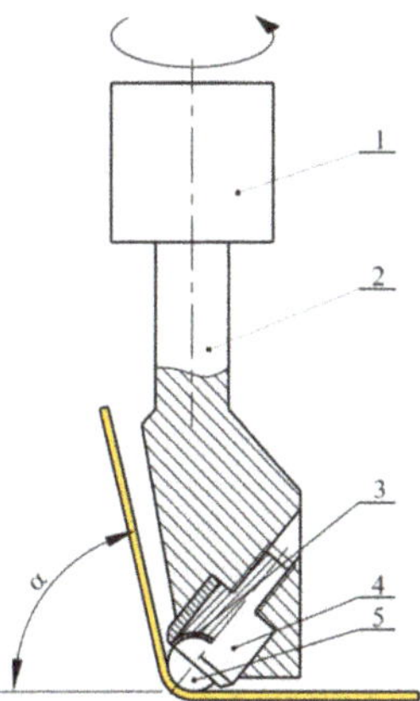

Figure 29. Schematic representation of the ORB test: c—CNC spindle, 2—tool arm, 3—ball bearings, 4—ball cup, 5—roller ball; prepared based on [99].

Precise calculation of COF is not easy since the measured horizontal force contains not only the friction but also the forming force. According to the approach of Xu et al. [100], Lu et al. [99] defined friction indicator μ^* in the analysis to evaluate COF. This friction indicator is the ratio between the horizontal and vertical force components at the mid-position of the groove in each pass, according to Equation (37) [99]:

$$\mu^* = \frac{f_H}{|f_Z|} = \frac{friction + forming\ load}{|f_Z|} \tag{37}$$

where f_H is the horizontal (or in-plane) force resisting the motion of the forming tool in the horizontal plane, f_Z is the vertical force acting normal to the horizontal plane and flattens the sheet.

It should be noted that μ^* is not only the result of friction conditions but also related to other effects such as strain hardening phenomenon and geometry of the desired component. The experimental friction tests revealed that lower friction could be obtained using roller-type tools. This reduction is more apparent when the pressure between the tool and sheet is high up to a specific level [99]. Durante et al. [101] concluded that the forces in SPIF continue to increase in the initial few contours due to the dynamic equilibrium between sheet thinning and strain hardening. Therefore, the friction indicator should be calculated once the force has achieved a steady-state.

Wei et al. [102] evaluated the COF using Equation (38) using sliding tests on 20 mm wide specimens, fixed at the ends. As illustrated in Figure 30, a tool path alternates between a stroke, back and forth twice along an 80 mm straight line [101], similar to the straight groove test, was used by Kim and Park [103] and more recently by Ilyas et al. [104] to evaluate forming limit curves in ISF.

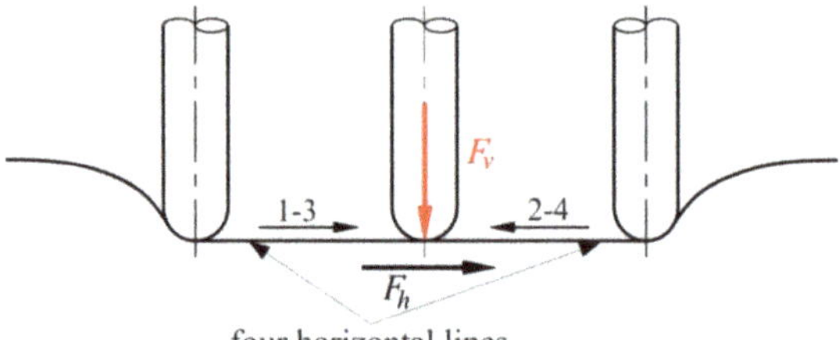

Figure 30. Sliding test path for the evaluation of COF (number of tool passes have been indicated by 1–3 and 2–4), prepared based on [101].

The formula for the determination of the COF value is [102]:

$$\mu = \left|\frac{F_h}{F_v}\right| \tag{38}$$

where F_h and F_v are horizontal and vertical components of the total force, respectively.

In order to evaluate the modulus of the equivalent force in the sheet plane, it is enough to calculate the force F_h [101]:

$$F_h = \sqrt{F_a^2 + F_b^2} \tag{39}$$

where F_a and F_b are values of the horizontal component forces evaluated in the central zones for sides a and b in Figure 31, respectively, and represent the moduli of the two components of the horizontal force.

Analysis of variance of the effect of some forming parameters (sheet thickness, tool diameter, forming angle, and step size) on the value of friction indicator μ^* conducted by Wei et al. [102] revealed that the response of friction indicator to parameter variations is relatively complex. During forming, the mechanical properties of the sheets change during the strain-hardening phenomenon [6]. Furthermore, the tool-sheet contact area changes when the forming is performed with an altered set of parameters [105]. These changes affect the values of force components, and hence, the friction indicator [102]. It should also be noted that the results of roughness findings of Wei et al. [102] are different from those presented in the ISF literature. The material type observed to be the major factor influencing the friction condition, and hence, the roughness. Recently, a number of researchers mainly studied the friction indicator as to the function of tool rotation. In many cases [100,101,105], it has been revealed that the friction indicator has a significant effect on the roughness.

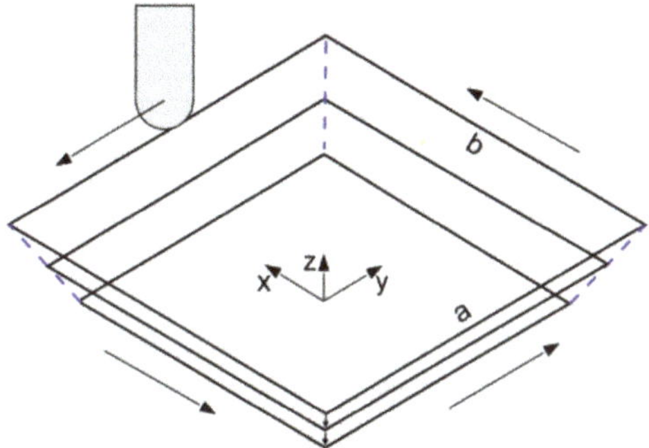

Figure 31. Sides of the tool path during the forming of pyramid frusta.

In the case of the pin-on-disc test, the tool contacts the workpiece over the same rounded loop repeatedly; whereas in the case of the SPIF, the forming tool never exactly touches the same workpiece area. Despite this, Petek et al. [106] used a conventional pin-on-disc test shown in Figure 21 to study the coating ability to prevent the adhesion of the workpiece at initial testing conditions during SPIF. Pin-on-disc test is also applied by Minutolo et al. [107] to determine COF value for a numerical model of the ISF process. The sliding speed was set equal to the punch feed rate in the motion of the groove. According to the authors' conclusion, results of finite element-based computations were observed to have good agreement with the experimental ones, showing that pin on disc test can be used for testing friction in the SPIF process. Hussain [108] successfully studied the different lubrication strategies during ISF of titanium sheets using pin-on-disc tribometer.

Saidi et al. [109] proposed an approach estimating the COF in SPIF based on the values of forming forces F_x, F_y, and F_z (Figure 32) while forming the truncated cones with wall angle α. The authors took the following assumptions: (i) value of COF varies from one zone to another, and (ii) COF is estimated on the basis of a uniform displacement of the tool with respect to the workpiece. Based on these assumptions, the COF was calculated by the following relation [109,110]:

$$\mu = \frac{\sqrt{F_x^2 + F_y^2}}{\sqrt{F_z^2}}$$

(40)

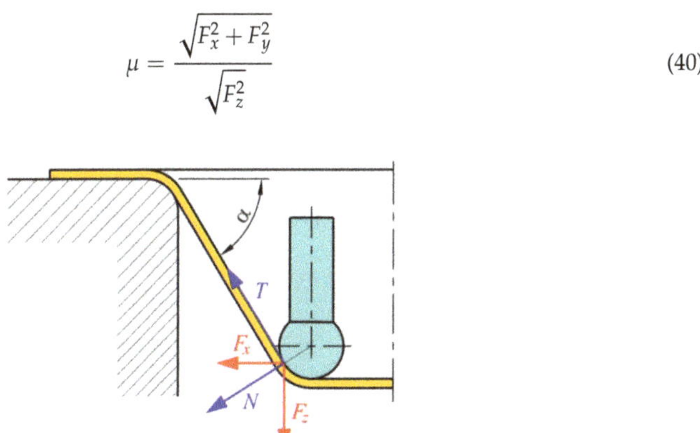

Figure 32. Relationship of the forming force components.

6. Conclusions

In the SMF occurs various zones in terms of stress, deformation, sliding velocity, and contact conditions. Therefore, over the years, a number of tribological tests to model the phenomenon of friction in individual regions of sheet metal to be formed have been developed. The aim of this review is (i) presentation of tribological tests aimed at determining the value of COF occurring in SMF and ISF, while maintaining the actual contact geometry and kinematics of the tool movement, and (ii) the discussion of the disadvantages and limitations of specific tests. Although most of the friction tests were developed many decades ago, they are still valid and used by many authors. Although the methods to determine friction resistance in conventional SMF at different temperature conditions have been fairly well understood [10,15,16,50], there are a very limited number of experimental tests modeling conditions in ISF [102,104,105,109]. This is due to the difficulty in separating the frictional resistance from the plastic deformation resistance in the process [110].

Based on the literature review, the tribological tests can be classified into direct and indirect measurement tests of the COF. In indirect methods of determination, the COF is determined on the basis of measuring other physical quantities, e.g., friction force and normal force [40,50–53,55,57,58,66]. Based on the adopted model, the COF is calculated. The disadvantage of this type of methods is that they allow the determination of the average value of COF, while they do not allow measuring and determining the real friction resistance. Most of the tribometers, which are typically used to measure the wear resistance, allows determining the instantaneous value of the COF. The rotational specimen causes that the length of the friction track is unlimited. This intensifies the wear processes and causes overestimation of the value of COF. Assigning the COF determined in a pin- and ball-on-disc tribotesters to the conditions arisen in conventional SMF should be limited. However, the pin-on-disc and ball-on-disc are able to determine the effect of directional topography on the wear and COF [111].

Most of the friction tests, are able to capture the COF values of tool-workpiece interfaces by controlling critical parameters, such as lubrication conditions [19,20,62,112], microgeometry of contact (surface roughness of sheet and tools surfaces) [50,53,91], contact pressures [40,90], temperature [42,46,47,81], tool material and coating applied [21,22,38,45], and sliding velocity [33]. The multitude of parameters and phenomena affecting frictional resistance means that the results obtained in various tribological tests are not comparable [35].

The BUT and DBS tests could not directly measure COF in such a way that their value has heavily relied on the material behavior of tested materials. During friction tests, the sheet is plastically deformed, which affects the value of process forces used to evaluate the value of the COF. In contrast, the plastic deformation of the sheet strip specimen in BUT tests allows determining the evolution of the value of the COF with the increased contact pressure.

In the classical Coulomb model commonly used in the solid mechanics, the friction force is proportional to the normal pressure. The ratio of the friction force to the contact pressure is handled as a material constant, named the COF [113]. In metal forming operations exist high local pressures which intensify the effects of adhesion and plowing in the friction resistance. In such conditions, the usage of the formula for the determination of the COF based on the Coulomb friction model (Equation (1)) is limited to small pressures.

In SDT, it was realized that in the presence of the cylindrical counter-samples, the relation of the normal force and the friction force is nonlinear [15,35]. It is a result of a non-proportional increase in real contact area with an increase in normal force. This phenomenon is observed by the non-linear decrease of the value of the COF (Equation (1)) with an increase of normal force [35]. In the case of lubricated conditions and tests with flat counter-samples, the value of pressure is crucial in the determination of the COF. As shown by Mousavi et al. [114], since the lubricant film is under relative high surface pressure, a uniform distribution of lubricant over the contact area rarely takes place. Therefore, lubricant pockets can provide the permanently reserved lubricant during the sheet metal forming.

The main problems observed in warm and hot forming processes are that high friction and wear occur because of high adhesion between the workpiece material and tool surface, and surface-initiated fatigue. The occurrence and prevention of galling mechanism are investigated by using tribometers for metal forming at elevated temperatures [12,27,63]. The typically used pin-on-disc tribometer to characterize frictional contact in hot stamping do not reproduce appropriate contact conditions. First, the cyclic contact of the pin with the specimen does not correspond to real contact conditions [58]. Second, the build-up in the front of pin causes plowing friction and causes overestimation of COF. Third, if the ball-shaped pin is used, the real contact area changes during the test [111]. In fact, according to the Coulomb law, the COF is independent of the contact area. However, the applicability of Coulomb friction model is very limited in the description of contact phenomena in hot SMF due to a high shear of adhesion in total contact resistance.

Author Contributions: T.T. had the initial idea for the manuscript and wrote the draft of an article; H.G.L. contributed in Funding acquisition, project administration, validation, and writing–review & editing. All authors have read and agreed to the published version of the manuscript.

Funding: This research received no external funding.

Acknowledgments: The authors gratefully acknowledge the financial support provided by the University of Stavanger under the activities of the COTech Program Area for Research.

Conflicts of Interest: The authors declare no conflict of interest.

References

1. Trzepieciński, T.; Lemu, H.G. Effect of computational parameters on springback prediction by numerical simulation. *Metals* **2017**, *7*, 380. [CrossRef]
2. Slota, J.; Jurcisin, M.; Lazarescu, L. Influence of technological parameters on the springback angle of high-strength steels. *Acta Metall. Slovaca* **2014**, *20*, 236–243. [CrossRef]
3. Altan, T.; Tekkaya, A.E. *Sheet Metal Forming Processes and Applications*; ASM International: Materials Park, OH, USA, 2012.
4. Figueiredo, L.; Ramalho, A.; Oliveira, M.C.; Menezes, L.F. Experimental study of friction in sheet metal forming. *Wear* **2011**, *271*, 1651–1657. [CrossRef]
5. Zhang, R.; Shao, Z.; Lin, J. A review on modelling techniques for formability prediction of sheet metal forming. *Int. J. Lightweight Mater. Manuf.* **2018**, *1*, 115–125. [CrossRef]
6. Shi, X.; Gao, L.; Khalatbari, H.; Xu, Y.; Wang, H.; Jin, L.L. Electric hot incremental forming of low carbon steel sheet: Accuracy improvement. *Int. J. Adv. Manuf. Technol.* **2013**, *68*, 241–247. [CrossRef]
7. Li, Y.; Chen, X.; Liu, Z.; Sun, J.; Li, F.; Li, J.; Zhao, G. A review on the recent development of incremental sheet-forming process. *Int. J. Adv. Manuf. Technol.* **2017**, *92*, 2439–2462. [CrossRef]
8. Peng, W.; Ou, H.; Becker, A. Double-sided incremental forming: A review. *J. Manuf. Sci. Eng.* **2019**, *141*, 1–12. [CrossRef]
9. Kirkhorn, L.; Bushlya, V.; Andersson, M.; Ståhl, J.E. The influence of tool steel microstructure on friction in sheet metal forming. *Wear* **2013**, *302*, 1268–1278. [CrossRef]
10. Kim, H.; Kades, N. Friction and lubrication. In *Sheet Metal Forming: Fundamentals*; Altan, T., Tekkaya, A.E., Eds.; ASM International: Materials Park, OH, USA, 2012; pp. 89–104.
11. Bay, N.; Azushima, A.; Groche, P.; Ishibashi, I.; Merklein, M.; Morishita, M.; Nakamura, T.; Schmid, S.; Yoshida, M. Environmentally benign tribo-systems for metal forming. *CIRP Ann.* **2010**, *59*, 760–780. [CrossRef]
12. Pelcastre, L. *Hot Forming Tribology: Galling of Tools and Associated Problems*; Luleå University of Technology: Lulea, Sweden, 2011.
13. Evin, E.; Tomáš, M.; Výrostek, M. Verification the numerical simulation of the strip drawing test by its physical model. *Acta Mech. Slovaca* **2016**, *20*, 14–21. [CrossRef]
14. Masters, L.G.; Williams, D.K.; Roy, R. Friction behaviour in strip draw test of pre-stretched high strength automotive aluminium alloys. *Int. J. Mach. Tools Manuf.* **2013**, *73*, 17–24. [CrossRef]
15. Trzepieciński, T.; Fejkiel, R. On the influence of deformation of deep drawing quality steel sheet on surface topography and friction. *Tribol. Int.* **2017**, *115*, 78–88. [CrossRef]

16. Bay, N.; Olsson, D.D.; Andreasen, J.L. Lubricant test methods for sheet metal forming. *Tribol. Int.* **2008**, *41*, 844–853. [CrossRef]

17. Menezes, P.L.; Kumar, K.K.; Kailas, S.V. Influence of friction during forming processes—A study using a numerical simulation technique. *Int. J. Adv. Manuf. Technol.* **2009**, *40*, 1067–1076. [CrossRef]

18. Jadhav, S.; Schoiswoh, M.; Buchmayr, B. Tribology test methods and simulations of the effect of friction on the formability of automotive steel sheets. *Metall. Ital.* **2018**, *9*, 56–63.

19. Lovell, M.R.; Deng, Z. Characterization of interfacial friction in coated sheet steels: Influence of stamping process parameters and wear mechanisms. *Tribol. Int.* **2002**, *35*, 85–95. [CrossRef]

20. Deng, Z.; Lovell, M.R. Effects of lubrication and die radius on the friction behavior of Pb-coated sheet steels. *Wear* **2000**, *244*, 42–52. [CrossRef]

21. Sulaiman, M.H.; Farahana, R.N.; Bienk, K.; Nielsenc, C.V.; Bay, N. Effects of DLC/TiAlN-coated die on friction and wear in sheet-metal forming under dry and oil-lubricated conditions: Experimental and numerical studiem. *Wear* **2019**, *438*, 203040. [CrossRef]

22. Kim, A.; Han, B.; Yan, Q.; Altan, T. Evaluation of tool materials, coatings and lubricants in forming galvanized advanced high strength steels (AHSS). *CIRP Ann. Manuf. Technol.* **2008**, *57*, 299–304. [CrossRef]

23. Carlsson, P. Surface Engineering in Sheet Metal Forming. Ph.D. Thesis, Acta Universitatis Upsaliensis, Upsalla, Sweden, 18 February 2005.

24. Hol, J.; Meinders, V.T.; de Rooij, M.B.; van den Boogaard, A.H. Multi-scale friction modeling for sheet metal forming: The boundary lubrication regime. *Tribol. Int.* **2015**, *81*, 112–128. [CrossRef]

25. Sigvant, M.; Pilthammar, J.; Hol, J.; Wiebenga, J.H.; Chezan, T.; Carleer, B.; van den Boogard, T. Friction in sheet metal forming: Influence of surface roughness and strain rate on sheet metal forming simulation results. *Procedia Manuf.* **2019**, *29*, 512–519. [CrossRef]

26. Li, G.; Long, X.; Yang, P.; Liang, Z. Advance on friction of stamping forming. *Int. J. Adv. Manuf. Technol.* **2018**, *96*, 21–38. [CrossRef]

27. Lu, J.; Song, Y.; Hua, L.; Zhou, P.; Xie, G. Effect of temperature on friction and galling behavior of 7075 aluminum alloy sheet based on ball-on-plate sliding test. *Tribol. Int.* **2019**, *140*, 105872. [CrossRef]

28. Littlewood, M.; Wallace, J.F. The effect of surface finish and lubrication on the fictional variation involved in the sheet-metal-forming process. *Sheet Met. Ind.* **1964**, *41*, 925–930.

29. Hutchings, I.M. Leonardo da Vinci's studies of friction. *Wear* **2016**, *360*, 51–66. [CrossRef]

30. Wang, D.; Yang, H.; Li, H. Advance and trend of friction study in plastic forming. *Trans. Nonferr. Met. Soc. China* **2014**, *24*, 1263–1272. [CrossRef]

31. Dohda, K.; Boher, C.; Reza-Aria, F.; Mahayotsanun, N. Tribology in metal forming at elevated temperatures. *Friction* **2015**, *3*, 1–27. [CrossRef]

32. Trzepieciński, T.; Lemu, H.G. Proposal for an experimental-numerical method for friction description in sheet metal forming. *Stroj. Vestn. J. Mech. Eng.* **2015**, *61*, 383–391. [CrossRef]

33. Üstünyagiz, E.; Nielsen, C.V.; Christiansen, P.; Martins, P.A.F.; Bay, N. Continuous strip reduction test simulating tribological conditions in ironing. *Procedia Eng.* **2017**, *207*, 2286–2291. [CrossRef]

34. De Souza, J.H.C.; Liewald, M. Analysis of the tribological behaviour of polymer composite tool materials for sheet metal forming. *Wear* **2010**, *268*, 241–248. [CrossRef]

35. Trzepieciński, T.; Bazan, A.; Lemu, H.G. Frictional characteristics of steel sheets used in automotive industry. *Int. J. Automot. Technol.* **2015**, *16*, 849–863. [CrossRef]

36. Vollertsen, F.; Hu, Z. Tribological size effects in sheet metal forming measured by a strip drawing test. *CIRP Ann.* **2006**, *55*, 291–294. [CrossRef]

37. Roizard, X.; von Stebut, J. Surface asperity flattening in sheet metal forming—A 3D relocation stylus profilometric study. *Int. J. Mach. Tools Manuf.* **1995**, *35*, 169–175. [CrossRef]

38. Guillon, O.; Roizard, X.; Belliard, P. Experimental methodology to study tribological aspects of deep drawing—Application to aluminium alloy sheets and tool coatings. *Tribol. Int.* **2001**, *34*, 757–766. [CrossRef]

39. Ter Haar, R. *Friction in Sheet Metal Forming, the Influence of (Local) Contact Conditions and Deformation*; Universiteit Twente: Enschede, The Netherlands, 1996.

40. Payen, G.R.; Felder, E.; Repoux, M.; Mataigne, J.M. Influence of contact pressure and boundary films on the frictional behaviour and on the roughness changes of galvanized steel sheets. *Wear* **2012**, *276*, 48–52. [CrossRef]

41. Coello, J.; Miguel, V.; Martínez, A.; Avellaneda, F.J.; Calatayud, A. Friction behaviour evaluation of an EBT zinc-coated trip 700 steel sheet through flat friction tests. *Wear* **2013**, *305*, 129–139. [CrossRef]

42. Venema, J.; Matthews, D.T.A.; Hazrati, J.; Wörmann, J.; van den Boogaard, A.H. Friction and wear mechanisms during hot stamping of AlSi coated press hardening steel. *Wear* **2017**, *380*, 137–145. [CrossRef]

43. Xue, H.; Nanxiang, K.; Zhu, H. The friction coefficient testing method of metallic sheet and strip. *Adv. Mater. Res.* **2012**, *629*, 75–78. [CrossRef]

44. Recklin, V.; Dietrich, F.; Groche, P. Influence of test stand and contact size sensitivity on the friction coefficient in sheet metal forming. *Lubricants* **2018**, *6*, 41. [CrossRef]

45. Kondratiuk, J.; Kuhn, P. Tribological investigations on friction and wear behaviour of coatings for hot sheet metal forming. *Wear* **2011**, *270*, 839–849. [CrossRef]

46. Uda, K.; Azushima, A.; Yanagida, A. Development of new lubricants for hot stamping of Al-coated 22MnB5 steel. *J. Mater. Process. Technol.* **2016**, *228*, 112–116. [CrossRef]

47. Tokita, Y.; Nakagaito, T.; Tamai, Y.; Urabe, T. Stretch formability of high strength steel sheets in warm forming. *J. Mater. Process. Technol.* **2017**, *246*, 77–84. [CrossRef]

48. Shi, Z.; Wang, L.; Mohamed, M.; Balint, D.S.; Lin, J.; Stanton, M.; Watson, D.; Dean, T.A. A new design of friction test rig and determination of friction coefficient when warm forming an aluminium alloy. *Procedia Eng.* **2017**, *207*, 2274–2279. [CrossRef]

49. Merklein, M.; Andreas, K.; Steiner, J. Influence of tool surface on tribological conditions in conventional and dry sheet metal forming. *Int. J. Precis. Eng. Manuf. Green Technol.* **2015**, *2*, 131–137. [CrossRef]

50. Sedlaček, M.; Podgornik, B.; Vižintin, J. Influence of surface preparation on roughness parameters, friction and wear. *Wear* **2009**, *266*, 482–487. [CrossRef]

51. Trzepieciński, T. 3D elasto-plastic FEM analysis of the sheet drawing of anisotropic steel sheet. *Arch. Civ. Mech. Eng.* **2010**, *10*, 95–106. [CrossRef]

52. Wang, L.; He, Y.; Zhou, J.; Duszczyk, J. Effect of temperature on the frictional behaviour of an aluminium alloy sliding against steel during ball-on-disc tests. *Tribol. Int.* **2010**, *43*, 299–306. [CrossRef]

53. Ajayi, O.O.; Erck, R.A.; Lorenzo-Martin, C.; Fenske, G.R. Frictional anisotropy under boundary lubrication: Effect of surface texture. *Wear* **2009**, *267*, 1214–1219. [CrossRef]

54. Altan, T. R&D Update: Twist-Compression Testing to Evaluate Friction, Lubrication. 2017. Available online: https://www.thefabricator.com/article/stamping/r-d-update-twist-compression-testing-to-evaluate-friction-lubrication (accessed on 28 June 2019).

55. Dou, S.; Xia, J. Analysis of sheet metal forming (stamping process): A study of the variable friction coefficient on 5052 aluminum alloy. *Metals* **2019**, *9*, 853. [CrossRef]

56. Singh, R.; Melkote, S.N.; Hashimoto, F. Frictional response of precision finished surfaces in pure sliding. *Wear* **2005**, *258*, 1500–1509. [CrossRef]

57. Masuko, M.; Aoki, S.; Suzuki, A. Influence of lubricant additive and surface texture on sliding friction characteristics of steel under varying speeds ranging from ultralow to moderate. *Tribol. Trans.* **2005**, *48*, 289–298. [CrossRef]

58. Wu, C.; Qu, P.; Zhang, L.; Li, S.; Jiang, Z. A numerical and experimental study on the interface friction of ball-on-disc test under high temperature. *Wear* **2017**, *376*, 433–442. [CrossRef]

59. Ghiotti, A.; Bruschi, S.; Borsetto, F. Tribological characteristics of high strength steel sheets under hot stamping conditions. *J. Mater. Process. Technol.* **2011**, *211*, 1694–1700. [CrossRef]

60. Carlsson, P.; Bexell, U.; Olsson, M. Tribological behaviour of thin organic permanent coatings deposited on hot-dip coated steel sheet—A laboratory study. *Surf. Coat. Technol.* **2000**, *132*, 169–180. [CrossRef]

61. Kirkhorn, L.; Frogner, K.; Andersson, M.; Ståhl, J.E. Improved tribotesting for sheet metal forming. *Procedia CIRP* **2012**, *3*, 507–512. [CrossRef]

62. Sulaiman, M.H.; Christiansen, P.; Bay, N. The influence of tool texture on friction and lubrication in strip reduction. *Procedia Eng.* **2017**, *207*, 2263–2268. [CrossRef]

63. Moghadam, M.; Christiansen, P.; Bay, N. Detection of the onset of galling in strip reduction testing using acoustic emission. *Procedia Eng.* **2017**, *183*, 59–64. [CrossRef]

64. Fukui, S.; Ohi, T.; Kudo, H.; Takita, I.; Seino, J. Some aspects of friction in metal-strip drawing. *Int. J. Mech. Sci.* **1992**, *4*, 297–312. [CrossRef]

65. Matuszak, A. Determination of the frictional properties of coated steel sheets. *J. Mater. Process. Technol.* **2000**, *106*, 107–111. [CrossRef]

66. Le, H.R.; Sutcliffe, M.P.F. Measurements of friction in strip drawing under thin film lubrication. *Tribol. Int.* **2002**, *35*, 123–128. [CrossRef]

67. Aleksandrović, S.; Dordević, M.; Stefanović, M.; Lazić, V.; Adamović, D.; Arsić, D. Different ways of friction coefficient determination in stripe ironing test. *Tribol. Ind.* **2014**, *36*, 293–299.

68. Andreasen, J.L.; Bay, N.; Andersen, M.; Christensen, E.; Bjerrum, N. Screening the performance of lubricants for the ironing of stainless steel with a strip reduction test. *Wear* **1997**, *207*, 1–5. [CrossRef]

69. Dohda, K.; Kawai, N. Compatibility between tool materials and workpiece in sheet-metal ironing process. *J. Tribol.* **1990**, *112*, 275–281. [CrossRef]

70. Zhang, S.; Hodgson, P.D.; Duncan, J.L.; Cardew-Hall, M.J.; Kalyanasundaram, S. Effect of membrane stress on surface roughness changes in sheet forming. *Wear* **2002**, *253*, 610–617. [CrossRef]

71. Weinmann, K.J.; Kernosky, S.K. Friction studies in sheet metal forming based on a unique die shoulder force transducer. *CIRP Ann.* **1996**, *45*, 269–272. [CrossRef]

72. Dilmec, M.; Arap, M. Effect of geometrical and process parameters on coefficient of friction in deep drawing process at the flange and the radius regions. *Int. J. Adv. Manuf. Technol.* **2016**, *86*, 747–759. [CrossRef]

73. Lemu, H.G.; Trzepieciński, T. Numerical and experimental study of frictional behavior in bending under tension test. *Stroj. Vestn. J. Mech. Eng.* **2013**, *59*, 41–49. [CrossRef]

74. Bay, N.; Tribo-Systems for Sheet Metal Forming. Metallurgiske Processer i Danmark. Available online: http://orbit.dtu.dk/ws/files/128041777/vm2009_a9.pdf (accessed on 12 September 2019).

75. Wenzloff, G.J.; Hylton, T.A.; Matlock, D.K. A new test procedure for the bending-under-tension friction test. *J. Mater. Eng. Perform.* **1992**, *1*, 609–613. [CrossRef]

76. Sniekers, R.J.J.M.; Smits, H.A.A. Experimental set-up and data processing of the radial strip-drawing friction test. *J. Mater. Process. Technol.* **1997**, *66*, 216–223. [CrossRef]

77. Andreasen, J.L.; Olsson, D.D.; Chodnikiewicz, K.; Bay, N. Bending under tension test with direct friction measurement. *Proc. Inst. Mech. Eng. Part B J. Eng. Manuf.* **2006**, *220*, 73–80. [CrossRef]

78. Sube, D. Possibilities and Results of Numeric and Experimental Simulation of Sheet Metal Forming Processes. In Proceedings of the 3 Conferência Nacional de Conformação de Chapas, Porto Alegre, Brasil, 8 October 2000.

79. Hassan, M.A.; Tan, C.J.; Yamaguchi, K. A developed friction test for sheet metal stretch forming process. *Int. J. Surf. Sci. Eng.* **2013**, *7*, 152–170. [CrossRef]

80. Wagoner, R.H.; Wang, W.; Sriram, S. Development of OSU formability test and OSU friction test. *J. Mater. Process. Technol.* **1994**, *45*, 13–18. [CrossRef]

81. Gali, O.A.; Riahi, A.R.; Alpas, A.T. The tribological behaviour of AA5083 alloy plastically deformed at warm forming temperatures. *Wear* **2016**, *302*, 1257–1267. [CrossRef]

82. Das, S.; Morales, A.T.; Riahi, A.R.; Meng-Burany, X.; Alpas, A.T. Role of plastic deformation on elevated temperature tribological behaviour of an Al-Mg alloy (AA5083): A friction mapping approach. *Metall. Mater. Trans. A* **2011**, *42*, 2384–2401. [CrossRef]

83. Jurkovic, M.; Jurkovic, Z.; Buljan, S. The tribological state test in metal forming processes using experiment and modelling. *J. Achiev. Mater. Manuf. Eng.* **2006**, *18*, 383–386.

84. Nine, H.D. Drawbead forces in sheet metal forming. In *Mechanics of Sheet Metal Forming*, 1st ed.; Koistinen, D.P., Wang, N.M., Eds.; Springer: Boston, MA, USA, 1978; pp. 179–211.

85. Samuel, M. Influence of drawbead geometry on sheet metal forming. *J. Mater. Process. Technol.* **2002**, *122*, 94–103. [CrossRef]

86. Skarpelos, P.; Morris, J.W., Jr. The effect of surface morphology on friction during forming of electrogalvanized sheet steel. *Wear* **1997**, *212*, 165–172. [CrossRef]

87. Nanayakkara, N.K.B.M.P.; Kelly, G.L.; Hodgson, P.D. Determination of the coefficient of friction in partially penetrated draw beads. *Steel Grips* **2004**, *2*, 677–680.

88. Meinders, V.T. Developments in Numerical Simulations of the Real-Life Deep Drawing Process. Ph.D. Thesis, University of Twenty, Enschede, The Netherlands, 11 February 2000.

89. Sanchez, L.R. Characterisation of a measurement system for reproducible friction testing on sheet metal under plane strain. *Tribol. Int.* **1999**, *32*, 575–586. [CrossRef]

90. Olsson, D.D.; Bay, N.; Andreasen, J.L. Direct Friction Measurement in Draw Bead Testing. In Proceedings of the 8th International Conference on Technology of Plasticity, Verona, Italy, 9–13 October 2005.

91. Vallance, D.W.; Matlock, D.K. Application of the bending-under-tension friction test to coated sheet steels. *J. Mater. Eng. Perform.* **1992**, *1*, 685–693. [CrossRef]

92. Kim, Y.S.; Jain, M.K.; Metzger, D.R. Determination of pressure-dependent friction coefficient from draw-bend test and its application to cup drawing. *Int. J. Mach. Tools Manuf.* **2012**, *56*, 69–78. [CrossRef]

93. Duncan, J.L.; Shabel, B.S.; Filho, J.G. *A Tensile Strip Test for Evaluating Friction in Sheet Metal Forming*; SAE Technical Paper 780391; SAE: Warrendale, PA, USA, 1978; p. 8.

94. Uyyuru, R.K.; Simjith, M.; Kailas, S.V. Friction in metal forming. In *Metal Forming Technology and Process Modelling*, 1st ed.; Dixit, U.S., Narayanan, R.G., Eds.; McGraw Hill Education: New Delhi, India, 2013; pp. 156–198.

95. Lovell, M.; Higgs, C.F.; Deshmukha, P.; Mobley, A. Increasing formability in sheet metal stamping operations using environmentally friendly lubricants. *J. Mater. Process. Technol.* **2006**, *177*, 87–90. [CrossRef]

96. Hao, S.; Klamecki, B.E.; Ramalingam, S. Friction measurement apparatus for sheet metal forming. *Wear* **1999**, *224*, 1–7. [CrossRef]

97. Azushima, A.; Zhu, J. Tribological Behavior at the Interface in Tension-Bending Type Test for Sheet Metal Forming. In Proceedings of the 7th International Conference on Technology of Plasticity, Tokyo, Japan, 27 October–1 November 2002.

98. Azushima, A.; Sakurmoto, M.; Inagaki, M. Direct Observation of Interface Between Tool and Workpiece in Sliding Condition Under Tension Bending. In Proceeding of the 55th Japanese Joint Conference for the Technology of Plasticity, Tokyo, Japan, 26–28 May 2006.

99. Lu, B.; Fang, Y.; Xu, D.K.; Chen, J.; Oub, H.; Moser, N.H.; Cao, J. Mechanism investigation of friction-related effects in single point incremental forming using a developer oblique roller-ball tool. *Int. J. Mach. Tools Manuf.* **2014**, *85*, 14–29. [CrossRef]

100. Xu, D.; Wu, W.; Malhotra, R.; Chen, J.; Lu, B.; Cao, J. Mechanism investigation for the influence of tool rotation and laser surface texturing (LST) on formability in single point incremental forming. *Int. J. Mach. Tools Manuf.* **2013**, *73*, 37–46. [CrossRef]

101. Durante, M.; Formisano, A.; Langella, A.; Memola, F.M.C. The influence of tool rotation on an incremental forming process. *J. Mater. Process. Technol.* **2009**, *209*, 4621–4626. [CrossRef]

102. Wei, H.; Hussain, G.; Iqbal, A.; Zhang, Z. Surface roughness as the function of friction indicator and an import ant parameters-combination having controlling influence on the roughness: Recent results in incremental forming. *Int. J. Adv. Manuf. Technol.* **2019**, *101*, 2533–2545. [CrossRef]

103. Kim, Y.H.; Park, J.J. Effect of process parameters on formability in incremental forming of sheet metal. *J. Mater. Process. Technol.* **2002**, *130*, 42–46. [CrossRef]

104. Ilyas, M.; Hussain, G.; Espinosa, C. Failure and strain gradient analyses in incremental forming using GTN model. *Int. J. Lightweight Mater. Manuf.* **2019**, *2*, 177–185. [CrossRef]

105. Durante, M.; Formisano, A.; Langella, A. Observations on the influence of tool-sheet contact conditions on an incremental forming process. *J. Mater. Eng. Perform.* **2011**, *20*, 941–946. [CrossRef]

106. Petek, A.; Podgornik, B.; Kuzman, K.; Čekada, M.; Waldhauser, W.; Vižintin, J. The analysis of complex tribological system of single point incremental sheet metal forming—SPIF. *Stroj. Vestn. J. Mech. Eng.* **2008**, *54*, 266–273.

107. Minutolo, F.C.; Durante, M.; Formisano, A.; Langella, A. Evaluation of the maximum slope angle of simple geometries carried out by incremental forming process. *J. Mater. Process. Technol.* **2007**, *194*, 145–150. [CrossRef]

108. Hussain, G. Microstructure and mechanical and tribological properties of a lubricant coating for incremental forming of a Ti sheet. *Trans. Mech. Eng.* **2014**, *38*, 423–429.

109. Saidi, B.; Boulila, A.; Ayadi, M.; Nasri, R. Prediction of the Friction Coefficient of the Incremental Sheet Forming SPIF. In Proceedings of the 6th International Congress Design and Modelling of Mechanical Systems CMSM'2015, Hammamet, Tunisia, 23–25 March 2015.

110. Azevedo, N.G.; Farias, J.S.; Bastos, R.P.; Teixeira, P.; Davim, J.P.; Jose, R.; Alves de Sousa, R.J. Lubrication aspects during single point incremental forming for steel and aluminum materials. *Int. J. Precis. Eng. Manuf.* **2015**, *16*, 1–7. [CrossRef]

111. Trzepieciński, T.; Trytek, A.; Lemu, H.G. Study of frictional properties of AMS nickel-chromium alloys. *Key Eng. Mater.* **2016**, *674*, 244–249. [CrossRef]

112. Oleksik, V. Influence of geometrical parameters, wall angle and part shape on thickness reduction of single point incremental forming. *Procedia Eng.* **2014**, *81*, 2280–2285. [CrossRef]

113. Popov, V.L. Coulomb's law of friction. In *Contact Mechanics and Friction, Physical Principles and Applications*; Springer-Verlag: Berlin/Heidelberg, Germany, 2010; pp. 133–154.
114. Mousavi, A.; Sperk, T.; Gietzelt, T.; Kunze, T.; Lasagni, A.F.; Brosius, A. Effect of contact area on friction force in sheet metal forming operations. *Key Eng. Mater.* **2018**, *767*, 77–84. [CrossRef]

MDPI

St. Alban-Anlage 66

4052 Basel

Switzerland

Tel. +41 61 683 77 34

Fax +41 61 302 89 18

www.mdpi.com

Metals Editorial Office

E-mail: metals@mdpi.com

www.mdpi.com/journal/metals